COHESION AND STRUCTURE
OF SURFACES

COHESION AND STRUCTURE

Volume 4

Series editors

F.R. de Boer

University of Amsterdam

D.G. Pettifor

University of Oxford

ELSEVIER

Amsterdam – Lausanne – New York – Oxford – Shannon – Tokyo

COHESION AND STRUCTURE
OF SURFACES

Contributors

K. Binder
Johannes Gutenberg-Universität Mainz

M. Bowker
*University of Reading
and
University of Liverpool*

J.E. Inglesfield
University of Nijmegen

P.J. Rous
University of Maryland Baltimore County

1995

ELSEVIER
Amsterdam – Lausanne – New York – Oxford – Shannon – Tokyo

ISBN: 0 444 89829 8

North-Holland
Elsevier Science B.V.
P.O. Box 211
1000 AE Amsterdam
The Netherlands

Printed on acid-free paper
Printed in The Netherlands

PREFACE

One of the principal aims of this series of books is to collate and order up-to-date experimental data bases on cohesion and structure in order to reveal underlying trends that subsequent theoretical chapters might help elucidate. In this volume we consider the cohesion and structure of surfaces. During the past fifteen years there has been a dramatic increase in the number of different surfaces whose structures have been determined experimentally. For example, whereas in 1979 there were only 25 recorded adsorption structures, to date there are more than 250. In Chapter I Philip Rous presents a timely compilation of this structural data base on surfaces within a series of tables that allows easy direct comparison of structural parameters for related systems. Experimental structural trends amongst both clean surfaces and adsorbate systems are highlighted and discussed.

The past fifteen years has witnessed an equally dramatic development in the ability of theory to understand structure and phase transitions at surfaces. In Chapter II John Inglesfield outlines the successes of local density functional theory in predicting the relaxations and reconstructions of clean metal and semiconductor surfaces, and the behaviour of adsorbates such as hydrogen, oxygen and alkali elements on metal surfaces, thereby explaining some of the experimental trends observed within the database. These *ab initio* density functional calculations are of ground state properties at the absolute zero of temperature. In Chapter III Kurt Binder introduces finite temperature effects in a pedagogical review of current statistical mechanical treatments of phase transitions at surfaces, many of which display the prominent rôle of fluctuations or non-mean-field behaviour. He considers in detail not only phase transitions and ordering phenomena within adsorbed *two-dimensional* monolayers on a substrate, but also phase transitions such as surface roughening and surface melting that occur locally at the boundary of *semi-infinite* bulk materials. In the final chapter Michael Bowker discusses the relationship of the reactivity of a surface to its morphology and composition, which is particularly relevant to a fundamental understanding of catalysis.

Any multi-authored book relies upon the individual authors to meet the publishing deadlines. Kurt Binder submitted his manuscript on time in May 1993. As editor I apologize for the subsequent delays that mean his list of references will be two years out of date at publication.

D.G. Pettifor

v

CONTENTS

Chapter I

SURFACE CRYSTALLOGRAPHY: THE EXPERIMENTAL DATA BASE

P.J. ROUS

*Department of Physics, University of Maryland Baltimore County,
Baltimore, MD 21228-5398, USA*

Contents

Cohesion and Structure of Surfaces
Edited by D.G. Pettifor
© 1995 Elsevier Science B.V. All rights reserved.

Abstract

The surface crystallography of crystalline solids is reviewed and compiled as a sequence of tables that allows the direct comparison of structural parameters of related structures. The presentation of structural data is organized according to the concept of chemical periodicity; the surface structures of elements or adsorbates belonging to the same group of the periodic table are considered together. Evidence for the existence of structural trends for clean surfaces and adsorbate systems is extracted and discussed. Structural information for a total of over four hundred surface structures, including 257 adsorption systems, is presented.

1. Introduction

In 1979, Michel Van Hove made one of the first attempts to bring together the results of surface crystallography and to extract structural trends in surface chemical bonding (Van Hove, 1979). Van Hove's contribution, which appeared in a book entitled "The Nature of The Surface Chemical Bond" (Rhodin and Ertl, 1979), anticipated advances in surface crystallography that would allow an understanding of chemical bonding at the level achieved for molecules and bulk solids, as exemplified by Pauling's famous monograph; "The Nature of the Chemical Bond" (Pauling, 1960). Fifteen years later, it seems appropriate to revisit this theme.

In the mid-1970s, whilst some structural trends were evident, detailed knowledge of surface bonding was limited. This was for two reasons. First, virtually the only technique that was capable of retrieving structural information of crystallographic quality was low-energy electron diffraction (LEED). Further, the quoted accuracy of LEED determinations at that time was usually no better than ± 0.1 Å. This implied adsorption induced bond-length changes of less than ± 0.1 Å were hidden from the surface crystallographer's view. Second, the number of determined surface structures was relatively small. In fact, Van Hove's 1979 survey listed only 25 determined adsorption structures. This represented a small "data-base" from which to attempt to extract meaningful structural trends.

During the intervening fifteen years, great progress has been made in overcoming these limitations. Many different and complementary surface structural techniques have been developed and the quality of LEED determinations has been improved significantly. This means that the contemporary surface crystallographer can bring to bear several different techniques to determine the structure of a surface. Now, many structural parameters can

3

be determined with an accuracy that often exceeds a few hundredths of an angstrom. In addition, the simple progress of time has allowed the accumulation of many more distinct structural determinations. In this chapter, we report the structural parameters for almost 400 different surfaces including 257 adsorption structures; a number which is almost an order of magnitude larger than that available in 1979.

In this chapter, we provide a compilation of determined surface structures as of December 1993. The foundation of this contribution is a sequence of tables that collect together the relevant structural information concerning closely related surface structures. Given the diversity of surface structure, we needed to select an appropriate method of organizing and classifying the structures. We have chosen to take an approach based upon the concept of periodicity by which surface structures of elements or adsorbates belonging to the same group of the periodic table are considered together.

Whilst the tables provide the primary means of conveying structural information, we have provided a brief commentary that discusses the structural trends for each group of related surfaces. For a more extensive interpretation, the reader is referred to the other chapters of this volume or to the original publications. Whilst the presentation of structural data in the form of tables is convenient, in some cases the diversity of surface structures makes the tabular format inappropriate. This is especially true of extensively reconstructed surfaces such as semiconductors and compounds. In these cases, the structural details are described in the text.

For a given surface structure, it is common to find many repeat determinations of the same system. For example, $Ni(100)c(2 \times 2)$–O has been the subject of no fewer than twenty independent determinations. Rather than list the raw results of all the determinations, we have taken a more discriminating, but less comprehensive, approach and have attempted to select just one, representative, determination for each distinct surface structure. The primary criteria used in selecting a particular determination were that it be recent, of relatively high accuracy, and represent a consensus of several contemporary determinations. Inevitably this means that some currently controversial, but correct, surface structures may have been omitted from this survey. For this reason, the reader wishing to obtain detailed information about one particular surface structure is encouraged to consult one of the surface structural data-bases or review articles listed in the selected bibliography found at the end of this chapter.

This chapter is organized as follows. First, in sect. 2, we consider the surfaces of metals. In sect. 2.1 we describe the structure of unreconstructed clean metal surfaces and then proceed, in sect. 2.2, to consider the reconstructed surfaces. The surface structure of ordered and disordered metallic alloys is described in sect. 2.3. In sect. 2.4 we describe the surface structures associated with atomic adsorption on metals and in sect. 2.5 we consider molecular adsorption on metals. The structure of semiconductor surfaces is

discussed in sect. 3. The surfaces of elemental and compound semiconductors are considered separately in sections 3.1 and 3.2, as is atomic adsorption
on these surfaces: sections 3.1.2 and 3.2.2. The surfaces of graphite and
diamond are considered in sect. 4 followed by a discussion of the surface
structures of carbides (5.1), silicides (5.2), oxides (5.3.1) and disulfides/
diselenides (5.3.2). Finally, in sect. 6, we give a selected bibliography of
contemporary reviews of surface structure.

2. Metals

2.1. Unreconstructed surfaces of metals

The unreconstructed low Miller index surfaces of fcc, bcc and hcp metals are
illustrated schematically in fig. 1. The primary structural feature associated

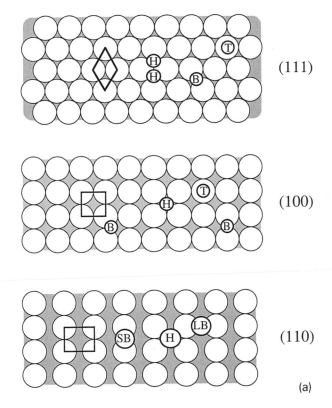

Fig. 1. (a) Top view of the fcc(111), (100) and (110) surfaces showing the surface unit cell
(bold lines) and possible high symmetry adsorption sites. The adsorption sites are: B: bridge
site, LB: long-bridge site, SB: short-bridge site, T: top site. On the (111) surface H denotes
one of two possible three-fold hollow sites; the fcc-hollow or the hcp-hollow. On the (100)
and (110) surfaces H denotes the four-fold and two-fold hollow sites respectively.

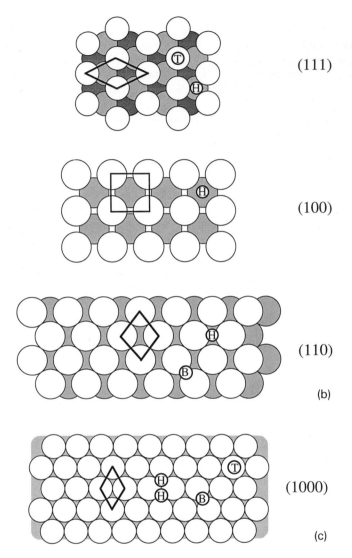

(111)

(100)

(110)

(b)

(1000)

(c)

Fig. 1 (contd.) (b) Top view of the bcc(111), (100) and (110) surfaces showing the surface unit cell (bold lines) and possible high symmetry adsorption sites. The adsorption sites are: B: bridge site, H denotes the hollow site. On both the (111) and (100) surfaces the preferred hollow site is the one in which the adatom sits directly above the second layer substrate atom. (c) Top view of the hcp(1000) surface showing the surface unit cell (bold lines) and possible high symmetry adsorption sites. The adsorption sites are: B: bridge site, H denotes the hollow site.

with unreconstructed surfaces of metals is the relaxation of the atomic planes at the selvedge. The relaxations of determined metallic surface structures are compiled in table 1.

Table 1

Structure of unreconstructed metal surfaces. ∂d_{12} is the relaxation of the first interplanar spacing expressed as a percentage of the bulk interplanar spacing. A positive value implies an expansion of the first interplanar spacing, a negative value implies a contraction. ∂d_{23}, ∂d_{23} and ∂d_{34} are the corresponding relaxations of deeper layers.

Element	Miller index	∂d_{12} (%)	∂d_{23} (%)	∂d_{34} (%)	∂d_{45} (%)	Reference
Divalent metals (IIA and IIB)						
Be	(1000)	$+5.8 \pm 0.4$	-0.2 ± 0.5	$+0.2 \pm 0.5$		Davis, 1992
						Feibelman, 1992
Zn	(1000)	-2.0				Unertl and Thapliyal, 1975
Cd	(100)	0.0				Shih et al., 1977a
Trivalent metals (IIIA and IIIB)						
Al	(111)	$+1.0 \pm 0.5$				Nielsen and Adams, 1982
	(100)	$+1.5$				Masud et al., 1983
		$+1.5$				Noonan and Davis, 1993
		$+1.2 \pm 0.4$	$+0.2 \pm 0.4$	-0.1 ± 0.4		Bohnen and Ho, 1988
	(110)	-8.5 ± 1.0	$+5.6 \pm 1.2$	$+2.3 \pm 1.3$	$+1.7 \pm 1.5$	Noonan and Davis, 1984
		-9.1 ± 1.0	$+4.9 \pm 1.0$	-1.7 ± 1.2	$+0.0 \pm 1.3$	Andersen et al., 1984
	(311)	-8.7 ± 0.8	$+8.8 \pm 1.6$			Noonan et al., 1985
	(331)	$-12.0 \pm 2.$	-4.0 ± 3.0	$+10.4 \pm 3.$	-5.1 ± 4.0	Adams and Sorensen, 1986
Sc	(1000)	-2.0				Tougaard et al., 1982
Transition metals (IVB)						
Ti	(1000)	-2.0 ± 0.8				Shih et al., 1977a
Zr	(1000)	-1.0 ± 2.0				Moore et al., 1979
Transition metals (VB)						
V	(100)	-6.6 ± 0.7	$+1.3 \pm 0.7$			Jensen et al., 1982
	(110)	-0.5 ± 0.5				Adams and Nielsen, 1981
						Adams and Nielsen, 1982
Ta	(100)	-11.0 ± 2.0	$+1.0$			Titov and Moritz, 1982
Transition metals (VIB)						
Mo	(100)	-9.5 ± 2.0	$+1.0 \pm 2.0$			Clarke, 1980
	(110)	-1.5 ± 2.0				Morales et al., 1981
W	(110)	0.0 ± 4.0				Van Hove et al., 1976
Transition metals (VIIB)						
Re	(1010)	-16.0	0.0			Davis and Zehner, 1980
Transition metals (VIII)						
Fe	(100)	-1.6 ± 2.8				Legg et al., 1977
	(110)	$+0.5 \pm 2.0$				Shih et al., 1980
	(111)	-16.1 ± 3.0	-9.3 ± 3.0	$+4.0 \pm 3.6$	-2.1 ± 3.6	Sokolov et al., 1986a
	(210)	-21.9 ± 4.6	-10.9 ± 4.6	-4.7 ± 4.6	0.0 ± 4.6	Sokolov et al., 1985
	(211)	-10.3 ± 2.6	$+5.0 \pm 2.6$	-1.7 ± 3.4		Sokolov et al., 1984a
	(310)	-16.1 ± 3.3	$+12.6 \pm 3.3$	-4.0 ± 4.4		Sokolov et al., 1984b
Ru	(1000)	-2.0 ± 1.0				Michalk et al., 1983
Co	(100)	-4.0				Maglietta et al., 1977
	(111)	0.0 ± 2.5				Lee et al., 1978
	(1000)	0.0 ± 2.5				Lee et al., 1978

Table 1 (contd.)

Ele-ment	Miller index	∂d_{12} (%)	∂d_{23} (%)	∂d_{34} (%)	∂d_{45} (%)	Reference
Transition metals (VIII) (contd.)						
Co	(1120)	-9.0 ± 3.0				Welz et al., 1978
Rh	(111)	0.0 ± 4.5				Van Hove and Koestner, 1984
	(100)	$+0.5 \pm 1.0$	0.0 ± 1.5			Oed et al., 1988c
	(110)	-6.9 ± 1.0	$+1.9 \pm 1.0$			Nitchl, 1987
	(311)	-14.5 ± 2.0	$+4.9 \pm 2.0$	-1.9 ± 2.0		Liepold et al., 1990
Ir	(111)	-2.6 ± 4.5				Chan et al., 1977
Ni	(111)	-1.2 ± 1.2				Demuth et al., 1975b
	(100)	-1.1 ± 2.0				Oed et al., 1989b
	(110)	-8.6 ± 0.5	$+3.5 \pm 0.5$	-0.4 ± 0.7		Adams et al. 1985a
		-9.8 ± 1.6	$+3.8 \pm 1.6$			Xu and Tong, 1985
		-9.0 ± 1.0	$+3.5 \pm 1.5$			Yalisove, 1986
	(311)	-15.9 ± 1.0				Adams et al., 1985b
Pd	(111)	-0.9 ± 1.3	-3.5 ± 1.3			Ohtani et al., 1987
	(100)	$+3.0 \pm 1.5$	-1.0 ± 1.5			Quinn et al., 1990
	(110)	-5.8 ± 2.0	$+0.7 \pm 2.0$			Barnes et al., 1985
Pt	(111)	0.0 ± 2.2				Hayek et al., 1985
		$+1.1 \pm 4.4$				Adams et al., 1979
		$+1.4 \pm 1.0$				Van der Veen, 1979
Noble metals						
Cu	(111)	-0.7 ± 1.0				Lindgren et al., 1984
	(100)	-1.1	$+1.7$	$+1.5$		Davis and Noonan, 1983
	(110)	-9.2	$+2.4$			Davis and Noonan, 1983
	(110)	-5.3 ± 1.5	$+3.3 \pm 1.5$			Stensgaard et al., 1983
	(110)	-8.5 ± 0.6	$+2.3 \pm 0.7$			Adams et al., 1983
	(311)	-9.2	$+2.4$			Streater et al., 1978
Ag	(111)	0.0 ± 5.0				Culberston et al., 1981
	(110)	-7.6 ± 3.0	$+4.2 \pm 3.0$			Kuk and Feldman, 1984

The majority of clean metal surfaces display a contraction of the spacing between the first and second atomic planes. However, there are several surfaces that exhibit the opposite behavior; an expansion of the first interplanar spacing. Table 1 allows us to identify seven surfaces for which an expansion is implicated; Be(1000), Al(111), Al(100), Fe(110), Rh(100), Pd(100), Pt(111). Of these seven surfaces, the error bars for the determinations of the first interlayer spacing of Rh(100), Fe(110) and Pt(111) do not allow us to conclusively ascribe an expansion to these surfaces. The authors of the Pd(100) determination note that the surface could be contaminated with hydrogen. This leaves only the group II and III metals Be and Al displaying a reproducible expansion of the top layer spacing.

The relaxation of deeper layers displays oscillatory behavior. Although many low Miller index surfaces exhibit strictly alternating relaxations (i.e. a contraction of the top layer spacing followed by an expansion of the second layer spacing, etc.), this behavior is not a general feature of surface structures of metals. More complex oscillatory behavior is observed for higher Miller index (stepped) surfaces which may also feature lateral motions of atom chains parallel to the surface.

Surface structures of clean metals, being the most studied of all surface structures, nicely illustrate the degree of reproducibility and accuracy achievable by modern surface crystallography. For example, three independent determinations of the first and second interlayer spacings of Ni(110) (see table 1) agree within 1.2% of the bulk interlayer spacing or 0.015 Å. Similarly, three independent determinations of the first and second interlayer spacings of Cu(110) (see table 1) agree within 3.9% of the bulk interlayer spacing or 0.05 Å.

2.2. Reconstructed surfaces of metals

The (110) surfaces of the transition metals Au, Ir, Pt display both a stable (1×2) and a metastable (1×3) reconstruction. The structural details of these reconstructions are listed in table 2 and the (1×2) reconstructed surface is illustrated in fig. 2. Both the (1×2) and (1×3) reconstructions are of the missing-row type which involve the "removal" of every second (1×2) or third (1×3) row of atoms from the top atomic plane of the bulk termination. The removal of this row is accompanied by significant atomic relaxations of at least the first three atomic planes perpendicular to the surface, see table 2. Both Pt(110)(1×2), Pt(110)(1×3) and Au(110)(1×2) have relaxations of a similar magnitude but the relaxations of Ir(110)(1×2) are significantly smaller. In addition to the planar relaxations, all of these surfaces exhibit lateral motions of the atoms within the second atomic plane out towards the valleys left by the missing rows. In addition, the removal of the atomic row causes a buckling of the third atomic layer which conforms with the "hill" and "valley" structure of the missing-row surface.

The missing row reconstruction may be induced in ordinarily unreconstructed fcc metals by driving electrons into the surface region, either electrochemically or by alkali-metal adsorption. For example, a (1×2) missing row reconstruction of Cu(110) and Pd(110) may be created by K and Cs adsorption (Barnes et al., 1985; Hu et al., 1990). The structural parameters of these surfaces are included in table 2 and show the smaller normal relaxations that are qualitatively similar to the stable missing-row forms of Ir(110).

The Ir(100)(1×5) reconstruction is caused by a lateral distortion of the top layer of Ir atoms along the (10) direction (Lang et al., 1983). This distortion allows the top layer of Ir atoms to form a quasi-hexagonal two-dimensional lattice which is commensurate with the underlying (100) plane formed by the

Table 2

Structural parameters determined for fcc(110)(1×2) and (1×3) missing row type reconstructions. ∂d_{12z}, ∂d_{23z} and ∂d_{34z} are the relaxations of the corresponding interlayer spacings expressed as a percentage of the bulk interplanar spacing of the unreconstructed surface. Δl_{2x} is the lateral displacement of the second layer atoms towards the missing row (+ is towards the missing row). b_3 is the normal buckling amplitude of the third atomic layer. The Cu and Pd reconstructions are induced by alkali metal adsorption and are not the stable structures of these surfaces.

Surface	Sym.	∂d_{12z} (%)	∂d_{23z} (%)	∂d_{34z} (%)	Δl_{2x} (Å)	b_3 (Å)	Reference
Ir	(1×2)	-13.0 ± 5.0	-12.0 ± 5.0	$+3.0 \pm 5.0$	$+0.02$	0.23 ± 0.07	Chan and Van Hove, 1986
	(1×3)	-8.0	0.0		$+0.04 \pm 0.01$		Shi et al., 1990
Pt	(1×2)	-21.0 ± 3.5	-6.0 ± 3.5		$+0.02 \pm 0.05$	0.03 ± 0.05	Fery et al., 1988
	(1×3)	-21.0 ± 3.5	-5.0 ± 3.5		$+0.03 \pm 0.04$	0.18 ± 0.05	Fery et al., 1988
Au	(1×2)	-20.0 ± 3.5	-6.3 ± 3.5	$+2.0 \pm 3.5$	$+0.03 \pm 0.03$	0.24 ± 0.05	Moritz and Wolf, 1985
Pd (ind.)	(1×2)	-5.0 ± 2.0					Barnes et al., 1985
Cu (ind.)	(1×2)	-12.0 ± 4.0	0.0 ± 4.0		$+0.05$		Hu et al., 1990

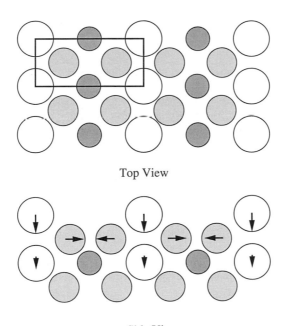

Top View

Side View

Fig. 2. Schematic illustration of the fcc(110) missing-row reconstruction. The bold rectangle identifies the (1×2) unit cell. The arrows indicate the direction of the atomic relaxations relative to the bulk termination.

second layer atoms. Since each top layer Ir atom in the (1×5) surface unit cell cannot occupy the continuation (hollow) site of the underlying lattice, the top Ir layer is buckled significantly with an amplitude of 0.48 Å. A similar quasi-(1×5) reconstruction occurs for Pt(100) although this surface has not be the subject of a complete structural analysis (Van Hove et al., 1981).

Below room temperature, the W(100)c(2×2) reconstructed surface is created by lateral movements of the W atoms in first W layer which propagate into at least the second layer of the surface (fig. 3). Alternate atoms move along the (011) direction to form zig-zag chains. A LEED structural study (Pendry et al., 1988) determines the amplitude of the lateral movements to be 0.24 ± 0.04 Å in the top W layer and 0.028 ± 0.007 Å in the second W layer. The top layer relaxes into the surface by $-7.0 \pm 2.0\%$ of the bulk interlayer spacing, the second layer spacing expands by $+1.2 \pm 2.0\%$. These structural parameters are in reasonable agreement with a recent X-Ray diffraction (XRD) determination (Altmann et al., 1988) which finds that the amplitude of the lateral movements is 0.24 ± 0.05 Å in the top W layer and 0.10 ± 0.05 Å in the second W layer. By XRD, the top layer is found to relax into the surface by $-4.0 \pm 1.0\%$ of the bulk interlayer spacing.

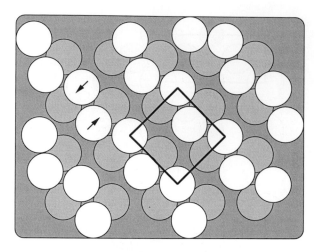

Fig. 3. Schematic illustration of the W(100)-c(2×2) reconstruction. Note the formation of zigzag rows by the atomic displacements indicated by the arrows.

2.3. Surfaces of metallic alloys

Metallic alloys may be divided into two types; those which form ordered bulk phases (such as NiAl) and those which are substitutionally disordered in the bulk. Table 3 presents the structural information for the surfaces of alloys which are disordered in the bulk. In addition to the relaxation of the atomic planes observed in clean monatomic metal surfaces, alloys possess an additional degree of structural freedom; the segregation profile at the selvedge. Table 3 shows that the majority of alloy surfaces display a significant deviation from the bulk composition in the first three or four atomic layers. For example, the surfaces of the PtNi alloys display segregation of Pt into the first atomic layer. Table 4 presents the analogous structural information for alloys which form ordered bulk phases.

2.4. Atomic adsorption on metals

2.4.1. Hydrogen

The adsorption geometry of hydrogen chemisorbed on metal surfaces has been determined primarily by LEED and high-resolution electron energy-loss spectroscopy (HREELS) and is tabulated in table 5. An extensive review of the interaction of hydrogen with solid surfaces has been given by Christmann (1988). The small scattering cross section of hydrogen makes the determination of the hydrogen position by LEED difficult; many LEED studies ignore the hydrogen scattering in the calculation of LEED IV spectra

Table 3

Compilation of structural parameters for the surfaces of unreconstructed disordered metallic alloys. C1–C4 are the percentage of atom type A in the corresponding layer of the bulk alloy AB. Atom type A is the first element listed in the alloy column of the table. ∂d_{12} is the change in the first interplanar spacing expressed as a percentage of the (mean) bulk interplanar spacing of the disordered alloy. ∂d_{23} and ∂d_{34} are the equivalent quantities for deeper layers.

Alloy	Miller index	C1 (%)	C2 (%)	C3 (%)	C4 (%)	∂d_{12} (%)	∂d_{23} (%)	∂d_{34} (%)	Reference
$Pt_{78}Ni_{22}$	(111)	99±1	30±5	87±10		−1.8±1.0	−1.8±1.0		Gauthier et al., 1985
$Pt_{50}Ni_{50}$	(111)	88±2	9±5	65±10		−2.0±1.0	−2.0±1.0		Gauthier et al., 1985
	(100)	86±10	24±10	65±10		−4.6±1.0	−9.0±3.0		Gauthier et al., 1985
	(110)	0±6	95±4	17±7	48±13	−19.2±0.6	10.5±1.0		Gauthier et al., 1987
$Pt_{10}Ni_{90}$	(111)	30±4	4±3			0.0±0.2	−0.8±0.9		Baudoing et al., 1986
	(100)	24±3	6=3			+2.0±1.0	−1.2±1.0	+1.61.0	Gauthier et al., 1990a
	(110)	6±4	52=2	10±10		−4.5±0.7	−3.6±1.1	+0.20.6	Gauthier et al., 1989
$Pt_{80}Fe_{20}$	(111)	96±2	88=3	84±10		+0.3±0.5	−0.6±1.1		Beccat et al., 1990
	(110)[1]	82	84	68	81	−13.0±7.0	+10.7±7.0		Baudoing-Savois et al. 1991
$Cu_{84}Al_{16}$	(111)[2]	66	84	84					Baird et al., 1986
Cu_3Au	(100)[3]	50	65	75					Stuck et al., 1991
$Cu_{85}Pd_{15}$	(110)[4]	70	50	100	85	−4.7	0.8		Lindroos et al., 1991

[1] This surface displays a missing row (2×1) reconstruction the structural details of which resemble Pt(110)(1×2).

[2] This surface displays a ($\sqrt{3}\times\sqrt{3}$)R30° pattern and has an ordered first bilayer.

[3] This is an ordered bulk alloy with a disordered surface segregation profile.

[4] This surface has an ordered second layer giving rise to a (2×2) LEED pattern.

Table 4

Compilation of structural parameters for the surfaces of ordered metallic alloys AB. Atom type A is the first element listed in the alloy column of the table. ∂d_{12} is the change in the first interplanar spacing expressed as a percentage of the (mean) bulk interlayer spacing of the disordered alloy. ∂d_{23} and ∂d_{34} are the equivalent quantities for deeper layers. b_1 and b_2 are the buckling amplitudes in the first and second layers respectively; a positive value implies that atom A moves out of its plane towards the surface.

Alloy	Miller index	∂d_{12} (%)	∂d_{23} (%)	∂d_{34} (%)	b_1 (Å)	b_2 (Å)	Reference
NiAl	(111)[1]	−50.0 ± 6.0	+15 ± 6				Noonan and Davis, 1987
	(111)[2]	−5.0 ± 6.0	+5.0 ± 6				Noonan and Davis, 1987
	(100)[3]	−8.5 ± 3.5	+4.0 ± 3.5				Davis and Noonan, 1988
	(110)	−4.6 ± 1.0	+1.0 ± 1.0				Davis et al., 1988
Ni$_3$Al	(111)	−0.5 ± 1.5			−0.20 ± 0.02	−0.02 ± 0.02	Sondericker et al., 1986a
	(100)[4]	−2.8 ± 1.7			−0.06 ± 0.03		Sondericker et al., 1986b
	(110)[4]	−12.0 ± 2.5	+3.0 ± 2.5		−0.02 ± 0.03		Sondericker et al., 1986c

[1] NiAl(111) consists of Ni and Al terminated domains. This structure is for the Ni terminated surface.
[2] NiAl(111) consists of Ni and Al terminated domains. This structure is for the Al terminated surface.
[3] NiAl(100) consists of a stack of alternating Ni and Al planes. The Al termination is favored.
[4] Top layer is NiAl, second 100% Ni; third NiAl etc.

Table 5

Compilation of structural parameters for H chemisorption systems. Δd_{12} is the percentage change in the first interlayer spacing of the substrate computed with respect to the bulk interplanar spacing normal to the surface. The H radius is computed by subtracting the metallic radius of the substrate atom from the derived M–H bond length.

Substrate	Overlayer	Site	Adsorption height (Å)	M–H bond length (Å)	Δd_{12} (%)	H-radius (Å)	Reference
Fe(110)	p(2×1)	3-fold	0.90±0.10	1.75±0.05		0.49±0.05	Moritz et al., 1985
	(3×1)–2H	3-fold	0.90±0.10	1.75±0.05		0.49±0.05	Kleinle et al., 1987
Rh(110)	(1×1)–2H	3-fold	0.78±0.10		−1.9±0.10		Oed et al., 1988b
	p(1×2)–3H	3-fold	0.71±0.10 (H1)	1.87±0.10	−3.8±1.0	0.53±0.10	Michl et al., 1989
			1.00±0.10 (H2)	1.93±0.10		0.59±0.10	
			1.15±0.10 (H3)	1.90±0.10		0.56±0.10	
Ru(0001)	p(2×1)–H	4-fold hol.	1.34±0.20	2.00±0.20	+1.0±3.0	0.66±0.20	Held et al., 1992
		4-fold hol.	0.90±0.15	1.91±0.15	−2.0	0.57±0.15	Lindroos et al., 1987
Ni(111)	c(2×2)	3-fold	1.15±0.05	1.84±0.06		0.59±0.06	Christmann et al., 1979
Ni(110)	p(2×1)–2H	3-fold	0.41±0.10	1.72±0.10	−4.5±1.5	0.48±0.10	Reimer et al., 1987
Pd(110)	(2×1)p2mg	3-fold	0.60±0.05	2.00±0.10		0.63±0.10	Skottke et al., 1988
Pt(111)	(1×1)	3-fold	1.00	1.90		0.52	Batra et al., 1984
	(1×1)	3-fold	0.71	1.76		0.38	Baro et al., 1979

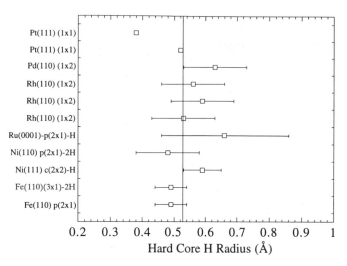

Fig. 4. Plot of the hard core radius of the H atom determined for various chemisorption systems (see table 5). The H radius is obtained by subtracting the metallic radius of the substrate atom from the derived metal-hydrogen bond length.

and therefore do not determine the adsorption geometry of the adsorbate.

In the systems examined, the hydrogen atom tends to occupy sites with high local coordination to the substrate atoms of low Miller index surfaces (see fig. 1). The H–metal bond length obtained from these studies ranges from 1.72 ± 0.10 Å for Ni(110)–H to 2.00 ± 0.20 Å for Ru(0001)–H, although the error bars of these analyses do not allow us to establish a definite trend for metal–H binding at surfaces. The hard-core radius of the hydrogen atom, also tabulated in table 5 and plotted in fig. 4, can be obtained by subtracting the radius of the metal from the determined bond length. Within the error bars, the hydrogen radius is found to be close to the Bohr radius of 0.529 Å. This implies that there is little charge transfer involved in hydrogen chemisorption.

A common feature of hydrogen adsorption is the reduction of the clean surface relaxation of the top layer spacing, especially for the more open fcc (110) surfaces. For example, the top layer relaxation of the clean surface is reduced from -8.5% to -4.5% in Ni(110)(2×1)–2H, from -8.0% to 0% in Cu(110)(1×1)–H, from -6.0% to 2% in Pd(110)(2×1)–2H and from -6% to -2% in W(100)(1×1)–2H. A tabular summary of H-induced relaxations can be found in the review article by Van Hove and Somorjai (1989)

2.4.2. Alkali metals

The adsorption geometry of alkali-metal atoms on metal surfaces has been the subject of study since the earliest days of quantitative surface crystallography. Despite the apparent maturity of this field, alkali-metal adsorption is of considerable current interest. The origin of this interest is twofold.

The first focus of interest is the surface extended X-ray adsorption fine structure (SEXAFS) study of the Ag(111)–Cs system by Lamble and coworkers (Lamble et al., 1988) which was observed to display a coverage dependent bond-length change as a function of coverage. This result was interpreted as an incremental change of the bond character from ionic towards metallic as the density of alkali-atom adatoms was increased. Subsequent, coverage dependent studies of two other alkali-metal adsorbates, Ru(0001)–K and Al(111)–Rb (Kerkar et al., 1992b), failed to demonstrate any coverage dependent bond-length change within the accuracy of the measurement.

Prior to the early 1990s, all structural studies of alkali-metal chemisorption found the adatom located at high coordination sites at which the alkali-metal atom is bound in three- or four-fold hollow sites. A comprehensive survey of alkali-metal adsorption studies prior to 1988 may be found in the book edited by Bonzel (Bonzel et al., 1989). Several more recent LEED, SEXAFS and X-ray studies have implicated low coordination (top) sites, as in the case of Cu(111)p(2×2)–Cs, or substitutional behavior. These results may signal that the current understanding of the alkali-metal bonding at surfaces is incomplete.

In table 6, we list the structural results obtained for the adsorption geometry of alkali-metal atoms at metal surfaces. In this table we have listed the effective radius of the adsorbed alkali-metal atom calculated by subtracting the metallic radius of the substrate atoms from the determined bond length. The result is expressed as a fraction of the metallic radius of the alkali-metal atom. Such a procedure gives an indication of the radius of the alkali-metal atom which allows comparison between alkali-metal adsorption on different substrates. At best, such a number is semi-quantitative because the effective radius of the substrate atoms depends upon the nature of the bond formed with the adsorbate. Nevertheless, with the exception of Al(111)($\sqrt{3} \times \sqrt{3}$)R30°–K, there is a general tendency for the effective radii of adsorbed alkali-metal atoms to be significantly smaller than the metallic radius. Further, alkali-metals atoms which occupy top sites appear to have significantly smaller radii than those systems in which the adatom occupies a high-coordination site. The radii of alkali metals adsorbed at top sites are, in fact, close to their respective ionic radii, which are approximately 50–60% of the metallic radius.

Table 6

The adsorption geometry of alkali-metal atoms chemisorbed on metal surfaces. The alkali metal to substrate bond length is derived from the determined coordinates. The adatom radius is obtained by subtracting the metallic radius of the substrate atom from the determined bond length. The adatom radius is expressed as the ratio of the adatom radius to the metallic radius of the adatom.

Substrate	Overlayer	Site	Adsorption height (Å)	M–A bond length (Å)	adatom radius (units of r metallic)	Reference.
Sodium (Na)						
Al(100)	c(2×2)	4-fold hollow	2.03±0.10	2.89±0.08	0.77±0.04	Hutchins et al., 1976 Van Hove et al., 1976
Al(111)	(√3×√3)R30°	3-fold substit.	1.67±0.03	3.31±0.03	0.99±0.02	Schmalz, 1991
Ni(100)	c(2×2)	4-fold hollow	2.23±0.10	2.83±0.08	0.83±0.05	Demuth et al., 1975a
Potassium (K)						
Ni(111)	p(2×2)	3-fold top	2.82±0.04	2.82±0.04	0.66±0.02	Fisher et al., 1992
Ru(0001)	(√3×√3)R30°	hcp hollow	2.94±0.03	3.29±0.05	0.83±0.02	Gierer et al., 1991
	p(2×2)	fcc hollow	2.90±0.03	3.25±0.05	0.81±0.02	Gierer et al., 1992
Co(10$\bar{1}$0)	c(2×2)	4-fold hollow	2.44±0.05	3.12±0.05	0.79±0.02	Barnes et al., 1991
Au(110)	c(2×2)	2-fold substit.	1.05±0.15	3.07±0.07	0.68±0.03	Haberle and Gustafsson, 1989
Ni(100)	c(4×2)	4-fold hollow	2.68±0.05	3.20±0.05	0.83±0.02	Muschiol et al., 1992
Rubidium (Rb)						
Al(111)	(2×2) to (√3×√3)R30°	3-fold top	3.13±0.10	3.13±0.10	0.70±0.04	Kerkar et al., 1992b
Caesium (Cs)						
Cu(111)	p(2×2)	3-fold top	3.01±0.05	3.01±0.05	0.63±0.02	Lindgren et al., 1983
Ag(111)	dis. 0.3 ml	3-fold hollow	3.07±0.03	3.50±0.03	0.77±0.01	Lamble et al., 1988
	dis. 0.15 ml	3-fold hollow	2.73±0.03	3.20±0.03	0.66±0.01	Lamble et al., 1988
Rh(100)	c(4×2)	4-fold hollow	2.87±0.06	3.44±0.06	0.77±0.02	von Eggling et al., 1989

2.4.3.   Group IVA chemisorption on metals

2.4.3.1.   Carbon

Carbon chemisorption on metals has been studied on the substrates Ni(100), Mo(100) and Zr(1000). Although studies of atomic C chemisorption are few, the resulting structures are diverse and are tabulated in table 7.

Of these three systems, C on Ni(100) forms a p4g-c(2×2)–2C overlayer by inducing an unusual reconstruction of the substrate (see fig. 5). In the Ni(100)p4g-c(2×2)–2C structure, the C adatoms occupy equivalent hollow sites in which the adjacent 4 Ni atoms in the top substrate layer undergo a clockwise rotation about the adsorbate. The second Ni layer is buckled by 0.15 Å with the Ni atoms directly below the C adatom being pulled out of the surface. There is a significant expansion of the top layer spacing induced by C adsorption. There are two relevant Ni–C bond lengths in this structure. The Ni–C distance between the adatom and the top layer Ni atoms is 1.82 ± 0.03 Å, the distance between the adatom and the second layer Ni atom directly underneath is 1.95 ± 0.03 Å. This latter distance may be compared to the sum of the covalent radii of Ni and C which is 1.92 Å.

C adsorption on the open Mo(100) surface results in conventional hollow site adsorption in which the shortest C–Mo distance is between the adatom and the second layer Mo atom directly below the adsorption site. This C–Mo bond length, 1.99 ± 0.05 Å, is significantly shorter than the C–Mo distance between the adatom and top-layer Mo atoms; 2.27 ± 0.03 Å. The sum of the covalent radii of C and Mo is 2.07 Å.

C adsorption on Zr(1000) involves the occupation by C of octahedral interstitial sites halfway between the first and second Zr layers. The resulting structure resembles that of bulk ZrC which involves the insertion of the C

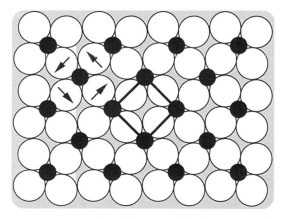

Fig. 5. Schematic illustration of the Ni(100)-p4g-c(2×2)–2C reconstruction. Note the rotational reconstruction of the substrate indicated by the arrows.

Table 7

Structural parameters for C and Si chemisorption on metals. δd_{12} is the relaxation of the first substrate interlayer spacing expressed as a percentage of the bulk interplanar spacing. The bond length is derived from the determined coordinates and is presented with the sum of the covalent radii of the two species.

Substrate	Overlayer	Site	Adsorption height (Å)	δd_{12} (%)	Bond length (Å)	Sum of cov. radii (Å)	Reference
Carbon							
Ni(100)	p4g-c(2×2)	hollow	0.12±0.04	+8.0±4.0[1]	1.82±0.03	1.92	Gauthier et al., 1991 / Kilcoyne et al., 1991
Mo(100)	c(2×2)	4-fold hollow	0.43±0.05	-1.0±3.0	1.99±0.05	2.07	Rous et al., 1991
Zr(1000)	(1×1)	interstitial	-1.33±0.10	+0.7±4.0	2.22±0.07	2.29	Wong and Mitchell, 1988
Silicon							
Mo(100)	(1×1)	4-fold hollow	1.61		2.51	2.47	Ignatiev et al., 1975a

[1] The second Ni layer is bucklec (see text). This interlayer expansion is measured from the center-of-mass plane of the second Ni layer.

Table 8

Structural parameters for N chemisorption on metals. δd_{12} is the relaxation of the first substrate interlayer spacing expressed as a percentage of the bulk interplanar spacing. The bond length is derived from the determined coordinates and is compared to the sum of the covalent radii of the adatom and metal atom.

Substrate	Overlayer	Site	Adsorption height (Å)	δd_{12} (%)	Bond length (Å)	Sum of cov. radii (Å)	Reference
Ti(1000)	(1×1)	interstitial	-1.22±0.05	+4.3±2.0	2.10±0.05	2.02	Shih et al., 1976
Zr(1000)	(1×1)	interstitial	-1.30±0.05	-1.5±3.7	2.27±0.05	2.15	Wong and Mitchell, 1987
Cr(100)	(1×1)	4-fold hollow	0.22±0.02	+25.0±0.5	2.04±0.02	1.92	Joly et al., 1989
Mo(100)	c(2×2)	4-fold hollow	1.02		2.45	2.00	Ignatiev et al., 1975b
W(100)	c(2×2)	4-fold hollow	0.49±0.06	+1.3	2.09±0.06	2.00	Griffiths et al., 1982
Fe(100)	c(2×2)	4-fold hollow	0.27±0.05	+7.7±3.5	1.81	1.87	Imbihl et al., 1982
Ni(100)	p4g-(2×2)	hollow + recon.	0.10±0.12	+7.3	1.85	1.85	Kilcoyne et al. 1991
Cu(100)	c(2×2)	4-fold hollow	0.06	+8.0	1.81	1.87	Zeng and Mitchell, 1989

atoms into the interstices of the close packed metallic lattice. The Zr–C bond distance is 2.29 ± 0.07 Å; the sum of the covalent radii of Zr and C is 2.22 Å

2.4.3.2.   Silicon

Silicon chemisorption on metals has been studied in only one case; an early LEED study of a (1×1)–Si overlayer on Mo(100) (see table 7). The Si occupies 4-fold hollow sites in which the shortest Si–Mo distance, 2.51 Å, is between the adatom and the Mo atoms in the top layer. This agrees well with the sum of the covalent radii; 2.49 Å. It is interesting to compare Si adsorption on Mo(100) to C adsorption on the same substrate. Since the C adatom is significantly smaller than Si (the covalent radii are 0.77 Å and 1.17 Å respectively), the C adatom is able to sit much deeper in the hollow site, actually forming a bond to the second layer Mo atom. This is apparent from the relative adsorption heights of the two species which are 0.12 Å and 1.61 Å for C and Si, respectively.

2.4.4.   Group VA chemisorption on metals: nitrogen

The structural parameters for N chemisorption on metals are compiled in table 8. The structural aspects of N chemisorption on metal surfaces is dominated by the small size of the N atom. The covalent radius of N is 0.70 Å which allows the N atom to penetrate into the hollow sites of the lower Miller index surfaces. For adsorption on two fcc (100) surfaces, Cu and Ni, the shortest N–M bond length is formed between the top layer substrate atoms and the adatom (Kilcoyne et al., 1991; Zeng and Mitchell, 1989). For Cr(100)(1×1)–N, the N–Cr distance between the N and Cu atoms in the first layer and the second layer is almost identical; 2.04 Å and 2.07 Å respectively (Joly et al., 1989). On the more open bcc(100) surfaces of Mo, Fe, and W, the N atom is able to sit directly above the second layer substrate atom with which the shortest bond is formed. In all cases (except interstitial site occupation and Mo(100)c(2×2)–N, the metal–N bond length is in good agreement with the simple sum of covalent radii. N adsorption on Mo(100) is an exception, with a determined bond length from an early LEED study (Ignatiev et al., 1975b) which is significantly larger.

The Cr(100)(1×1)–N structure involves an anomalous expansion of the first Cr interplanar spacing by +25%, although bucklings of the second Cr layer (as are found in N adsorption on Cu(100)) were not investigated. The Cu(100)c(2×2)–N overlayer structure displayed adsorbate induced buckling of the second Cu layer with an amplitude of 0.09 ± 0.02 Å. The Cu atoms directly below the N adatom are pushed into the surface. The shortest Cu–N bond length, 1.81 Å, is between the adatom and the top layer Cu atoms. The bond distance between the adatom and the second layer Cu atom directly below it is 2.00 Å. The sum of the covalent radii of N and Cu is 1.87 Å, intermediate between these two values.

Atomic nitrogen adsorbed on Ni(100) induces a rotational reconstruction qualitatively identical to the Ni(100)p 4 g-c(2×2)–2C surface structure. The N adatoms occupy equivalent hollow sites in which the adjacent 4 Ni atoms in the top substrate layer undergo a clockwise rotation about the adsorbate by 0.55 Å. There is a significant expansion of the top layer spacing, +7.3%, induced by N adsorption. There are two relevant Ni–N bond lengths in this structure. The Ni–N distance between the adatom and the top layer Ni atoms is 1.85 Å, the distance between the adatom and the second layer Ni atom directly underneath is 1.99 Å. The former distance may be compared to the sum of the covalent radii of Ni and C which is 1.85 Å.

N adsorption on Ti(1000) and Zr(1000) involves the formation of an intercalation compound by the occupation by N of octahedral interstitial sites halfway between the first and second substrate layers. The resulting structures resembles that of the bulk nitrides TiN/ZrN and involves the insertion of the N atoms into the interstices of the close packed metallic lattice. For Ti(1000)–N, the Ti–N bond distance is 2.10 ± 0.05 Å; the sum of the covalent radii of Ti and N is 2.02 Å. For Zr(1000)–N, the Zr–N bond distance is 2.27 ± 0.05 Å; the sum of the covalent radii of Zr and N is 2.15 Å. These structures are analogous to the Zr(1000)(1×1)–C adsorption system (Wong et al., 1988).

2.4.5. Group VIA (chalcogen) chemisorption on metals

2.4.5.1. Oxygen

Oxygen is the most extensively studied of all atomic adsorbates. In table 9 the structural results obtained for oxygen adsorption on metals are summarized.

During the early 1980s, there were at least four independent determinations of the c(2×2) overlayer phase on Cu(100). However, more recent investigations suggest that this is not the thermodynamically stable phase and is probably a metastable, disordered, version of the stable Cu(100)(2√2×√2)R45° structure. The structure of this phase is tabulated in table 9 and is described here. The Cu(100)(2√2×√2)R45° surface is a missing-row structure in which one of the four nearest-neighbor Cu atoms in the top layer is removed. Lateral motions of the Cu atoms and a 0.1-Å amplitude buckling in both the first and second substrate layers is observed. The Cu atom below the adsorption site is pulled out of the surface. Cu(110)p(2×1)–O and Fe(211)p(2×1)–O are also missing row structures in which the O atom is adsorbed at the long-bridge sites. The study of the adsorption of atomic oxygen on the stepped surface Cu(410) indicates that the adatom adsorbs at the step a distance of 0.39 ± 0.20 Å above the (100) terraces.

Oxygen adsorption on Ni(100) is one of the most extensively studied of all surface systems with almost twenty distinct structure determinations using a virtually every surface structure technique. For the latter part of the 1980s, the lateral position of the O atom in the c(2×2) overlayer was the origin of some controversy. However, the apparently artificial source of the controversial pseudo-bridge site was elucidated in a recent, extensive, LEED study in which substrate buckling (0.35 Å) was found in a detailed structural survey. The Ni(100)p(2×2)–O and -c(2×2)–O overlayer systems are of interest because the determined Ni–O bond lengths of 1.92 ± 0.01 Å and 1.93 ± 0.03 Å are significantly larger than the value of 1.80 ± 0.02 Å observed in other Ni/O adsorption systems. This appears to be a reproducible trend, since all structural determinations of Ni(100)c(2×2)–O and -p(2×2)–O (except the one leading to the pseudo-bridge model) yield a long Ni–O bond length of greater than 1.92 Å.

A shorter Ni–O bond length is found in the Ni(110)p(2×1)–O system which is also found to be a missing row structure analogous to Cu(110)p(2×1)–O. The adatom occupies the long-bridge site. On the Ni substrate, the O atom position in not quite equidistant from the nearest neighbor Ni atoms leading to a pair of Ni–O bond lengths; 1.86 Å and 1.77 Å. Ni(111)($\sqrt{3} \times \sqrt{3}$)R30°–O appears to be a fairly conventional adsorption structure in which the adatom is located at a hollow site with no lateral relaxations or buckling of the substrate. Oxygen adsorption in the lower coverage p(2×2) phase on Ni(111) induces a buckling of the top Ni layer and a small amplitude (0.12 ± 0.06 Å) rotational reconstruction centered upon each of the adatoms.

The p(2×2)–O overlayers on Rh(100) and Rh(111) may be compared to the same oxygen overlayers on Ni(100). The Rh–O bond length found in the Rh(100)p(2×2)–O structure is significantly longer (2.13 ± 0.03 Å) than that found on the Rh(111) substrate (1.98 ± 0.06 Å). This seems to reproduce the trend seen for oxygen adsorption on Ni(100) and Ni(111) surfaces. Like Ni(100)p(2×2)–O and -c(2×2)–O, the Rh(100)p(2×2)–O structure exhibits substrate buckling in the second layer, although the amplitude, only 0.01 Å, is significantly smaller in magnitude than that seen in Ni(100) adsorption structures.

On Ru(1000), the p(2×1)– and p(2×2)–oxygen overlayers have been studied. The O overlayers induce a buckling of both the first and second layers of the substrate. The amplitude of the buckling in the top Ru layer is 0.07 ± 0.04 Å for both the p(2×1)–O and p(2×2)–O overlayers. The amplitude of the buckling of the second Ru layer is 0.01 ± 0.04 and 0.08 ± 0.04 Å for p(2×1) and p(2×2) overlayers, respectively. The Ru–O bond lengths found in the two overlayers are identical within the error bars; the largest difference being the much more extensive buckling of the second substrate layer observed for the lower coverage phase. This decrease of the

Table 9

Structural parameters for group VIA (chalcogen) chemisorption on metals. δd_{12} is the relaxation of the first substrate interlayer spacing expressed as a percentage of the bulk interplanar spacing. The bond length is derived from the determined coordinates and is compared to the sum of the covalent radii of the adatom and metal atom.

Substrate	Overlayer	Site	Adsorption height (Å)	$\delta d_{12}\%$	Bond length (Å)	Sum of cov. radii (Å)	Reference
Oxygen							
Al(111)	(1×1)	fcc hollow	0.70±0.08		1.79±0.05	1.91	Martinez et al., 1983
							Kerkar et al., 1992a
Co(100)	c(2×2)	4-fold hollow	0.80		1.93	1.82	Maglietta et al., 1978
Cu(100)	(2√2×√2)R45°	4-fold	0.15 [1]	+11.0	1.82	1.83	Zeng and Mitchell, 1990
							Asenio et al., 1990
Cu(110)	p(2×1)	long-bridge	0.04±0.03	+16.0±2.0	1.81±0.02	1.83	Parkin et al. 1990
Cu(410)	O and 2O	quasi 4-fold	0.39±0.20		1.85±0.10	1.83	Thompson and Fadley, 1984
Fe(100)	(1×1)	4-fold hollow	0.45±0.04	+8.2±2.8	2.00±0.04	1.83	Jona and Marcus, 1987
Fe(211)	p(2×1)	long bridge	0.26±0.05	-7.0±3.0	2.05±0.05	1.83	Sokolov et al., 1986b
Ir(110)	c(2×2)	short bridge	1.37±0.05	-2.0±5.0	1.93±0.04	1.93	Chan et al., 1978
Ir(111)	p(2×2)	fcc-hollow	1.30±0.05		2.04±0.04	1.93	Chan and Weinberg, 1979
Ni(100)	c(2×2)	4-fold hollow	0.77±0.02	+5.7±1.1	1.92±0.01	1.81	Oed et al., 1989a
Ni(100)	p(2×2)	4-fold hollow	0.80±0.05	+1.0±2.8	1.94±0.03	1.81	Oed et al., 1990
Ni(110)	p(2×1)	long-bridge	0.20	+4.3	1.77 / 1.86	1.81	Kleinle et al., 1990
Ni(111)	(√3×√3)R30°	fcc hollow	1.08±0.02	+0.7±1.0	1.80±0.02	1.81	Mendez et al., 1991
Ni(111)	p(2×2)	fcc hollow	1.15±0.03 [2]	-1.3±1.5	1.80±0.02	1.81	Grimsby et al., 1990
Rh(100)	p(2×2)	4-fold hollow	0.95±0.04	-0.3±2.1	2.13±0.03	1.91	Oed et al. 1988a
Rh(111)	p(2×2)	fcc hollow	1.23±0.09		1.98±0.07	1.91	Wong et al., 1986
Ru(1000)	p(2×1)	hcp hollow	1.25±0.02	-0.8±1.0	2.02±0.02	1.91	Pfnür et al., 1989
Ru(1000)	p(2×2)	hcp hollow	1.21±0.03	-2.2±1.5	2.03±0.02	1.91	Lindroos et al. 1989
Ta(100)	p(3×1)	interstitial	-0.43		1.95	2.00	Titov and Jagodzinski, 1985
W(100)	disordered	4-fold hollow	0.59		2.10	1.96	Rous et al., 1986
W(100)	p(2×1) dis.	top	0.06		2.00	1.96	Mullins and Overbury, 1989
W(110)	p(2×1)	3-fold hollow	1.25±0.03		2.08±0.02	1.96	Van Hove and Tong, 1975a
Zr(1000)	(2×2)	interstitial	-1.37±0.05	0.0±0.05	2.31±0.03	2.11	Hui et al., 1985

Sulfur

Co(100)	c(2×2)	4-fold hollow	1.30		2.20	2.20	Maglietta, 1982
Cr(100)	c(2×2)	4-fold hollow	1.17±0.02		2.35±0.02	2.22	Terminello et al., 1988
Cu(100)	p(2×2)	4-fold hollow	1.28±0.03		2.25±0.02	2.21	Shih et al., 1981
Fe(100)	c(2×2)	4-fold hollow	1.10±0.02	0.0±3.0	2.30±0.01	2.21	Zeng et al., 1990
Fe(110)	p(2×2)	hollow	1.43		2.17	2.21	Zhang et al., 1988
Ge(100)	p(2×1)	bridge	1.08	0.0	2.36	2.26	Leiung et al., 1988
Ge(111)	p(2×2)	bridge	1.03±0.05	−10.56	2.11±0.02	2.26	Robey et al., 1987
Ir(110)	p(2×2)−2S	fcc hollow	0.94	−3.3	2.38	2.31	Chan and Van Hove, 1987
Ir(111)	(√3×√3)R30°	fcc hollow	1.65±0.07		2.28	2.31	Chan and Weinberg, 1979
Mo(100)	c(2×2)	4-fold hollow	1.00±0.06	−7.3±3.0	2.38	2.34	Rous et al., 1991
Ni(100)	c(2×2)	4-fo.d hollow	1.30±0.02	+2.0±1.0	2.19	2.19	Starke et al., 1990
Ni(100)	p(2×2)	4-fold hollow	1.25±0.02	+0.5±1.0	2.16	2.19	Oed et al., 1990b
Pd(100)	c(2×2)	4-fold hollow	1.30±0.05		2.34	2.32	Berndt et al., 1982
Pd(111)	(√3×√3)R30°	fcc hollow	1.53±0.05		2.20±0.03	2.32	Máca et al., 1985
Pt(111)	(√3×√3)R30°	fcc hollow	1.62±0.05		2.28±0.03	2.34	Hayek et al. 1985
Rh(100)	p(2×2)	4-fold hollow	1.29		2.29	2.29	Hengrasmee et al., 1979
Rh(110)	c(2×2)	center	0.77		2.12	2.29	Hengrasmee et al., 1980
Rh(111)	(√3×√3)R30°	fcc-hollow	1.53		2.18	2.29	Wong et al., 1985

Selenium (Se)

Ag(100)	c(2×2)	4-fold hollow	1.91±0.04		2.80	2.51	Ignatiev et al., 1973
Ni(100)	c(2×2)	4-fold hollow	1.55		2.35	2.32	Rosenblatt et al., 1982a
Ni(100)	p(2×2)	4-fold hollow	1.55±0.10		2.35±0.06	2.32	Van Hove and Tong, 1975b
Ni(110)	c(2×2)	2-fold hollow	1.10±0.04		2.35±0.02	2.32	Rosenblatt et al., 1982a
Ni(111)	p(2×2)	fcc hollow	1.80±0.04		2.30±0.02	2.32	Rosenblatt et al., 1982b

Tellurium (Te)

Cu(100)	p(2×2)	4-fold hollow	1.90		2.62	2.54	Comin et al., 1982
Ni(100)	c(2×2)	4-fold hollow	1.90±0.10		2.59±0.06	2.52	Demuth et al., 1973
Ni(100)	p(2×2)	4-fold hollow	1.80±0.10		2.52±0.06	2.52	Van Hove and Tong, 1975c

[1] Measured relative to the c.o.m plane of the buckled top Cu layer.
[2] Measured relative to the c.o.m. plane of the buckled top Ni layer.

second layer buckling amplitude with coverage is the same trend exhibited by oxygen overlayers on Ni(100).

The open surface of bcc W(100) allows the penetration of O into the hollow site to bond to W atoms within the second W layer. The disordered O overlayer at 120K reconstructs the top W layer such that the four adjacent W atoms move towards the adatom. At elevated temperatures the p(2×1) phase can be prepared and is found to be a missing row structure in which the adatom adsorption geometry is similar to that of the disordered phase.

In the Ta(100)p(3×1)–O structure, the oxygen atom penetrates the top layer to sit at a sub-surface long-bridge site. The interstitial O atom causes a 0.1 Å buckling of the top Ta layer. Another interstitial oxygen adsorption structure is Zr(1000)p(2×2)–O in which the O atoms occupy a p(2×2) underlayer situated halfway between the Zr layers. These oxygen adsorption systems may be compared to the C and N intercalated structures of Zr and Ti.

Finally, we note that for oxygen adsorption on all of the metallic substrates shown in table 9 (except the bcc Fe(100) and W(100) surfaces and the interstitials), the shortest O–metal bond is formed between the top layer metal atom and the adatom

2.4.5.2.   Sulfur

The Ni(100)c(2×2)–S and -p(2×2)–S systems have been extensively studied and are of interest primarily because they are analogous to the oxygen overlayers of the same symmetry. The Ni(100)c(2×2)–S system involves an almost insignificant buckling of the second Ni layer by 0.01 ± 0.03 Å whilst the p(2×2) overlayer induces a significant buckling with an amplitude of 0.07 ± 0.05 Å. This decrease of the second layer buckling amplitude with coverage is the same trend as seen for oxygen overlayers on Ni(100) and Ru(1000). In contrast to oxygen adsorbed on Ni(100), the Ni–S bond lengths are close to the sum of the covalent radii and not significantly larger, a trend which is observed for Ni–O bonds in O on Ni(100). Sulfur adsorption on Cu(100) resembles that of oxygen and sulfur adsorption on Ni(100) rather than O adsorption on Cu(100). The p(2×2) structure is similar to the Ni(100)p(2×2)–S system but with a smaller second layer buckling amplitude of 0.02 Å instead of 0.07 Å.

On reconstructed Ir(110), S adsorbs on the fcc-hollow sites to form a p(2×2) overlayer with two S atoms in the unit cell. The sulfur atom is adsorbed equidistant (2.38 Å) from three Ir atoms; two in the top layer and one in the second. Ir(111)($\sqrt{3} \times \sqrt{3}$)R30°–S appears to be a straightforward adsorption structure in which the S atom occupies the fcc-hollow site.

Sulfur adsorption onto the surfaces of bcc metals is qualitatively different to that on fcc substrates. Sulfur adsorbed onto Fe(110) induces lateral distortions of the Fe substrate. These movements produce a "pseudo" four-fold site at the Fe(110) surface in which the S adatoms bond to Fe atoms within the top layer. The open bcc Mo(100) surface allows the S atom to sit deep within the hollow sites to form a bond to the second layer Mo atoms. In the Mo(100)c(2×2)–S structure, the second Mo layer is buckled by 0.16 Å with the atom directly underneath the S atom being pulled out of the surface.

Finally, we note that sulfur adsorption on all the low Miller index surfaces tabulated involves bonding of the adatom to a metal atom in the top layer. There are two exceptions, Rh(110) and Ir(110), surfaces which are sufficiently open for the S atom to bond directly to a second layer metal atom.

2.4.5.3.　Selenium and tellurium

There are a few, relatively early, studies of Se and Te adsorption on metals. Selenium is found to adsorb at the high coordination (hollow) sites on the low Miller index surfaces Ni(100) and Ag(100). On the most open surface to have been examined, Ni(110), the bond distance to the Ni atom in the second substrate layer (2.35 Å) is slightly shorter than that to the top layer (2.42 Å), suggesting the formation of a Ni–Se bond to the second substrate layer. However, it should be noted that the LEED studies of Se adsorption on metals originate before 1975, whilst more recent studies (1982) were by photo-electron diffraction only. Consequently, detailed substrate distortions, of the type seen in more recent studies of O and S adsorption on metals, have not been searched for.

Like Se, Te is found to adsorb at the high coordination (hollow) sites on the low Miller index surfaces of Cu and Ni(100). Again, detailed substrate distortions, of the type seen in more recent studies of O and S adsorption, have not been considered.

2.4.6.　Group VIIA (halogen) chemisorption on metals

Table 10 gives a tabulation of the determined surface structures for chemisorption of chlorine, bromine and iodine on metal surfaces. In all of the halogen adsorption systems which have been examined, the adatoms form simple overlayer structures in which the adatom is located at the high-coordination site and the halogen–metal bond length is (within the measurement error) identical to the bond length of the corresponding bulk solid. For example, the Cl–Cu bond length determined for Cu(100)($\sqrt{3}\times\sqrt{3}$)R30°–Cl is within $2\pm1\%$ of the bond length in bulk CuCl.

2.4.7.　Atomic adsorption of metals on metals

Tables 11–13 tabulate the structural results for the elements which form metallic solids adsorbed on metal surfaces. In addition to the transition

Table 10

Structural parameters for group VIIA (halogen) chemisorption on metals. δd_{12} is the relaxation of the first substrate interlayer spacing expressed as a percentage of the bulk interplanar spacing. The bond length is derived from the determined coordinates and is compared to the bond length found the the corresponsing bulk solid (e.g., AgBr, AgCl).

Substrate	Overlayer	Site	Adsorption height (Å)	δd_{12} (%)	Bond length (Å)	d/d_{bulk}	Reference
Chlorine							
Ag(100)	c(2×2)	4-fold hollow	1.62±0.10	0.0±5.0	2.61±0.06		Jona and Marcus, 1983
	c(2×2)	4-fold hollow	1.61±0.04		2.60±0.03		Chang and Winograd, 1990
	c(2×2)	4-fold hollow	1.75±0.05		2.69±0.03		Lamble et al., 1987
	c(2×2)	4-fold hollow	1.96±0.20		2.83±0.12		Cardillo et al., 1983
Ag(110)					2.56		Winograd and Chang, 1989
Ag(110)							Holmes et al., 1987
Ag(111)	(√3×√3)R30°–Cl and –2Cl	fcc hollow	2.12±0.01		2.70±0.01		Lamble et al., 1986
Cu(111)	(√3×√3)R30°	fcc hollow	1.88±0.03	−3.3±1.0	2.39±0.02	1.02±0.01	Crapper et al., 1987
Cu(100)	c(2×2)	4-fold hollow	1.60±0.03		2.41±0.02	1.03±0.01	Jona et al., 1983
			1.59±0.02		2.37±0.02	1.01±0.01	Citrin et al., 1982b
			1.53±0.02	+3.9±2.2	2.37±0.02	1.01±0.01	Patel et al., 1989
			1.604±0.005	+0.4±1.2	2.410±0.005	1.030±0.002	Wang et al., 1991a
Ni(111)	(√3×√3)R30°	fcc hollow	1.837±0.001	−2.5	2.33±0.005		Wang et al., 1991b
			1.83±0.05	−0.9	2.33±0.02		Funabashi et al., 1990
							Takata et al., 1992
Ni(100)	c(2×2)	4-fold hollow	1.60±0.02	+11.4	2.38±0.02		Yokoyama et al., 1989
			1.58±0.02	4.5	2.35±0.02		Sette et al., 1988
			1.60±0.02		2.38±0.02		Yokoyama et al., 1990
Bromine							
Ni(100)	c(2×2)	4-fold hollow	1.51±0.03		2.25±0.02	0.98±0.01[1]	Lairson et al., 1985

Iodine

Ag(111)	(√3×√3)R30°	mix. fcc and hcp hollow	2.28±0.08	−3.0±2.5	2.83±0.06	1.01±0.02	Maglietta et al., 1981
Cu(100)	p(2×2)	3-fold hollow	2.34±0.02		2.87±0.03	1.02±0.01[2]	Citrin et al., 1980
Cu(111)	(√3×√3)R30°	4-fold hollow	1.98±0.02		2.68±0.02	1.03±0.01	Citrin et al., 1980
		3-fold hcp	2.21±0.02		2.69±0.02	1.03±0.01	Citrin et al., 1980
Ni(100)	c(2×2)	4-fcld hollow	2.15±0.02		2.78±0.02	1.00±0.01	Jones et al., 1987

[1] Bulk bond length from sum of covalent radii (Ni=1.15 Å, Br=1.14 Å).
[2] Phys. Rev. Lett., 47 (1981) 1567.

Table 11

Structural parameters for metal adsorption on transition metals involving the formation of monolayer alloys. The given bond length is the shortest distance between the substitutional atom and the host atom. d_{12} is the spacing between the first and second layers of the substituted surface, b_1 is the amplitude of the buckling in the mixed top layer.

Element	Substrate	Overlayer	Site	d_{12} (Å)	b_1 (Å)	Bond length (Å)	Reference
Mn	Pd(100)	c(2×2)	subs	1.84±0.05	0.20±0.05	2.76±0.03 / 2.68±0.03	Tian et al., 1990
Au	Cu(100)	c(2×2)	subs	1.88±0.05	0.10±0.05	2.56±0.03	Wang et al., 1987
Pd	Cu(100)	c(2×2)	subs	1.81±0.03	0.02±0.03	2.56±0.02	Wu et al., 1988
Sn	Pt(111)	(2×2)	subs		0.20±0.05	2.78±0.04	Overbury et al., 1991
		(√3×√3)R30°	subs		0.22±0.05	2.78±0.04	Overbury et al., 1991

Table 12

Structural parameters for transition metal adsorption on transition metals; epitaxial structures. The given bond length is the shortest distance between the adatom and a substrate atom. d_{12} is the spacing between the first and second layers of the epitaxial system, d_{23} is the spacing between the second and third layers, etc. The adatom–adatom and adatom–substrate bond lengths are derived from the determined structural paramaters.

Element	Substrate	No. of layers	d_{12} (Å)	d_{23} (Å)	d_{34} (Å)	A–S Bond length (Å)	A–A bond length (Å)	Reference
Fe	Ag(100)	25	1.45±0.03	1.45±0.03	1.42±0.03		2.89±0.02 2.51±0.02	Li et al., 1990
	Cu(100)	1	1.78±0.02	0.0±2.0%		2.53±0.02	2.55	Clarke et al., 1987a
	Cu(110)	1	1.25±0.03	−0.8±2.0%		2.52±0.02 2.54±0.02	2.56	Marcano et al., 1989 Marcano et al., 1989
	Cu(111)	1	2.02±0.02	−1.0±1.0%	0.0±1.0%	2.50±0.02	2.55	Darici et al., 1988
	Ni(100)	1	1.85±0.05	−0.4±2.8%		2.55±0.02	2.49	Lu et al., 1989
	Ni(100)	2	1.90±0.05	1.75±0.05%		2.48±0.03	2.49 2.59±0.03	Lu et al., 1989
Ni	Ru(0001)	1	2.05±0.05			2.58±0.05	2.71	Tian et al., 1991
	Cu(100)	1	1.80±0.02	−1.7±1.1%	0.0±1.1%	2.55±0.02	2.55	Abu-Joudeh et al., 1986
Co	Cu(100)	1	1.70±0.02	−6.0±2.0%		2.48±0.02	2.55	Clarke et al., 1987b
	Cu(111)	1	1.98±0.03	0.0±1.0%		2.47±0.03	2.55	Chandesris et al., 1986
Au	Pd(111)	1	2.25±0.20			2.75±0.15	2.75	Kuk et al., 1983
Cu	Ni(100)	1	1.80±0.03			2.52±0.03	2.50	Abu-Joudeh et al., 1984
Cd	Ti(0001)	1	2.6±0.1			3.08±0.07	2.95	Shih et al., 1977a

Table 13

Structural parameters for group IVA and group VA metal adsorption on transition metals. The given bond length is the shortest distance between the adatom and a substrate atom. d_{12} is the spacing between the adatom and the first substrate atomic plane, d_{23} is the first interlayer spacing of the substrate. The adatom–substrate bond lengths are derived from the determined structural parameters.

Element	Substrate	Overlayer	Site	d_{12}[1] (Å)	$d_{23}\%$	A–S bond length (Å)	d_a/d_{met}[2]	Reference
Pb	Cu(100)	c(2×2)	4-fold hollow	2.29 ± 0.04	1.81 ± 0.03	3.62 ± 0.03	1.33	Hoesler et al., 1986
		$(5\sqrt{2}\times\sqrt{2})$R45°	hollow and pseudo hollow	2.31 ± 0.10	1.81 ± 0.10	2.70 ± 0.04	0.81	Hoesler and Moritz, 1986
						2.93 ± 0.04	0.94	
Bi	Ni(100)	c(2×2)	4-fold hollow	2.05		2.70	0.85	Klink et al., 1991

[1] Spacing from second substrate layer to the subplane in the first atomic plane which contains the substrate atoms.
[2] d_a/d_{met} is the ratio of the derived adatom radius to the metallic radius.

metals, we have also included in this section (table 13) the group IVA and group VA atoms which form metallic solids (chemisorption of other atoms belonging to these groups are treated separately in sections 2.4.3 and 2.4.4).

Transition metal adsorption results in one of two types of behavior; alloy formation and the initial growth of epitaxial structures. Mn on Pd(100), Au and Pd on Cu(100) and Sn on Pt(111) can result in substitutional structures in which the "adatom" replaces an host atom in the top layer of the substrate. This results in the formation of an ordered alloy which is confined to the first atomic layer of the surface. Because of the size difference between the adatom and host atom, this substitution can result in a buckling of the alloy monolayer (see table 11).

In principle, epitaxial structures involve the deposition and growth of complete monolayers of the adatom which occupy the continuation sites of the substrate. Because of lattice mismatch, the epitaxial structure is subject to lateral stress which is relieved by defect formation. In reality, the distinction between alloy formation and epitaxy is often a function of the surface growth conditions and preparation. If grown at elevated temperatures, seemingly epitaxial structures can exhibit significant amounts of interdiffusion across the interfacial region.

2.5. Molecular adsorption on metals

2.5.1. Carbon monoxide (CO) and NO

The adsorption of carbon monoxide on transition metals surfaces has been more extensively studied than any other molecule–surface system (see table 14). In all cases, the molecule bonds upright, or almost upright, with the carbon atom closest to the surface. The adsorption site appears to vary with the identity of the substrate and as a function of the coverage. At low coverages (<0.33 ml), CO adsorbs at the top site on Cu(100), Ni(100) and Ru(1000). On (111) surfaces, the bridge sites are occupied at low coverages, but the top sites can be populated at higher coverages. An exception appears to be CO adsorption on Pd; CO adsorbs at bridge sites on Pd(100) (Behm et al., 1980) and at fcc-hollow sites on Pd(111) (Ohtani et al., 1987). At monolayer coverage on Ni(100)(p(2×1)–2CO), the molecules are found to tilt by 17 degrees from the surface normal (Hannaman and Passler, 1988).

In all of these cases, no significant change in the C–O bond length is observed upon adsorption, within the errors of each determination (the bond length for gas-phase CO is 1.13 Å). Figure 6 shows the observed C–substrate bond length plotted as a function of the radius of the metal atom. There is a discernible and expected trend to longer C–metal bonds as the effective metal radius increases, although there is considerable scatter in the data. The one detailed structural study of NO chemisorption for

Table 14
CO and NO chemisorption structures on metals. M–C/N denotes the shortest substrate atom to carbon/nitrogen distance.

Substrate	Overlayer	Coverage	Site	Adsorption height (Å)	Molecular bond length (Å)	M–C/N bond length (Å)	Reference
Carbon monoxide							
Cu(100)	c(2×2)	0.5 ml	top (C down)	1.92 ± 0.05	1.13	1.92 ± 0.05	McConville et al., 1986
Ni(100)	c(2×2)	0.5 ml	top (C-down)	1.80 ± 0.04	1.13 ± 0.10	1.80 ± 0.04	Kevan et al., 1981
Pd(100)	(2√2×√2)R45°-2CO	0.5 ml	bridge (C-down)	1.36 ± 0.10	1.15 ± 0.10	1.93 ± 0.06	Behm et al., 1980
Ni(110)	p(2×1)	1.0 ml	short-bridge tilted (C down)	1.34 ± 0.10	1.12 ± 0.10	1.95 ± 0.06	Hannaman and Passler, 1988
Ni(111)	(√3×√3)R30°	0.33 ml	bridge (C-down)	1.27 ± 0.10	1.13 ± 0.10	1.78 ± 0.06	Kevan et al., 1981
Pd(111)	(√3×√3)R30°	0.33 ml	fcc hollow (C-down)	1.29 ± 0.05	1.15 ± 0.05	2.04 ± 0.06	Ohtani et al., 1987
Pt(111)	c(4×2)-2CO	0.5 ml	top and bridge	2.26 ± 0.025 1.85 ± 0.025	1.15 ± 0.05 1.15 ± 0.05	2.26 ± 0.025	Ogletree et al., 1986
Pt(111)	disordered	0.33 ml	top (12%) bridge (88%)	2.26 1.85	1.15 1.15	2.26	Blackman et al., 1988
Rh(111)	(√3×√3)R30°	0.33 ml	top (C-down)	1.95 ± 0.10	1.07 ± 0.10	1.95 ± 0.10	Koestner et al., 1981
Rh(111)	(2×2)-3CO	0.75 ml	2 top 1 bridge	2.19 1.84	1.15	2.03	Van Hove et al., 1983
Ru(1000)	(√3×√3R)30°	0.33 ml	top (C-down)	2.00 ± 0.10	1.09 ± 0.10	2.00 ± 0.10	Michalk et al., 1983
Ru(1000)	disordered	0.05 to 0.20 ml	top (C-down)	2.10 ± 0.15	1.10 ± 0.10	2.00	Piercy et al., 1989
NO							
Rh(111)	(2×2)-3NO	0.75 ml	2 top 1 bridge	2.15 ± 0.10 1.55 ± 0.10	1.15	2.05 ± 0.10	Kao et al., 1989

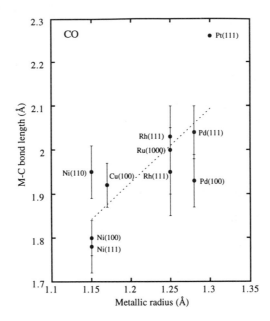

Fig. 6. The determined C-metal bond length for CO adsorption on metallic surfaces plotted as a function of the covalent radius of the substrate atom. The dashed line is a guide to the eye through the data, the error bars are those of the determination. Note that although the covalent radius has been used to compile this figure, for all metals shown the metallic radius is a uniform 0.09 Å larger.

Rh(111)(2×2)–3NO (Kao et al., 1989) shows that the adsorption geometry is almost identical to that of the equivalent CO system; Rh(111)(2×2)–3CO (Van Hove et al., 1983).

2.5.2. Adsorption of organic molecules on metals

The simplest adsorbed organic molecule to have been studied is acetylene, C_2H_2, on Ni(111) and Cu(100). When adsorbed onto Ni(111), the molecule remains intact and becomes oriented parallel to the surface with C–C bond lying across a bridge site (Casalone et al., 1982). This orientation locates one C atom approximately in the fcc-hollow site and the other in the hcp-hollow (see fig. 7). Since the C–C bond length was fixed at 1.21 Å in this study, the intramolecular bond lengths were not determined. On Cu(100), the molecule also lies parallel to the surface with the C atoms close to the bridge sites (see fig. 7). A C–C bond length of 1.42 ± 0.05 Å (Arvanitis et al., 1987) and C–Cu distance of 1.73 Å was found. Ethylene, C_2H_4, adsorbs intact on Cu(100) with the C atoms close to the bridge sites and with the molecule centered over the 4-fold hollow (Tang et al., 1991). This adsorption geometry

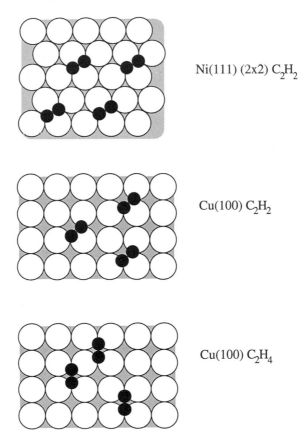

Ni(111) (2x2) C_2H_2

Cu(100) C_2H_2

Cu(100) C_2H_4

Fig. 7. Schematic illustration of the adsorption geometry for Ni(111)-(2×2)-C_2H_2 and the disordered overlayers of Cu(100)–C_2H_2 and Cu(100)–C_2H_4. The hydrogen positions were not determined and therefore only the carbon atom locations are shown.

is similar to that of acetylene on Cu(100) (Arvanitis et al., 1987). The C–C bond length is found to be 1.47 Å, the (single) C–Cu distance is 1.90 Å

The exposure of Rh(111) and Pt(111) to ethylene, C_2H_4, produces the ethylidyne radical, C_2H_3, which adsorbs with the C–C bond axis perpendicular to the surface. In a (2×2) overlayer on Pt(111) (Kesmodel et al., 1979), the lower C atom is found to be located in fcc-hollow site and is attached to a methyl group (the presence of H is inferred from HREELS). The C–C bond length is 1.50 ± 0.05 Å and the shortest C–Pt distance is 2.00 ± 0.02 Å. In the same (2×2) ethylidyne adsorption structure on Rh(111) (Wander et al., 1991a), the lower C atom is found to be located in the *hcp-hollow* site. The C–C bond length is 1.48 ± 0.10 Å, the shortest C–Rh distance is 2.06 ± 0.06 Å and an adsorbate induced buckling of the uppermost two Rh layers is found.

Ethylidyne adsorption on Rh(111) has also been studied in the form of a c(4×2) overlayer produced by coadsorption with either NO or CO (Blackman et al., 1988b). When coadsorbed with NO, the NO molecule is located at the fcc-hollow site and the C_2H_3 species remains oriented upright above the hcp-hollows. When coadsorbed with CO, the C_2H_3 species is displaced to the fcc-hollow sites by the CO molecule which then occupies the hcp-hollow site. In both structures, the local adsorption geometry of the ethylidyne species is identical, within the quoted error bars of the determination. The C–C bond length of the ethylidyne species is found to be 1.45 ± 0.05 Å and the C–Rh distance is 2.03 ± 0.04 Å. The determined adsorption geometry is also identical as to that of the (2×2) overlayer discussed earlier (Wander et al., 1991a).

The exposure of Cu(100) to methanol produces the adsorbed methoxy radical, CH_3O. A near-edge extended X-ray adsorption fine structure (NEX-AFS) study finds that the methoxy species is adsorbed with the O atom closest to the surface above a bridge site (Lindner et al., 1988). The C–O bond axis is oriented within 10 degrees of the surface normal and the C–O bond length is determined to be 1.43 Å, the O–Cu distance is 2.37 Å. The adsorption geometry of the formate species, HCO_2, has been studied on Cu(100) and Cu(110). On Cu(100) the formate radical is adsorbed with the O–C–O plane perpendicular to the surface. The O–C–O trimer is centered above a bridge site with the O atoms closest to the Cu substrate atoms. The O–C bond lengths are 1.25 Å and the O–Cu distance is 1.98 Å. On Cu(110), the local adsorption geometry of the oxygen atoms is similar to that on Cu(100) with the O–C–O group centred above a top site (see fig. 8). On Cu(110), the O–C bond lengths are 1.25 Å and the O–Cu distance is 1.98 Å; identical to the local adsorption geometry of the formate species on Cu(100).

The adsorption structures of benzene C_6H_6 have been studied on Rh(111) and Pt(111) with and without the coadsorption of CO (fig. 9). All of these LEED studies find that the benzene ring lies parallel to the surface and that there are distortions of the C–C bonds within the benzene ring induced by adsorption. A disordered C_6H_6 overlayer on Pt(111) has been studied using diffuse LEED (Wander et al., 1991b). The distorted benzene molecule is found to be centered over bridge sites of the substrate. The C_6 ring is found to be buckled with two C–C bond lengths of 1.47 Å and 1.64 Å present. The ordered structure Pt(111)$(2\sqrt{3}\times4)$–$2C_6H_6$–4CO (Ogletree et al., 1987) contains C_6H_6 centered over the bridge sites, as in the disordered overlayer. However the molecule is rotated 30 degrees from its orientation in the disordered phase and the ring distortions give somewhat different C–C bond lengths of (2×) 1.64 Å and (4×) 1.76 Å.

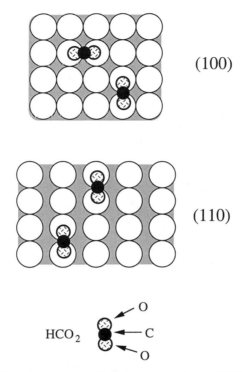

(100)

(110)

HCO_2

Fig. 8. The determined adsorption geometry of the formate species on Cu(100) and Cu(110). The hydrogen atom is not shown but is assumed to be directly above the C atom.

On the (111) surfaces of Pd and Rh benzene forms a (3×3)–C_6H_6–2CO overlayer when coadsorbed with CO. The Pd(111)(3×3)–C_6H_6–2CO structure (Ohtani et al., 1988) contains a flat benzene molecule centered over the fcc-hollow site. C–C bond lengths of 1.40 ± 0.1 Å and 1.46 ± 0.1 Å are determined. In Rh(111)(3×3)–C_6H_6–2CO (Van Hove et al., 1987) the molecule is centered over the hcp-hollow site with C–C bond lengths of 1.56 Å and 1.45 Å. On Rh(111), the Rh(111)$(2\sqrt{3}\times4)$–C_6H_6–CO structure has also been studied (Van Hove et al., 1986) and shows a similar adsorption geometry to the Rh(111)(3×3)–C_6H_6–2CO system with the benzene adsorbed over hcp-hollow sites with C–C bond lengths of 1.81 Å and 1.34 Å.

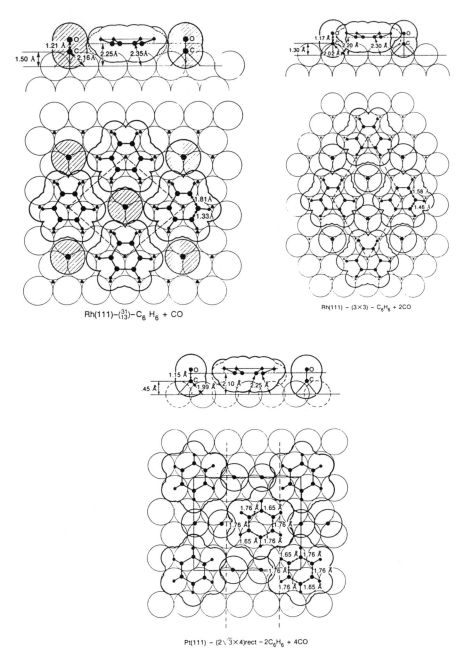

Fig. 9. Schematic illustrations of a selection of the benzene adsorption structures on transition metal surfaces. (After Maclaren et al., 1987.)

3.   Semiconductors

3.1.   Elemental semiconductor surfaces

3.1.1.   Clean surfaces of elemental semiconductors

3.1.1.1.   Clean (100) surfaces

Both the Si(100) and Ge(100) surfaces have stable (2×1) reconstructions which involve the saturation of dangling bonds by the formation of dimers between the atoms in the top layer of the bulk termination of the solid. In the case of Si(100)(2×1), although the existence of surface dimers is no longer controversial, there have been contradictory reports of dimers oriented parallel to the surface (the symmetric dimer model) or tilted in the plane perpendicular to the surface (the asymmetric dimer model), see fig. 10.

Recent studies which employ diffraction techniques, including medium energy ion scattering (MEIS) (Tromp et al., 1983), LEED (Holland et al., 1984) and grazing incidence X-ray scattering (Jedrecy et al., 1990), favor the asymmetric dimer model in which the top layer dimer is tilted by between 13.3 and 7.6 degrees. However, a kinematic LEED study (Zhao et al., 1991), photoemission studies (Johansson et al., 1990; Uhrberg and Hansson, 1991)

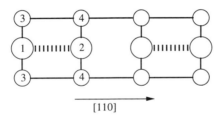

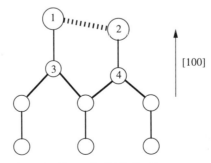

Fig. 10. Schematic illustration of the Si(100)- or Ge(100)-(2×1) reconstructed surface showing the formation of tilted dimers in the top layer. Shown is the asymmetric dimer model, in the symmetric dimer model the atoms labeled 1 and 2 lie in the same plane parallel to the surface.

and STM images (Tromp et al., 1985; Hamers et al., 1986, 1987) appeared
to demonstrate the presence of both asymmetric and symmetric dimers in
a mixture of c(4×2) and p(2×1) domains, the tilted dimers being stabilized
by vacancies produced by missing dimers. The KLEED study claims that the
nominally "(2×1)" surface actually comprises of only 75% (2×1) domains
in which the dimers are tilted by less than 4 degrees. The buckled dimers
are found in the coexisting c(4×2) phase which covers the remaining 25%
of the surface and have a much larger tilt angle of 13.6 degrees. Recently,
temperature dependent STM measurements have shown that the density of
asymmetric dimers increases as the temperature is reduced, an indication
that the buckled dimer is the true ground state (Wolkow, 1992). This
conclusion is supported by a recent measurement of the optical properties
of the surface which, by comparison to a tight-binding calculation, strongly
support the existence of tilted dimers. The structural parameters for several
determinations of the Si(100)(2×1), and c(4×2) reconstructed surfaces are
given in table 15.

The Ge(100)p(2×1) surface has been the subject of fewer studies that
the analogous Si surface. In particular, the symmetric *vs* asymmetric dimer
question remains unresolved for this surface. Photoemission (Kevan, 1985),
He-atom scattering (Lambert et al., 1987), STM (Kubby et al., 1987) and
an earlier XRD study (Eisenberger and Marra, 1981) indicate that the
asymmetric is the majority species at the surface. However a recent, more
detailed, X-ray diffraction study favors symmetric, or almost symmetric,
dimer formation (Grey et al., 1988).The structural parameters for two relatively complete XRD determinations of the Ge(100)(2×1), reconstructed
surfaces are given in table 15.

Table 15

Structural parameters for Si(100) and Ge(100) (2×1) and c(4×2) surfaces. z_{12}, z_{23} and z_{34}
denote the distances between the atoms numbered as in fig. 10, measured perpendicular to
the surface. d_{12} is the Si–Si or Ge–Ge bond length of the top layer dimer. The tilt angle is
measured relative to the surface plane and is derived from the coordinates.

Structural parameter	Si(100) (2×1)				Si(100) c(4×2)	Ge(100) (2×1)	
Tilt angle (°)	13.3	7.6	8.3 ±4.6	3.4	13.6	27.0 ±1.0	0.0
z_{12} (Å)	0.56	0.31	0.36±0.20	0.14	0.56	1.28±0.04	0.00
z_{23} (Å)	0.83	0.99	0.69±0.20	1.19	0.82	0.75±0.04	1.17
z_{34} (Å)	0.09	0.00	0.08	0.05	0.00		
d_{12} (Å)	2.36	2.32	2.47	2.38	2.31	2.51	2.33
Reference	Tromp et al., 1983	Jedrecy et al., 1990	Holland et al., 1984	Zhao et al., 1991	Zhao et al., 1991	Eisenberger and Marra, 1981	Grey et al., 1988

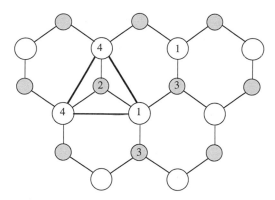

Fig. 11. Schematic illustration of the bulk termination of the Si(111) and Ge(111) surfaces. The numbers label the atoms for comparison to fig. 12. The bold lines show the surface unit cell. The actual, stable, surfaces are reconstructed.

3.1.1.2.   Clean (111) surfaces

The bulk termination of the (111) surfaces of Si and Ge has a surface atom bonded to three neighbors in the second atomic plane (see fig. 11). This surface atom has one dangling bond. The Si(111) surface displays a variety of reconstructions depending upon the treatment of the surface. At low temperatures the Si(111)(2×1) structure is formed which may be irreversibly converted to Si(111)(1×1) by heating to 360°C. The Si(111)(1×1) structure may also be stabilized by impurities. At 400°C the Si(111)(7×7) is formed. This structure may be converted to a metastable (1×1) structure by heating beyond 900°C. Laser-annealing the (7×7) surface results is a variety of structures exhibiting c(4×2), (2×2), (5×5) and (9×9) reconstructions.

The determination of Si(111)(1×1) structure produced by laser-annealing the Si(111)(7×7) remains inconclusive. A detailed LEED study (Jones and Holland, 1985) found two candidate structures. In both structures, the surface atom sinks into the surface whilst the second layer spacing is expanded. In the first candidate structure, the top layer spacing contracts by 0.2 Å from its bulk terminated value of 0.78 Å, whilst the second layer spacing increases by 0.1 Å from its bulk value of 2.35 Å. The second structure comprises of much more extensive relaxations in which the top layer spacing contracts by 0.70 ± 0.02 Å whilst the second layer spacing increases by 0.6 Å This produces an almost graphitic surface in which the separation between the top two Si planes is less than 0.05 Å.

The Si(111)(2×1) surface has been the subject of a pair of relatively complete structure determinations (Himpsel et al., 1984; Smit et al., 1985) both of which indicate that the so called Pandey π-bonded chain model describes the essential structural features of this system (Pandey, 1981). This

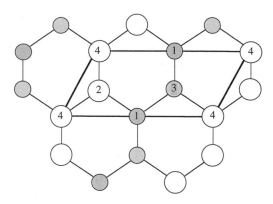

Fig. 12. Schematic illustration of the (2×1) reconstruction of the Si(111) surface. The numbers label the atoms in the bulk termination (see fig. 11). The bold lines show the surface unit cell.

structure is shown in fig. 12. In this structure, the second layer atoms in the bulk termination move up towards the top layer, breaking half of their bonds with the third layer. Half of the top layer atoms then bond with the third layer atoms. Although, the net change in the number of dangling bonds is zero, this reconstruction allows the dangling bonds to move closer together to form a π-bond. As is shown in fig. 12, the first and second layers of the reconstructed surface consist of zig-zag chains of almost coplanar atoms. The most recent medium-energy ion scattering (MEIS) study (Smit et al., 1985) concludes that the Si–Si bond length in the uppermost zig-zag chain is 2.3 Å and that this chain is corrugated by 0.30 ± 0.10 Å normal to the surface. At a distance of 1.0 Å below the upper chain, the lower chain has Si–Si bond lengths of 2.40 Å and is corrugated by 0.15 ± 0.10 Å.

Like Si(111), Ge(111) exhibits a complex set of reconstructions depending upon the surface preparation. Ge(111) also forms a metastable (2×1) phase at low temperatures but there has been no structural determination to date. There is some indirect evidence from total energy calculations (Zhu and Louie, 1991), angle-resolved photoemission (ARPES) (Nicholls et al., 1984) and inverse-photoemission (Nicholls and Reihl, 1989), that the (2×1) structure is similar to the π-bonded chain model found for Si(111)(2×1).

Unlike Si(111), Ge(111) displays a c(2×8) reconstructed phase which, like the Si(111)(7×7) structure, appears to be the thermodynamically stable phase for this surface. Despite the size of the unit cell, the structural elements of this reconstruction are quite simple, being generated by an ordered array of Ge adatoms located on a relaxed bulk termination. A recent LEED study (Tong et al., 1990) determined that the surface unit cell contains four Ge adatoms all of which are situated on the T4 sites of

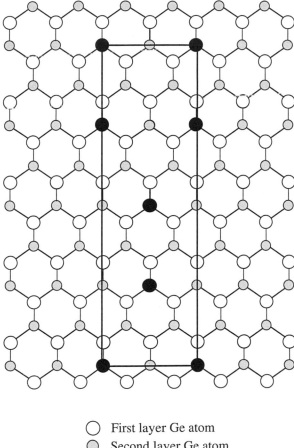

○   First layer Ge atom
◐   Second layer Ge atom
●   Ge adatom

Fig. 13. Schematic illustration of the (2×8) reconstruction of the Ge(111) surface which is formed from four Ge adatoms. The bold lines show the surface unit cell.

the "substrate". The unit cell contains two inequivalent Ge adatoms and is illustrated in fig. 13.

It is now generally accepted that the correct structural characterization of the complex Si(111) (7×7) structure is the dimer–adatom stacking fault model proposed by Takayanagi and coworkers (Takayanagi et al., 1985a, b). This structure consists of twelve Si adatoms located upon of a layer of Si dimers with vacancies at the vertices of the (7×7) unit cell. The physical origin of the tendency of this surface to form such a complex reconstruction is believed to be the result of the competition between two factors; the reduction of the total energy by saturation of dangling bonds and an increase in the total energy due to an increase in surface strain. The Si adatoms, released by the formation of vacancies, saturate three dangling bonds each. However, the creation of these vacancies increases the surface strain. The question of why Si(111) favors this complex reconstruction whilst Ge(111) favors the qualitatively much simpler (2×8) adatom reconstruction has been addressed by Vanderbilt (Vanderbilt, 1987)

3.1.2. Atomic adsorption on elemental semiconductors

3.1.2.1. Atomic adsorption on (100) surfaces

The adsorption structures of Si(100) and Ge(100) are tabulated in table 16. Alkali-metal adsorption of K and Na on Si(100)(2×1) does not remove the reconstruction present on the clean surface. For a 1-ml coverage of Na on Si(100)(2×1), the preferred model has the adatoms occupying the four-fold hollow sites between the two Si dimers (Wei et al., 1990). However, the LEED study of this system could not distinguish between this and another model in which the Na adatoms occupy two locations shown in fig. 14; the four-fold hollows (site III) or the valley bridge sites formed between the dimers (site II). In the Si(100)(2×1)–2K surface half of the adatoms also occupy the hollow sites between the dimers. The location of the remaining adatoms is uncertain, with the LEED study (Urano et al., 1991) slightly favoring the occupation of the adjacent hollows; site III in fig. 14. The determined Si–Na and Si–K bond lengths for adatoms in the four-fold hollow are 2.98 Å and 3.16 Å respectively.

The adsorption of 0.4 ML of the transition metal Co on Si(100) results in the removal of the (2×1) reconstruction with the adatom occupying the coplanar four-fold hollow site formed by four top layer Si atoms (Meyerheim et al., 1991). The Si–Co bond length is found to be 2.35 Å.

Chalcogen adsorption on Ge(100) has been investigated for the case of Ge(100)(2×1)–S (Leiung et al., 1988). Although the symmetry of the surface

Table 16

Structural parameters for atomic adsorption on (100) surfaces of Si and Ge. d_{12} is the adsorption height. The bond length is derived from the determined coordinates.

Substrate	Overlayer	Site(s)	d_{12} (Å)	Bond length (Å)	Reference.
Si(100)	(2×1)–Na	4-fold hollow (I)	1.85	2.98	Wei et al., 1990
	(2×1)–2K	4-fold hollow (I)	1.75	3.16	Urano et al., 1991
		+ hollow (III)	1.35	3.55	
	Co (0.4 ml)	4-fold hollow	0.00	2.35	Meyerheim et al., 1991
Ge(100)	p(2×1)–S	bridge	1.08	2.36	Leiung et al., 1988
		(continuation)			

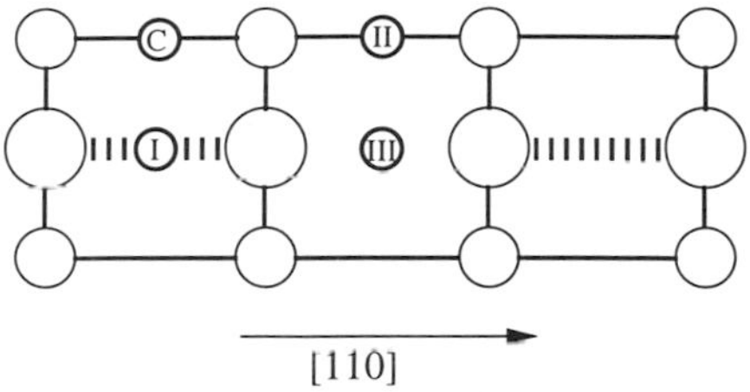

[110]

Fig. 14. Schematic illustration of the four high-coordination adsorption sites on the Si/Ge(100)-(2×1) reconstructed surface which preserve the (2×1) reconstruction.

is the same as the clean reconstructed surface, S adsorption disrupts the Si dimers by sitting in the bridging continuation sites; site C in fig. 14.

3.1.2.2.   Atomic adsorption on (111) surfaces

The majority of studies of atomic adsorption on Si(111) and Ge(111) surfaces have been of the $(\sqrt{3} \times \sqrt{3})R30°$ symmetry overlayer structures formed by group III, IV and V adsorbates. The structural results of these studies are tabulated in table 17.

With the exception of boron, group III and IV adatoms are located at the T4 site. This site, which is shown in fig. 15, is a three-fold symmetric site in which the adsorbed atom is located directly above a second layer atom of the substrate. The T4 site allows the adatom to bond to three Si atoms, whilst the $(\sqrt{3} \times \sqrt{3})R30°$ overlayer structure allows each Si atom to bond to just one adatom. In the case of group III adsorbates, all dangling bonds are saturated for a $(\sqrt{3} \times \sqrt{3})R30°$ overlayer, which presumably explains the stability of this structure. For group IV adsorbates, one dangling bond must remain which can be saturated only by back-bonding to the substrate. Group V adatoms are located at either substitutional sites in which a Si atom is replaced by an adatom or at T4 sites with considerable subsurface rearrangements (see fig. 15).

There have been two studies of chalcogen adsorption on elemental semiconductors. Sulfur forms a (2×2) super-structure on Ge(111) in which the adatom occupies a bridge site. On Si(111)(7×7), S also occupies a bridge site, although only the local adsorption structure has been determined by SEXAFS. A single study of a transition metal on Si(111) shows that Co adsorbs in the plane of top layer of Si(111). Halogen adatoms adsorb in a (1×1) array on both Si(111) and Ge(111) removing the clean surface reconstruction. The adatom is found to be located on the top site and the structural results for Cl, I and Br on Si(111) and Ge(111) are given in table 17.

Table 17

Structural parameters for atomic adsorption on (111) surfaces of Si and Ge. The bond length is derived from the determined coordinates.

Adsorbed atom	Sub-strate	Pattern	Site	Adsorption height (Å)	A–S Bond length (Å)	Reference
Group III						
B	Si(111)	$(\sqrt{3}\times\sqrt{3})R30°$	Exch. T4(Si) B5(B)	1.34 ± 0.10 2.19 ± 0.07	2.15 ± 0.07	Huang et al., 1990b
Al	Si(111)	$(\sqrt{3}\times\sqrt{3})R30°$	T4	1.39	2.49	Huang et al., 1990a
Ga	Si(111)	$(\sqrt{3}\times\sqrt{3})R30°$	T4	1.35	2.50	Kawazu and Sakama, 1988
Group IV						
Sn	Si(111)	$(\sqrt{3}\times\sqrt{3})R30°$	T4	1.59	2.56	Conway et al., 1989
Pb	Si(111)	$(\sqrt{3}\times\sqrt{3})R30°$	T4	1.43 ± 0.05	2.40	Doust and Tear, 1991
Group V						
As	Si(111)	(1×1)	Substi-tutional	0.96 ± 0.03	2.42 ± 0.02	Patel et al., 1987
Bi	Si(111)	$(\sqrt{3}\times\sqrt{3})R30°$	T4	1.11	2.39	Wan et al., 1991a
Bi	Ge(111)	$(\sqrt{3}\times\sqrt{3})R30°$	T4	1.32	2.09	Wan et al., 1991b
Group VI (Chalcogens)						
S	Ge(111)	(2×2)	Bridge	1.03 ± 0.05	2.26	Robey et al., 1987
Te	Si(111) (7×7)	(7×7)	Bridge	1.51	2.44	Citrin et al., 1982a
Group VIII						
Co	Si(111)	0.4 ml	Substi-tutional	0.00 ± 0.02	2.30 ± 0.05	Meyerheim et al., 1991
Group VIIA (Halogens)						
Cl	Ge(111)	(1×1)	top	2.07 ± 0.03	2.07 ± 0.03	Citrin et al., 1983
Cl	Si(111)	(1×1)	top	1.98 ± 0.04	1.98 ± 0.04	Citrin et al., 1983
Cl	Si(111) (7×7)	(1×1)	top	2.03 ± 0.03	2.03 ± 0.03	Citrin et al., 1983
Br	Si(111)	0.67 ml	top	2.14 ± 0.10	2.14 ± 0.10	Golovchenko et al., 1982
Br	Si(111)	0.25 ml	top	2.18 ± 0.06	2.18 ± 0.06	Materlik et al., 1984
I	Ge(111)	(1×1)	top	2.50 ± 0.04	2.50 ± 0.04	Bedzyk et al., 1989
I	Si(111) (7×7)	(1×1)	top	2.44 ± 0.03	2.44 ± 0.03	Citrin et al., 1982a

3.2. Compound semiconductor surfaces

3.2.1. Clean surfaces of compound semiconductors

The (110) surface of the zincblende structure compound semiconductors and the (10$\bar{1}$10) surface of the wurtzite structure compound semiconductors

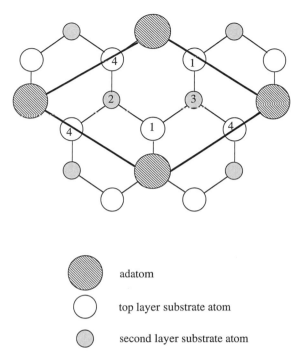

adatom

top layer substrate atom

second layer substrate atom

Fig. 15. Schematic illustration of the $\sqrt{3} \times \sqrt{3}R30°$ adsorption structure on Si(111) or Ge(111).

have an equal number of each atomic species in the top layer. The observed symmetry of these clean surfaces is (1×1), although substantial relaxations are present at the selvedge. The common feature of these relaxations, and the surface structure of compound semiconductors, is an approximately bond conserving rotation of the atom pair in the top layer with the cation sinking into the surface (Duke, 1990). This rotational distortion propagates to the second layer atom pair which rotates in the opposite direction in order to conserve the bond lengths in that layer (see fig. 16). The structural results for compound semiconductor surfaces are given in table 18 following the standard notation for the presentation of structural parameters put forward by Duke (Duke, 1990). Table 18 shows the basic similarity of the surface structures of compound semiconductors.

3.2.2. Atomic adsorption on compound semiconductors

Complete structure determinations for Al, Bi and Sb adsorption on GaAs(110) have been performed. Al adsorption on GaAs(110) leads to the substitution of Al atoms for Ga atoms in one or more layers (depending upon the coverage). This atomic exchange occurs because the As–Al bond energy is larger

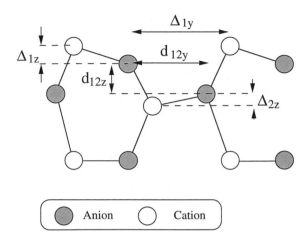

Fig. 16. Schematic illustration of the zinblende compound semiconductor surface structure (side view). The labeling convention is that of Duke (1990).

Table 18
Structural parameters for atomic adsorption on the (110) and (1010) surfaces of zincblende and wurtzite structure compound semiconductors. The bond length is that of the anion–cation dimer in the first layer. Ω is the tilt angle in the top layer. The other parameters are defined by fig. 16.

Surface	Δ_{1z}	Δ_{2z}	d_{12z}	d_{23z}	Ω (°)	Bond length (Å)	Reference
AlP	0.63	−0.07	1.33	1.96	25	2.22	Duke et al., 1983a
GaP	0.63	0.00	1.39	1.93	28	2.36	Duke et al., 1984
GaAs	0.69	−0.06	1.44	2.02	31	2.48	Ford et al., 1990
GaSb	0.77	0.00	1.62	2.16	30	2.65	Duke et al., 1983b
InP	0.69	0.00	1.55	2.08	30	2.56	Meyer et al., 1980
InAs	0.78	−0.15	1.50	2.21	36	2.50	Duke et al., 1983c
InSb	0.78	−0.10	1.54	2.34	29	2.81	Meyer et al., 1980
ZnS	0.59	0.00	1.40	1.91	28	2.29	Duke et al., 1984
ZnO	0.40	0.00			23	1.98	Duke et al., 1976
CdTe	0.81	−0.15	1.59	2.37	32	2.81	Duke et al., 1982a

than that of Ga–As. Kahn and coworkers (Kahn et al., 1981) have performed LEED determinations of 0.5–3.5 ml of Al on GaAs(110). At low coverages, the Al substitutes for the Ga atom in the second bilayer of the substrate. Increasing the coverage further forces the Al atoms to substitute into the third GaAs bilayer. Beyond 3.5 ml, the Al substitutes into the top layer and deeper bilayers of the substrate. The resulting Al substituted structures are the same as clean GaAs(110) with a slight (0.1 Å) reduction of the first interlayer spacing. The As–Al bond length was found to be 2.47 Å and is

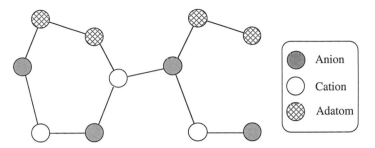

Fig. 17. Schematic illustration of the epitaxial continued lattice structure (ECLS) for adsorption on zinblende compound semiconductor surfaces structure (side view).

identical (within the measurement error) to the Ga–As bond length in the GaAs(110) surface.

Both Bi and Sb adatoms form a p(1×1) overlayer on GaAs(110) containing two adatoms in the surface unit cell. Dynamical LEED analyses of these adsorption structures (Ford et al., 1990; Duke et al., 1982b) have determined that these monolayers form an epitaxial continued layer structure (ECLS) shown in fig. 17. In the ECLS structure, the two adatoms per unit cell simply occupy the anion and cation continuation sites of the zincblende (110) lattice. The lower adatoms saturate the dangling bonds of the substrate and remove the bond rotation seen for the clean surface. Two of the four dangling bonds remaining on the group V adatom are saturated by bonding to neighboring adatoms to form zig-zag chains (see fig. 17).

4. Graphite and diamond surfaces

4.1. Graphite: C(1000)

The surface structure of the basal plane of graphite, C(1000), is unreconstructed and displays a small 1.5% relaxation of the top C sheet relative to the bulk interplanar spacing (Wu and Ignatiev, 1982). Alkali-metal adsorption on C(1000) has been studied for Cs and K adatoms. Cesium forms both a $(\sqrt{3}\times\sqrt{3})$R30° and a (2×2) phase upon adsorption onto C(1000) (Hu et al., 1989). In both cases, the adatom adsorbs on the six-fold hollow site. For the low coverage phase, the substrate remains as the bulk termination of the graphite lattice. However, the $(\sqrt{3}\times\sqrt{3})$R30° adsorption structure generates a stacking fault shift of the top C layer relative to the remainder of the substrate planes. K on C(1000) produces an intercalation structure in which the alkali-metal atom occupies the six-fold hollows between the C sheets. K intercalation generates a stacking fault in the graphitic layers and an increase of the C–C interplanar spacing from 3.35 to 5.35 Å (Wu and Ignatiev, 1983).

4.2. Diamond: C(111)

LEED (Yang et al., 1982) and MEIS (Derry et al., 1986) structural analyses of the C(111) diamond surface indicate that the surface is almost identical to the bulk termination of the diamond lattice. However, both of these studies note that the dangling bonds of the surface C atoms were probably saturated by adsorbed H. The MEIS investigation (Derry et al., 1986) suggests that the first interlayer spacing of C(111) is slightly contracted by $-1 \pm 1\%$ of the bulk interplanar spacing. The earlier LEED study saw no contraction within the error of the analysis.

5. Compounds

5.1. Carbides

Gruzalski and coworkers (Gruzalski et al., 1989) have performed a comparative LEED study of TaC(100) and HfC(100), the bulk crystal form of which is the NaCl structure. The bulk termination at the (100) surface consists of a stack of coplanar bilayers each containing one cation and an anion. In both cases a buckling of the top and second bilayer was found in which the C atom moves out of the surface. For TaC(100), the buckling amplitude was found to be 0.2 Å in the first bilayer and 0.04 Å in the second. For HfC(100), the buckling amplitudes were found to be 0.1 Å and 0.03 Å respectively. In both surfaces the top bilayer relaxes into the surface by $-10 \pm 4\%$ of the bulk inter-bilayer spacing for TaC(100) and $-4 \pm 4\%$ for HfC(100). This buckling behavior is similar to that observed in the (100) surfaces of NaCl structure oxides (eg. MgO(100), see sect. 5.3.1).

  Bulk silicon carbide has the zincblende crystal structure and has been studied, not in single crystal form, but as a micron depth thin film created by chemical vapor deposition on a Si(100) substrate (Powers et al., 1992). Two C-terminated $c(2 \times 2)$ structures have been studied by LEED, one with, and one without exposure, to C_2H_4 following cleaning. In both cases, the surface is terminated with coplanar C–C dimers which bridge the second layer Si sites. The Si rich surface terminates with an asymmetric Si dimer (Powers et al., 1992).

5.2. Silicides

Both $NiSi_2$ and $CoSi_2$ crystallize in the bulk fluorite crystal structure. The (111) surface of these solids is a stack of Si–Co–Si or Si–Ni–Si trilayers. Consequently, these surfaces are structurally similar to the disulfides and diselenides discussed in sect. 5.3.2. There have been several studies of the (111) surfaces of $NiSi_2$ and $CoSi_2$, although the overall structural trends exhibited by these surfaces remain unclear. In particular, the earlier LEED and later MEIS results do not agree on the extent of the relaxations in the surface trilayer.

An initial LEED study of $NiSi_2(111)$ (Yang et al., 1983) showed that the surface terminated at a complete Si–Ni–Si trilayer and the investigated the relaxation of the top Si–Ni interplanar spacing within this trilayer. It was found that the uppermost Si plane moves into the surface by -0.2 ± 0.1 Å. In contrast, a recent MEIS study (Vrijmoeth, 1991) finds that the internal interplanar spacings of the top trilayer are unrelaxed, but that the entire trilayer relaxes towards the bulk by -3% of the bulk inter-trilayer spacing. In the case of $CoSi_2(111)$ (Wu et al., 1986), which also terminates with a complete trilayer, the entire surface Si–Co–Si trilayer is compressed normal to the surface. The Si–Co interplanar spacings within the top trilayer are reduced from 0.77 Å to 0.73 Å. The entire trilayer also relaxes into the surface by -1.5% of the bulk inter-trilayer spacing.

5.3. Surface structure of chalcogenides

5.3.1. Oxides

There have been relatively few studies of oxide surfaces, primarily due to charging problems encountered when attempting to perform LEED measurements on these insulating surfaces. Because of the sparsity of structural information, it is difficult to establish any structural trends for oxide surfaces at present. However, a common feature of all determinations that have looked for it, is an approximately 0.1 Å buckling of the top metal–O layer in which the O atom moves out of the surface.

The only alkali-metal oxide that has been investigated is an early LEED study of fluorite structure $Na_2O(111)$. The (111) surface of the fluorite crystal structure consists of a stack of Na–O–Na trilayers offering the possibility of two terminations. The $Na_2O(111)$ surface is found to terminate at a complete Na–O–Na trilayer. A detailed structural study of surface relaxations was not performed (Andersson et al., 1977)

The bulk crystal form of the alkaline earth oxides is the NaCl structure. The bulk termination at the (100) surface consists of a stack of coplanar bilayers each containing one cation and an anion. CaO(100) has been the subject of a LEED study which found a contraction of top interlayer spacing by -1.2% (Prutton et al., 1979). Buckling of the top bilayer was not investigated. Buckling of the top bilayer was found in a recent LEED study of MgO(100) (Blanchard et al. 1990). In this case, the oxygen atom moves out of the surface by 0.05 ± 0.025 Å whilst the Mg atom sinks into the surface the same distance. The center-of-mass plane of the top MgO bilayer is unrelaxed. The (100) surface of transition metal oxide CoO, which also has the NaCl bulk structure, is found to be an unrelaxed bulk termination. As for CaO, buckling of the top bilayer was not considered. The (100) surface of TiO_2 reconstructs to form a surface with a (3×1) superstructure produced by removing rows of oxygen atoms along the (001) direction (Zschack, 1991).

The (100) surface of the perovskite structure oxide $SrTiO_3$(100) has been studied by LEED (Bickel et al., 1989). Parallel to the (100) plane, $SrTiO_3$ consists of alternate planes of O–Ti–O and Sr–O. The O–Ti–O trilayer terminated surface has a buckled top trilayer in which the O atoms move out of the surface by 0.04 ± 0.04 Å whilst the Ti atoms sink into the surface by the same distance. The position of the reference plane of the originally coplanar top layer is unchanged. The behavior resembles that of MgO(100) described above. The Sr–O terminated surface terminates in a buckled bilayer with the O atom moving out of the surface. The buckling amplitude is 0.16 ± 0.08 Å. The entire Sr–O bilayer relaxes towards the bulk by $-6 \pm 2\%$ of the bulk Sr–O/O–Ti–O interlayer spacing.

5.3.2. Disulfides and diselenides

The surfaces of the MS_2 and MSe_2 structures which have been investigated to date all have in common a repeat S–M–S (Se–M–Se) trilayer unit which is periodically repeated in a direction normal to the (1000) surface. MoS_2(1000) (Mrtsik et al., 1977; Van Hove et al., 1977), $TiSe_2$(1000) (Kasch et al. 1989) and $NbSe_2$(1000) (Mrtsik et al., 1977; Van Hove et al., 1977), terminate in a complete S–M–S (Se–M–Se) trilayer without stacking faults and with S/Se atomic plane outermost. In MoS_2(1000), the outermost plane of S atoms moves into the surface by -0.07 ± 0.01 Å. The entire S–Mo–S trilayer at the surface relaxes towards the bulk by $3 \pm 0.4\%$ of the bulk spacing of 2.96 Å. Similar behavior is observed for $NbSe_2$(1000); the outermost Se atom sinks into the surface by -0.02 ± 0.02 Å. The entire S–Mo–S trilayer at the surface relaxes toward the bulk by $-1 \pm 1\%$ of the bulk spacing of 2.91 Å.

The $TiSe_2$(1000) surface relaxations are qualitatively different to those of MoS_2(1000) and $NbSe_2$(1000). For this surface, both Se atoms in the top trilayer move away from the Ti plane by 0.4 Å. The entire Se–Ti–Se trilayer at the surface relaxes toward the bulk by $-1 \pm 1\%$ of the bulk spacing of 2.84 Å.

6.   Selected bibliography

There are a number of sources of information concerning the structure of surfaces. From a crystallographic viewpoint, perhaps the most comprehensive is the computerized database published by Watson and coworkers (Watson et al., 1993). This data-base allows the user to search for determinations of surface structures which were performed prior to 1992. The structural information is not tabulated by this data-base and therefore does not allow easy comparison of related structures. A printed copy of an earlier version of this data-base, which lists structures determined prior to 1986, is available (Maclaren et al., 1987).

Tabulations of some surface structures may be found in a review by Van Hove and coworkers (Van Hove et al., 1989) and the reviews by Watson that compare the results of surface structure determinations utilizing different crystallographic techniques (Watson, 1990, 1992). Van Hove has also published a recent review of crystal surface structure, without the tabular presentation of the structural data (Van Hove, 1992). Van Hove and Somorjai have reviewed surface structure from the point-of-view of adsorbate induced restructuring of surfaces (Van Hove and Somorjai, 1989). Ohtani and coworkers have listed all observed overlayer structures and surface symmetries, albeit without any reference to the detailed surface structure (Ohtani et al., 1987).

6.1.   Selected references

Maclaren, J.M., J.B. Pendry, P.J. Rous, D.K. Saldin, G.A. Somorjai, M.A. Van Hove and D.D. Vvedensky, 1987, Surface Crystallographic Information Service; A Handbook of Surface Structures (Reidel, Dordrecht).

Ohtani, H., C.T. Kao, M.A. Van Hove and G.A. Somorjai, 1987, Prog. Surface Sci. **23**, 155.

Van Hove, M.A., 1992, in: Structure of Solids, ed. V. Gerold (Springer-Verlag Chemie, Weinheim).

Van Hove, M.A. and G.A. Somorjai, 1989, Prog. Surf. Sci. **30**, 201.

Van Hove, M.A., S.W. Wang, D.F. Ogletree and G.A. Somorjai, 1989, Adv. Quant. Chem. **20**, 2.

Watson, P.R., 1990, J. Phys. Chem. Ref. Data **19**, 85.

Watson, P.R., 1992, J. Phys. Chem. Ref. Data **21**, 123.

Watson, P.R., M.A. Van Hove and K. Hermann, 1993, NIST Surface Structure Database Ver. 1.0, NIST Standard Reference Data Program, Gaitherburg, MD.

Acknowledgements

This work was supported by the Donors of the Petroleum Research Fund administered by the American Chemical Society and the Science Innovators Program of the Digital Equipment Corporation. I am grateful to John Pendry and Michel Van Hove for introducing me to surface crystallography.

References

Abu-Joudeh, M.A., P.P. Vaishnava and P. Montano, 1984, J. Phys. **C17**, 6899.

Abu-Joudeh, M.A., B.M. Davies and P.A. Montano, 1986, Surface Sci. **171**, 331.

Adams, D.L. and H.B. Nielsen, 1981, Surface Sci. **107**, 305.

Adams, D.L. and H.B. Nielsen, 1982, Surface Sci. **116**, 598.

Adams, D.L., and C.S. Sorensen, 1986, Surface Sci. **166**, 495.

Adams, D.L., H.B. Nielsen and M.A. Van Hove, 1979, Phys. Rev. **B20**, 4789.

Adams, D.L., H.B. Nielsen and J.N. Andersen, 1983, Surface Sci. **128**, 294.

Adams, D.L., L.E. Petersen and C.S. Sorensen, 1985a, J. Phys. **C18**, 1753.

Adams, D.L., W.T. Moore and K.A.R. Mitchell, 1985b, Surface Sci. **149**, 407.

Altmann, M.S., P.J. Estrup and I.K. Robinson, 1988, Phys. Rev. **B38**, 5211.

Andersen, J.N., H.B. Nielsen, L. Petersen and D.L. Adams, 1984, J. Phys. **C17**, 173.

Andersson, S., J.B. Pendry and P.M. Echenique, 1977, Surface Sci. **65**, 539.

Asensio, M.C., M.J. Ashwin, A.L.D. Kilcoyne, D.P. Woodruff, A.W. Robinson, Th. Lindner, J.S. Somers, D.E. Ricken and A.M. Bradshaw, 1990, Surface Sci. **236**, 1.

Arvanitis, D., L. Wenzel and K. Babeschke, 1987, Phys. Rev. Lett. **59**, 2435.

Baird, R.J., D.F. Ogletree, M.A. Van Hove and G.A. Somorjai, 1986, Surface Sci. **165**, 345.

Batra, I.P., J.A. Barker and D.J. Auerbach, 1984, J. Vac. Sci. Technol. **2**, 943.

Barnes, C.J., M.Q. Ding, M. Lindroos, R.D. Diehl and D.A. King, 1985, Surface Sci. **162**, 59.

Barnes, C.J., P. Hu, M. Lindroos and D.A. King, 1991, Surface Sci. **251/252**, 561.

Baro, A., H. Ibach and H.D. Bruchmann, 1979, Surface Sci. **88**, 384.

Baudoing, R., Y. Gauthier, Y. Joly, M. Lundberg and J. Rundgren, 1986, Phys. **C19**, 2825.

Baudoing-Savois, R., Y. Gauthier and W. Moritz, 1991, Phys. Rev. **B38**, 12977.

Bedzyk, M.J., Q. Shen, M.E. Keeffe and G. Navrotski, 1989, Surface Sci. **220**, 419.

Beccat, P., Y. Gauthier, R. Baudoing-Savois and J.C. Bertolini, 1990, Surface Sci. **238**, 105.

Behm, R.J., K. Christmann, G. Ertl and M.A. Van Hove, 1980, J. Chem. Phys. **73**, 2984.

Berndt, W., R. Hora and M. Scheffler, 1982, Surface Sci. **117**, 188.

Bickel, N., G. Schmidt, K. Heinz and K. Muller, 1989, Phys. Rev. Lett. **62**, 2009.

Blackman, G.S., M-L. Xu, M.A. Van Hove and G.A. Somorjai, 1988a, Phys. Rev. Lett. **61**, 2352.

Blackman, G.S., C.T. Kao, B.E. Bent, C.M. Mate, M.A. Van Hove and G.A. Somorjai, 1988b, Surface Sci. **207**, 66.

Blanchard, D.L., D.L. Lessor, J.P. LaFemina, D.R. Baer, W.K. Ford and T. Guo, 1990, J. Vac. Sci. Technol. **A9**, 1814.

Bohnen, H.P. and W. Ho, 1988, Surface Sci. **207**, 105.

Bonzel, H.P., A.M. Bradshaw and G. Ertl, eds., 1983, Physics and Chemistry of Alkali Metal Adsorption (Elsevier, Amsterdam).

Cardillo, M.J., G.E. Becker, D.R. Hamann, J.A. Serri, L. Whitman and L.F. Mattheiss, 1983, Phys. Rev. **B28**, 494.

Casalone, G., M.G. Cattania, F. Merati and M. Simonetta, 1982, Surface Sci. **120**, 171.

Chan, C-M. and M.A. Van Hove, 1986, Surface Sci. **171**, 226.

Chan, C-M. and M.A. Van Hove, 1987, Surface Sci. **183**, 303.

Chan, C-M. and W.H. Weinberg, 1979, J. Chem. Phys. **71**, 2788.

Chan, C-M., S.L. Cunningham, M.A. Van Hove, W.H. Weinberg and S.P. Withrow, 1977, Surface Sci. **66**, 394.

Chan, C-M., K.L. Luke, M.A. Van Hove, W.H. Weinberg and S.P. Withrow, 1978, Surface Sci. **78**, 386.

Chandesris, D., P. Roubin, G. Rossi and J. Lecante, 1986, Surface Sci. **169**, 57.

Chang, C-C. and N. Winograd, 1990, Surface Sci. **230**, 27.

Christmann, K., 1988, Surface Sci. Rep. **9**, 1.

Christmann, K., R.J. Behm, G. Ertl, M.A. Van Hove and W.H. Weinberg, 1979, J. Chem. Phys. **70**, 4168.

Citrin, P.H., P. Eisenberger and R.C. Hewitt, 1980, Phys. Rev. Lett. **45**, 1948.

Citrin, P.H., P. Eisenberger and J.E. Rowe, 1982a, Phys. Rev. Lett. **48**, 802.

Citrin, P.H., D.R. Hamann, L.F. Mattheiss and J.E. Rowe, 1982b, Phys. Rev. Lett. **49**, 1712.

Citrin, P.H., J.E. Rowe and P. Eisenberger, 1983, Phys. Rev. **B28**, 2299.

Clarke, A., P.J. Rous, M. Arnott, G. Jennings and R.F. Willis, 1987a, Surface Sci. **192**, L843.

Clarke, A., G. Jennings, R.F. Willis, P.J. Rous and J.B. Pendry, 1987b, Surface Sci. **187**, 327.

Clarke, L.J., 1980, Surface Sci. **91**, 131.

Comin, F., P.H. Citrin, P. Eisenberger and J.E. Rowe, 1982, Phys. Rev. **B26**, 7060.

Conway, K.M., J.E. MacDonald, C. Norris, E. Vlieg and J.F. van der Veen, 1989, Surface Sci. **215**, 555.

Crapper, M.D., C.E. Riley, P.J.J. Sweeney, C.F. McConville and D.P. Woodruff, 1987, Surface Sci. **182**, 213.

Culbertson, R.J., L.C. Feldman, P.J. Silverman and H. Boehm, 1981, Phys. Rev. Lett. **47**, 657.

Darici, Y., J. Marcano, H. Min and P.A. Montano, 1987, Surface Sci. **182**, 477.

Darici, Y., J. Marcano, H. Min and P.A. Montano, 1988, Surface Sci. **195**, 566.

Davis, H.L., 1992, Phys. Rev. Lett. **68**, 2632.

Davis, H.L. and D.M. Zehner, 1980, J. Vac. Sci. Technol. **17**, 190.

Davis, H.L. and J.R. Noonan, 1983, Surface Sci. **126**, 245.

Davis, H.L. and J.R. Noonan, 1988, in: The Structure of Surfaces II (Springer, Berlin) p. 152.

Debe, M.K. and D.A. King, 1982, J. Phys. **C15**, 2257.

Demuth, J.E., D.W. Jepsen and P.M. Marcus, 1973, J. Phys. **C6**, L307.

Demuth, J.E., D.W. Jepsen and P.M. Marcus, 1975a, J. Phys. **C8**, L25.

Demuth, J.E., P.M. Marcus and D.W. Jepsen, 1975b, Phys. Rev. **B13**, 1460.

Derry, T.E., L. Smit and J.F. van der Veen, 1986, Surface Sci. **167**, 502.

Doust, T.N. and S.P. Tear, 1991, Surface Sci. **251/252**, 568.

Duke, C.B., 1990, in: Reconstructions of Solid Surfaces, eds. K. Christmann and K. Heinz (Springer, Berlin).

Duke, C.B., A.R. Lubinsky, B.W. Lee and P. Mark, 1976, J. Vac. Sci. Technol. **13**, 761.

Duke, C.B., A. Paton, W.K. Ford, A. Kahn and G. Scott, 1982a, J. Vac. Sci. Technol. **20**, 778.

Duke, C.B., A. Paton, W.K. Ford, A. Kahn and J. Carelli, 1982b, Phys. Rev. **B26**, 803.

Duke, C.B., A. Kahn and C.R. Bonapace, 1983a, Phys. Rev. **B28**, 852.

Duke, C.B., A. Paton and A. Kahn, 1983b, Phys. Rev. **B27**, 3436.

Duke, C.B., A. Paton, A. Kahn and C.R. Bonapace, 1983c, Phys. Rev. **B27**, 6189.

Duke, C.B., A. Paton and A. Kahn, 1984, J. Vac. Sci. Technol. **A2**, 515.

Eisenberger, P., and W.C. Marra, 1981, Phys. Rev. Lett. **46**, 1081.

Feibelman, P., 1992, Phys. Rev. **B46**, 2532.

Fery, P., W. Moritz and D. Wolf, 1988, Phys. Rev. **B38**, 7275.

Fisher, D., S. Chandavarkar, I.R. Collins, R.D. Diehl, P. Kaukasoina and M. Lindroos, 1992, Phys. Rev. Lett. **68**, 2786.

Ford, W.K., T. Guo, D.L. Lessor and C.B. Duke, 1990, Phys. Rev. **B42**, 8952.

Funabashi, M., T. Yokoyama, Y. Takata, T. Ohta and Y. Kitajima, 1990, Surface Sci. **242**, 59.

Gauthier, Y., R. Baudoing, Y. Joly, J. Rundgren, J.C. Bertolini and J. Massardier, 1985, Surface Sci. **162**, 342.

Gauthier, Y., R. Baudoing, M. Lundberg and J. Rundgren, 1987, Phys. Rev. **B35**, 7867.

Gauthier, Y., R. Baudoing and J. Jupille, 1989, Phys. Rev. **B40**, 1500.

Gauthier, Y., W. Hoffmann and M. Wuttig, 1990a, Surface Sci. **233**, 239.

Gauthier, Y., Y. Joly, J. Rundgren, L.I. Johansson and P. Wincott, 1990b, Phys. Rev. **B42**, 9328.

Gauthier, Y., R. Baudoing-Savois, K. Heinz, and H. Landskron, 1991, Surface Sci. **251/252**, 493.

Gierer, M., H. Bludau, T. Hertel, H. Over, W. Moritz and G. Ertl, 1991, Surface Sci. **259**, 85.

Gierer, M., H. Bludau, T. Hertel, H. Over, W. Moritz and G. Ertl, 1992, Surface Sci. **279**, L170.

Golovchenko, J.A., J.R. Patel, D.R. Kaplan, P.L. Cowan and M.J. Bedzyk, 1982, Phys. Rev. Lett. **49**, 1560.

Gruzalski, G.R., D.M. Zehner, J.R. Noonan, H.L. Davis, R.A. DiDio and K. Muller, 1989, J. Vac. Sci. Technol. **A7**, 2054.

Grey, F., R.L. Johnson, J.S. Pederson, M. Nielsen and R. Feidenhans'l, 1988, in: The Structure of Surfaces II (Springer, Berlin) p. 292.

Grimsby, D.T., Y.K. Wu and K.A.R. Mitchell, 1990, Surface Sci. **232**, 51.

Griffiths, K., D.A. King, G.C. Aers and J.B. Pendry, 1982, J. Phys. **C15**, 4921.

Haberle, P., and T. Gustafsson, 1989, Phys. Rev. **B40**, 8218.

Haberle, P., P. Fenter and T. Gustafsson, 1989, Phys. Rev. **B39**, 5810.

Hamers, R.J., R.M. Tromp and J.E. Demuth, 1986, Phys. Rev. **B34**, 5345.

Hamers, R.J., R.M. Tromp and J.E. Demuth, 1987, Surface Sci. **181**, 346.

Hannaman, D.J., and M.A. Passler, 1988, Surface Sci. **203**, 449.

Hayek, K., H. Glassl, A. Gutmann, H. Leonhard, M. Prutton, S.P. Tear and M.R. Welton-Cook, 1985, Surface Sci. **152**, 419.

Held, G., H. Pfnür and D. Menzel, 1992, Surface Sci. **271**, 21.

Hengrasmee, S., P.R. Watson, D.C. Frost and K.A.R. Mitchell, 1979, Surface Sci **87**, L249.

Hengrasmee, S., P.R. Watson, D.C. Frost and K.A.R. Mitchell, 1980, Surface Sci. **92**, 71.

Himpsel, F.J., P.M. Marcus, R. Tromp, I.P. Batra, M.R. Cook, F. Jona and H. Liu, 1984, Phys. Rev. **B30**, 2257.

Hoesler, W., and W. Moritz, 1986, Surface Sci. **175**, 63.

Hoesler, W., W. Moritz, E. Tamura and R. Feder, 1986, Surface Sci. **171**, 55.

Holland, B.W., C.B. Duke and A. Paton, 1984, Surface Sci **140**, L269.

Holmes, D.J., N. Panagiotides, C.J. Barnes, R. Dus, D. Norman, G.M. Lamble, F. Della-Valle and D.A. King, 1987, J. Vac. Sci. Technol. **A5**, 703.

Hu, Z.P., J. Li, N.J. Wu and A. Ignatiev, 1989, Phys. Rev. **B39**, 13201.

Hu, Z.P., B.C. Pan, W.C. Fan and A. Ignatiev, 1990, Phys. Rev. **B41**, 9692.

Huang, H., S.Y. Tong, J. Quinn and F. Jona, 1990a, Phys. Rev. **B41**, 3276.

Huang, H., S.Y. Tong, W.S. Yang, H.D. Shih and F. Jona, 1990b, Phys. Rev. **B42**, 7483.

Hui, K.C., R.H. Milne, K.A.R. Mitchell, W.T. Moore and M.Y. Zhou, 1985, Solid State Commun. **56**, 83.

Hutchins, B.A., T.N. Rhodin and J.E. Demuth, 1976, Surface Sci. **54**, 419.

Ignatiev, A., F. Jona, D.W. Jepsen and P.M. Marcus, 1973, Surface Sci. **40**, 439.

Ignatiev, A., F. Jona, D.W. Jepsen and P.M. Marcus, 1975a, Phys. Rev. **B11**, 4780.

Ignatiev, A., F. Jona, D.W. Jepsen and P.M. Marcus, 1975b, Surface Sci. **49**, 189.

Imbihl, R., R.J. Behm, G. Ertl and M. Moritz, 1982, Surface Sci. **123**, 123.

Jedrecy, N., M. Sauvage-Simkin, R. Pinchaux, J. Massies, N. Greiser and V.H. Etgens, 1990, Surface Sci. **230**, 197.

Jensen, V., J.N. Andersen, H.B. Nielsen and D.L. Adams, 1982, Surface Sci. **116**, 66.

Johansson, L.S.O., R.I.G. Uhrberg, P. Mårtensson and G.V. Hansson, 1990, Phys. Rev. **B42**, 1305.

Joly, Y., Y. Gauthier and R. Baudoing, 1989, Phys. Rev. **B40**, 10119.

Jona, F., and P.M. Marcus, 1983, Phys. Rev. Lett. **53**, 1823.

Jona, F., and P.M. Marcus, 1987, Solid State Commun. **64**, 667.

Jona, F., D. Westphal, A. Goldman and P.M. Marcus, 1983, J. Phys. **C16**, 3001.

Jones, G.J.R., and B.W. Holland, 1985, Solid State Commun. **53**, 45.

Jones, R.G., S. Ainsworth, M.D. Crapper, C. Somerton and D.P. Woodruff, 1987, Surface Sci. **179**, 425.

Kahn, A., J. Carelli, D. Kanani, C.B. Duke, A. Paton and L. Brillson, 1981, J. Vac Sci. Technol. **19**, 331.

Kao, C.T., G.S. Blackman, M.A. Van Hove, G.A. Somorjai and C-M. Chan, 1989, Surface Sci. **224**, 77.

Kasch, M., E. Pehlke, W. Schattke, T. Kurberg, H.P. Barnscheidt, R. Manze and M. Skibowski, 1989, Surface Sci. **214**, 436.

Kawazu, A., and H. Sakama, 1988, Phys. Rev. **B37**, 2704.

Kerkar, M., D. Fisher, D.P. Woodruff and B. Cowie, 1992a, Surface Sci. **271**, 45.

Kerkar, M., D. Fisher, D.P. Woodruff, R.G. Jones, R.D. Diehl and B. Cowie, 1992b, Phys. Rev. Lett. **68**, 3204.

Kesmodel, L.L., L.H. Dubois and G.A. Somorjai, 1979, J. Chem. Phys. **70**, 2180.

Kevan, S.D., 1985, Phys. Rev. **B32**, 2344.

Kevan, S.D., R.F. Davis, D.H. Rosenblatt, J.G. Tobin, M.G. Mason, D.A. Shirley, C.H. Li and S.Y. Tong, 1981, Phys. Rev. Lett. **46**, 1629.

Kilcoyne, A.L.P, D.P. Woodruff, A.W. Robinson, Th. Lindner, J.S. Somers and A.M. Bradshaw, 1991, Surface Sci. **253**, 107.

Kleinle, G., M. Skottke, V. Penka, R.J. Behm, G. Ertl and W. Moritz, 1987, Surface Sci. **189/190**, 177.

Kleinle, G., J. Wintterlin, G. Ertl, R.J. Behm, F. Jona and W. Moritz, 1990, Surface Sci. **225**, 171.

Klink, C., M. Foss, I. Stensgaard and F. Besenbacher, 1991, Surface Sci. **251/252**, 841.

Koestner, R.J., M.A. Van Hove and G.A. Somorjai, 1981, Surface Sci. **107**, 439.

Kubby, J.A., J.E. Griffith, R.S. Becker and J.S. Vickers, 1987, Phys. Rev. **B36**, 6079.

Kuk, Y., and L.C. Feldman, 1984, Phys. Rev. **B30**, 5811.

Kuk, Y., L.C. Feldman and P.J. Silverman, 1983, Phys. Rev. Lett. **50**, 511.

Lairson, B., T.N. Rhodin and W. Ho, 1985, Solid State Commun. **55**, 925.

Lambert, R., R.L. Trevor, M.J. Cardillo, A. Sakai and D.R. Hamann, 1987, Phys. Rev. **B35**, 8055.

Lamble, G.M., R.S. Brooks, S. Ferrer, D.A. King and D. Norman, 1986, Phys. Rev. **B34**, 2975.

Lamble, G.M., R.S. Brooks, J.C. Campuzano and D.A. King, 1987, Phys. Rev. **B36**, 1796.

Lamble, G.M., R.S. Brooks, D.A. King and D. Norman, 1988, Phys. Rev. Lett. **61**, 1112.

Lang, E., K. Muller, K. Heinz, M.A. Van Hove, R.J. Koestner and G.A. Somorjai, 1983, Surface Sci. **127**, 347.

Lee, B.W., A. Alsenz, A. Ignatiev and M.A. Van Hove, 1978, Phys. Rev. **B17**, 1510.

Legg, K.O., F. Jona, D.W. Jepsen and P.M. Marcus, 1977, J. Phys. **C10**, 937.

Leiung, K.T., L.J. Terminello, Z. Hussain, X.S. Zhang, Y. Hayashi and D.A. Shirley, 1988, Phys. Rev. **B38**, 8241.

Li, H., Y.S. Li, J. Quinn, D. Tian, J. Sokolov, F. Jona and P.M. Marcus, 1990, Phys. Rev. **B42**, 9195.

Liepold, S., N. Elbel, M. Michl, W. Nichtl-Pecher, K. Heinz and K. Muller, 1990, Surface Sci. **240**, 81.

Lindgren, S.A., L. Wallden, J. Rundgren, P., Westrin and J. Neve, 1983, Phys. Rev. **B28**, 6707.

Lindgren, S.A., L. Wallden, J. Rundgren and P. Westrin, 1984, Phys. Rev. **B29**, 576.

Lindner, Th., J. Somers, A.M. Bradshaw, A.L. Kilcoyne and D.P. Woodruff, 1988, Surface Sci. **203**, 333.

Lindroos, M., H. Pfnür, P. Feulner and D. Menzel, Surface Sci., 1987, 180, 421.

Lindroos, M., H. Pfnür, G. Held and D. Menzel, 1989, Surface Sci. **222**, 451.

Lindroos, M., C.J. Barnes, M. Bowker and D.A. King, 1991, in: The Structure of Surfaces III (Springer, Berlin) p. 287.

Lu, S.H., Z.Q. Wang, D. Tian, Y.S. Li, F. Jona and P.M. Marcus, 1989, Surface Sci. **221**, 35.

Máca, F., M. Scheffler and W. Bernt, 1985, Surface Sci. **160**, 467.

Maclaren, J.M., J.B. Pendry, P.J. Rous, D.K. Saldin, G.A. Somorjai, M.A. Van Hove and D.D. Vvedensky, 1987, Surface Crystallographic Information Service; A Handbook of Surface Structures (Reidel, Dordrecht).

Maglietta, M., 1982, Solid State Commun 43, 43.

Maglietta, M., E. Zanazzi, F. Jona, D.W. Jepsen and P.M. Marcus, 1977, Appl. Phys. 15, 409.

Maglietta, M., E. Zanazzi, U. Bardi and F. Jona, 1978, Surface Sci. 77, 101.

Maglietta, M., E. Zanazzi, U. Bardi, D. Sondericker, F. Jona and P.M. Marcus, 1981, Phys. Rev. Lett. 47, 657.

Materlik, G., A. Frohm and M.J. Bedzyk, 1984, Phys. Rev. Lett. 52, 441.

Marcano, J., Y. Darici, H. Min, Y. Yin and P.A. Montano, 1989, Surface Sci. 217, 1.

Martinez, V., F. Soria, M.C. Munoz and J.L. Sacedon, 1983, Surface Sci. 128, 424.

Masud, N., R. Baudoing, D. Aberdam and C. Gaubert, 1983, Surface Sci. 133, 580.

McConville, C.F., D.P. Woodruff, K.C. Prince, G. Paolucci, V. Chab, M. Surman and A.M. Bradshaw, 1986, Surface Sci. 166, 221.

Mendez, M.A., W. Oed, A. Fricke, L. Hammer, K. Heinz and K. Muller, 1991, Surface Sci. 253, 99.

Meyer, R.J., C.B. Duke, A. Paton, J.C. Tsang, J.L. Yeh, A. Kahn and P. Mark, 1980, Phys. Rev. B22, 6171.

Meyerheim, H.L., U. Dobler and A. Puschmann, 1991, Phys. Rev. B44, 5738.

Michalk, G., W. Moritz, H. Pfnür and D. Menzel, 1983, Surface Sci. 129, 92.

Michl, M., W. Nichtl-Pecher, W. Oed, H. Landskron, K. Heinz and K. Muller, 1989, Surface Sci. 220, 59.

Moore, W.T., P.R. Watson, D.C. Frost and K.A.R. Mitchell, 1979, J. Phys. C12, L887.

Morales, L., D.O. Garza and L.J. Clarke, 1981, J. Phys. C14, 5391.

Moritz, W., and D. Wolf, 1985, Surface Sci. 163, L655.

Moritz, W., R. Imbihl, R.J. Behm, G. Ertl, and T. Matsushima, 1985, J. Chem. Phys. 83, 1959.

Mrtsik, B.J., R. Kaplan, T.L. Reinecke, M.A. Van Hove and S.Y. Tong, 1977, Phys. Rev. B15, 897.

Mullins, D.R. and S.H. Overbury, 1989, Surface Sci. 210, 481.

Muschiol, U., P. Bayer, K. Keinz, W. Oed and J.B. Pendry, 1992, Surface Sci. 275, 185.

Nicholls, J.M. and B. Reihl, 1989, Surface Sci. 218, 237.

Nicholls, J.M., G.V. Hansson, U.O. Karlsson, R.I.G. Uhrberg, R. Engelhart, K. Seki, S.A. Flodstrom and E.E. Kock, 1984, Phys. Rev. Lett. 52, 1555.

Nielsen, H.B. and D.L. Adams, 1982, J. Phys. C15, 615.

Nitchl, W., 1987, Surface Sci. 188, L729.

Noonan, J.R. and H.L. Davis, 1984, Phys. Rev. B29, 4349.

Noonan, J.R. and H.L. Davis, 1987, Phys. Rev. Lett. 59, 1714.

Noonan, J.R. and H.L. Davis, 1990, J. Vac. Sci. Technol. A8, 2671.

Noonan, J.R. and H.L. Davis, 1993, unpublished manuscript.

Noonan, J.R., H.L. Davis and W. Erley, 1985, Surface Sci. 152/153, 4349.

Oed, W., B. Doetsch, L. Hammer, K. Heinz and K. Mueller, 1988a, Surface Sci. 207, 207.

Oed, W., W. Puchta, N. Bickel, K. Heinz, W. Nitchl and K. Muller, 1988b, J. Phys. C21, 237.

Oed, W., B. Dotsch and L. Hammer, 1988c, Surface Sci. 240, 81.

Oed, W., H. Lindner, U. Starke, K. Heinz K. Mueller and J.B. Pendry, 1989a, Surface Sci. 224, 224.

Oed, W., H. Lindner, U. Starke, K. Heinz, K. Mueller and J.B. Pendry, 1989b, Surface Sci. 224, 179.

Oed, W., H. Lindner, U. Starke, K. Heinz, K. Mueller, D.K. Saldin, P.L. de Andres and J.B. Pendry, 1990a, Surface Sci. 225, 242.

Oed, W., U. Starke, F. Bothe and K. Heinz, 1990b, Surface Sci. 234, 72.

Ogletree, D.F., M.A. Van Hove and G.A. Somorjai, 1986, Surface Sci. 173, 351.

Ogletree, D.F., M.A. Van Hove and G.A. Somorjai, 1987, Surface Sci. **183**, 1.

Ohtani, H., M.A. Van Hove and G.A. Somorjai, 1987, Surface Sci. **187**, 372.

Ohtani, H., M.A. Van Hove and G.A. Somorjai, 1988, J. Phys. Chem. **92**, 3974.

Over, H., H. Bludau, M. Skottke-Klein, W. Moritz, C.T. Cambell and G. Ertl, 1992, Phys. Rev. **B63**, 8638.

Overbury, S.H., D.R. Mullins, M.F. Paffett and B.E. Koel, 1991, Surface Sci. **254**, 45.

Pandey, K.C., 1981, Phys. Rev. Lett. **47**, 1913.

Patel, J.R., J.A. Golovchenko, P.E. Freeland and H.J. Grossman, 1987, Phys. Rev. **B36**, 7715.

Patel, J.R., D.W. Berreman, F. Sette, P.H. Citrin, J.E. Rowe, P.L. Cowan, T. Jach and B. Karlin, 1989, Phys. Rev. **B40**, 1330.

Parkin, S.R., H.C. Zeng, M.Y. Zhou and K.A.R. Mitchell, 1990, Phys. Rev. **B41**, 5432.

Pauling, L., 1960, The Nature of the Chemical Bond (Cornell University Press, Ithaca, NY).

Pendry, J.B., K. Heinz, W. Oed, H. Landskron, K. Muller and G. Schmidtlein, 1988, Surface Sci. **193**, L1.

Pfnür, H., G. Held, M. Lindroos and D. Menzel, 1989, Surface Sci. **220**, 43.

Piercy, P., P.A. Heimann, G. Michalk and D. Menzel, 1989, Surface Sci. **219**, 189.

Powers, J.M., A. Wander, P.J. Rous, M.A. Van Hove and G.A. Somorjai, 1991, Phys. Rev. **B44**, 11159.

Powers, J.M., A. Wander, M.A. Van Hove and G.A. Somorjai, 1992, Surface Sci. **260**, L7.

Prutton, M., J.A. Ramsey, J.A. Walker and M.R. Welton-Cook, 1979, J. Phys. **C12**, 5271.

Quinn, J., Y.S. Li, D. Tian, F. Jona and P.M. Marcus, 1990, Phys. Rev. **B42**, 11348.

Reimer, W., V. Penka, M. Skottke, R.J. Behm, G. Ertl and W. Moritz, 1987, Surface Sci. **186**, 45.

Rhodin, T.N. and G. Ertl, eds., 1979, The Nature of The Surface Chemical Bond (North-Holland, Amsterdam).

Robey, S.W., C.C. Bahr, Z. Hussain, J.J. Barton, K.T. Leung, J. Lou, A.E. Schach von Wittenau and D.A. Shirley, 1987, Phys. Rev. **B35**, 5657.

Rosenblatt, D.H., S.D. Kevan, J.G. Tobin, R. Davis, M.G. Mason, D.A. Shirley, J.C. Tang and S.Y. Tong, 1982a, Phys. Rev. **B26**, 3181.

Rosenblatt, D.H., S.D. Kevan, J.G. Tobin, R. Davis, M.G. Mason, D.A. Shirley, J.C. Tang and S.Y. Tong, 1982b, Phys. Rev. **B26**, 1812.

Rous, P.J., J.B. Pendry, D.K. Saldin, K. Heinz, K. Muller and N. Bickel, 1986, Phys. Rev. Lett. **57**, 2951.

Rous, P.J., D. Jentz, D.G. Kelly, R.Q. Hwang, M.A. Van Hove and G.A. Somorjai, 1991, The Structure of Surfaces III (Springer, Berlin) p. 432.

Salwen, A. and J. Rundgren, 1975, Surface Sci. **53**, 523.

Schmalz, A., S. Aminpirooz, L. Becker, J. Haase, J. Neugebauer, M. Scheffler, D.R. Batchelor, D.L. Adams and E. Bøgh, 1991, Phys. Rev. Lett. **2163**, 2163.

Sette, F., T. Hashizume, F. Comin, A.A. MacDowell and P.H. Citrin, 1988, Phys. Rev. Lett. **61**, 1384.

Shi, M., H. Bu and J.W. Rabalais, 1990, Phys. Rev. **B42**, 2852.

Shih, H.D., F. Jona, D.W. Jepsen and P.M. Marcus, 1976, Surface Sci. **60**, 445.

Shih, H.D., F. Jona, D.W. Jepsen and P.M. Marcus, 1977a, Phys. Rev. **B15**, 5550.

Shih, H.D., F. Jona, D.W. Jepsen and P.M. Marcus, 1977b, J. Phys. **C9**, 1405.

Shih, H.D., F. Jona, U. Bardi and P.M. Marcus, 1980, J. Phys. **C13**, 3801.

Shih, H.D., F. Jona, D.W. Jepsen and P.M. Marcus, 1981, Phys. Rev. Lett. **46**, 731.

Skottke, M., R.J. Behm, G. Ertl, V. Penka and W. Moritz, 1988, J. Chem. Phys. **46**, 981.

Smit, L., R.M. Tromp and J.F. van der Veen, 1985, Surface Sci. **163**, 198.

Sokolov, J., H.D. Shih, U. Bardi, F. Jona and P.M. Marcus, 1984a, J. Phys. **C17**, 371.

Sokolov, J., F. Jona and P.M. Marcus, 1984b, Phys. Rev. **B29**, 5402.

Sokolov, J., F. Jona and P.M. Marcus, 1985, Phys. Rev. **B31**, 1929.

Sokolov, J., F. Jona and P.M. Marcus, 1986a, Phys. Rev. **B33**, 1397.

Sokolov, J., F. Jona and P.M. Marcus, 1986b, Europhys. Lett. **1**, 401.

Sondericker, D., F. Jona and P.M. Marcus, 1986a, Phys. Rev. **B34**, 6770.

Sondericker, D., F. Jona and P.M. Marcus, 1986b, Phys. Rev. **B33**, 900.

Sondericker, D., F. Jona and P.M. Marcus, 1986c, Phys. Rev. **B34**, 6775.

Starke, U., F. Bothe, W. Oed and K. Heinz, 1990, Surface Sci. **232**, 56.

Stensgaard, I., R. Feidenhans'l and J.E. Sorensen, 1983, Surface Sci. **128**, 281.

Streater, R.W., W.T. Moore, P.R. Watson, D.C. Frost and K.A.R. Mitchell, 1978, Surface Sci. **72**, 744.

Stuck, A., J. Osterwalder, L. Schlapbach and H.C. Poon, 1991, Surface Sci. **151/152**, 670.

Takata, Y., H. Sato, S. Yagi, T. Yokoyama, T. Ohta and Y. Kitajima, 1992, Surface Sci. **265**, 111.

Takayanagi, K., Y. Tanishiro, M. Takahashi and S. Takahashi, 1985a, J. Vac. Sci. Technol. **A3**, 1302.

Takayanagi, K., Y. Tanishiro, M. Takahashi and S. Takahashi, 1985b, Surface Sci. **164**, 367.

Tang, J.C., X.S. Feng, J.F. Shen, T. Fujikawa and T. Okazawa, 1991, Phys. Rev. **B44**, 13018.

Terminello, L.J., X.S. Zhang, Z.Q. Huang, S. Kim, A.E. Schach von Wittenau, K.T. Leung and D.A. Shirley, 1988, Phys. Rev. **B38**, 3879.

Thompson, K.A. and C.S. Fadley, 1984, Surface Sci. **146**, 281.

Tian, D., R.F. Lin, F. Jona and P.M. Marcus, 1990, Solid State Commun. **74**, 1017.

Tian, D., H. Li, F. Jona and P.M. Marcus, 1991, Solid State Commun. **80**, 783.

Titov, A., and W. Moritz, 1982, Surface Sci. **123**, L709.

Titov, A., and H. Jagodzinski, 1985, Surface Sci. **152/153**, 409.

Tong, S.Y., H. Huang and C.M. Wei, 1990, Chem. Phys. Solid. Surfaces VIII **22**, 395.

Tougaard, S., A. Ignatiev and D.L. Adams, 1982, Surface Sci. **115**, 270.

Tromp, R.M., R.G. Smeenk, F.W. Saris and D.J. Chadi, 1983, Surface Sci. **133**, 137.

Tromp, R.M., R.J. Hamers and J.E. Demuth, 1985, Phys. Rev. Lett. **55**, 1303.

Uhrberg, R.I.G., and G.V. Hansson, 1991, Crit. Rev. Solid State Mater. Sci. **17**, 133.

Unertl, W.N. and H.V. Thapliyal, 1975, J. Vac. Sci. Technol. **12**, 142.

Urano, T., Y. Uchida, S. Hongo and T. Kanaji, 1991, Surface Sci. **242**, 39.

Vanderbilt, D., 1987, Phys. Rev. **B36**, 6209.

van der Veen, J.F., R.G. Smeenk, R.M. Tromp and F.W. Saris, 1979, Surface Sci. **79**, 219.

Van Hove, M.A., 1979, in: The Nature of The Surface Chemical Bond, eds. T.N. Rhodin and G. Ertl (North-Holland, Amsterdam).

Van Hove, M.A. and R.J. Koestner, 1984, in: Determination of Surface Structure by LEED, eds. F. Jona and P.M. Marcus (Plenum, New York, NY), p. 357.

Van Hove, M.A. and G.A. Somorjai, 1989, Prog. Surface Sci. **30**, 201.

Van Hove, M.A. and S.Y. Tong, 1975a, Phys. Rev. Lett. **35**, 230.

Van Hove, M.A. and S.Y. Tong, 1975b, J. Vac. Sci. Technol. **12**, 1092.

Van Hove, M.A. and S.Y. Tong, 1975c, J. Vac. Sci. Technol. **12**, 230.

Van Hove, M.A., S.Y. Tong and N. Stoner, 1976, Surface Sci. **54**, 259.

Van Hove, M.A., B.J. Mrtsik, R. Kaplan, T.L. Reinecke and S.Y. Tong, 1977, Surface Sci. **64**, 85.

Van Hove, M.A., R.J. Koestner, P.C. Stair, J.P. Biberian, L.L. Kesmodel, I. Bartos and G.A. Somorjai, 1981, Surface Sci. **127**, 218.

Van Hove, M.A., R.J. Koestner, J.C. Frost and G.A. Somorjai, 1983, Surface Sci. **129**, 482.

Van Hove, M.A., R.F. Lin and G.A. Somorjai, 1986, J. Am. Chem. Soc. **108**, 2532.

Van Hove, M.A., R.F. Lin, G.S. Blackman and G.A. Somorjai, 1987, Acta Cryst. **B43**, 368.

von Eggeling, C., G. Schmidt, G. Besold, L. Hammer, K. Heinz and K. Muller, 1989, Surface Sci. **221**, 11.

Vrijmoeth, J., P.M. Zagwijn, J.W.M. Frenken and J.F. van der Veen, 1991, Phys. Rev. Lett. **67**, 1134.

Wan, K.J., W.K. Ford, G.J. Lapeyre and J.C. Hermanson, 1991a, Phys. Rev. **B44**, 3471.

Wan, K.J., W.K. Ford, G.J. Lapeyre and J.C. Hermanson, 1991b, Phys. Rev. **B44**, 6500.

Wander, A., M.A. Van Hove and G.A. Somorjai, 1991a, Phys. Rev. Lett. **67**, 626.

Wander, A., G. Held, R.Q. Hwang, G.S. Blackman, M.L. Xu, P.L. de Andres, M.A. Van Hove and G.A. Somorjai, 1991b, Surface Sci. **249**, 21.

Wang, L-Q., A.E. Schach von Wittenau, Z.G. Ji, L.S. Wang, Z.Q. Huang and D.A. Shirley, 1991a, Phys. Rev. **B44**, 1292.

Wang, L-Q., Z. Hussain, Z.Q. Huang, A.E. Schach von Wittenau, D.W. Lindle and D.A. Shirley, 1991b, Phys. Rev. **B44**, 1292.

Wang, Z.Q., Y.S. Li, C.K.C. Lok, J. Quinn, F. Jona and P.M. Marcus, 1987, Solid State Commun. **62**, 181.

Wei, C.M., H. Huang, S.Y. Tong, G.S. Glander and M.B. Webb, Phys. Rev., 1990, B42, 11284.

Welz, M., W. Moritz and D. Wolf, 1978, Surface Sci. **125**, 473.

Winograd, N. and C-C. Chang, 1989, Phys. Rev. Lett. **62**, 2568.

Wolkow, R.A., 1992, Phys. Rev. Lett. **68**, 2636.

Wong, P.C., and K.A.R. Mitchell, 1987, Surface Sci. **187**, L599.

Wong, P.C., M.Y. Zhou, K.C. Hui and K.A.R. Mitchell, 1985, Surface Sci. **163**, 172.

Wong, P.C., K.C. Hui, M.Y. Zhou and K.A.R. Mitchell, 1986, Surface Sci. **165**, L21.

Wong, P.C., J.R. Lou and K.A.R. Mitchell, 1988, Surface Sci. **206**, L913.

Wu, N.J., and A. Ignatiev, 1982, Phys. Rev. **B25**, 2983.

Wu, N.J., and A. Ignatiev, 1983, Phys. Rev. **B28**, 7288.

Wu, S.C., Z.Q. Wang, Y.S. Li, F. Jona and P.M. Marcus, 1986, Phys. Rev. **B33**, 2900.

Wu, Z.Q., S.H. Lu, Z.Q. Wang, C.K.C. Lok, J. Quinn, Y.S. Li, D. Tian, F. Jona and P.M. Marcus, 1988, Phys. Rev. **B38**, 5363.

Xu, M.L. and S.Y. Tong, 1985, Phys. Rev. **B31**, 6332.

Yasilove, S.M., 1986, Surface Sci. **171**, 400.

Yang, W.S., J. Sokolov, F. Jona and P.M. Marcus, 1982, Solid State Commun **41**, 191.

Yang, W.S., F. Jona and P.M. Marcus, 1983, Phys. Rev. **B28**, 7377.

Yokoyama, T., Y. Takata, M. Yoshiki, T. Ohta, M. Funabashi and Y. Kitajima, 1989, Jpn. J. Appl. Phys. **28**, L1637.

Yokoyama, T., Y. Takata, T. Ohta, M. Funabashi, Y. Kitajima and H. Kuroda, 1990, Phys. Rev. **B42**, 7000.

Zeng, H.C., and K.A.R. Mitchell, 1989, Langmuir **5**, 829.

Zeng, H.C., and K.A.R. Mitchell, 1990, Surface Sci. **239**, L571.

Zeng, H.C., R.A. McFarlane and K.A.R. Mitchell, 1990, Can. J. Phys. **68**, 353.

Zhang, X.S., L.J. Terminello, S. Kim, Z.Q. Huang, A.E. Schach von Wittenau and D.A. Shirley, 1988, J. Chem. Phys. **89**, 6538.

Zhao, R.G., J. Jia, Y. Li and W.S. Yang, 1991, Surface Sci. **24**, 517.

Zhu, X and S.G. Louie, 1991, Phys. Rev. **B43**, 12146.

Zschack, P., 1991, The Structure of Surfaces III (Springer, Berlin) p. 646.

Chapter II

THEORY OF SURFACE STRUCTURE AND BONDING

J.E. INGLESFIELD

*Institute for Theoretical Physics, University of Nijmegen,
NL-6525 ED Nijmegen, The Netherlands*

Contents

Cohesion and Structure of Surfaces
Edited by D.G. Pettifor
© 1995 Elsevier Science B.V. All rights reserved.

Abstract

This chapter describes theoretical and computational studies of surface structure, based on solving the electronic Schrödinger equation. This is done within the framework of density functional theory, in which the complicated many-body motion of all the electrons is replaced by an equivalent but simpler problem of each electron moving in an effective potential. The basis of density functional theory, the way that the Schrödinger equation is solved at surfaces, and how the equilibrium atomic structure is determined are presented. These are used to discuss the energetics, surface relaxation and surface reconstructions of metals, adsorbates on metals, and the surface reconstructions of semiconductors.

1. Introduction

The electrons are the glue which binds solids together, and the change in electronic wavefunctions at the surface causes the re-arrangement of atoms which we call surface relaxation and reconstruction. Advances in experimental techniques for determining surface atomic structure have been matched by advances in theory and computational methods for calculating surface electronic structure. It is now possible to *predict* the complicated reconstructions of semiconductor surfaces, with equal or better accuracy than experimental measurements. In fact these measurements often involve calculations of surface electronic structure in order to interpret the spectra in terms of atomic positions: in LEED the comparison of I/V curves with theoretical predictions plays an essential role in full surface structure determinations (Van Hove and Tong, 1979), in SEXAFS the X-ray absorption spectrum can only be interpreted in terms of the atomic structure if the scattering of the excited electron is calculated (Woodruff and Delcher, 1986), and even in STM a proper understanding of the image depends on knowing the energy distribution of the electronic states at the surface which can tunnel through to the tip (or *vice versa*) (Hörmandinger, 1994). The final link between surface electronic structure and experiment consists of experiments like photoemission and inverse photoemission which directly measure the energy distribution of states at the surface (Andrews et al., 1992).

There is a whole range of techniques designed to tackle the problem of bonding and surface electronic structure, ranging from density functional theory which has resulted in very accurate studies of the electronic charge density and ground-state energy, and the relationship between energy and atomic structure, through to simplified models which describe bonding in terms of the local environment of atoms together with interatomic potentials These techniques and results found using them form the subject of this chapter.

2. Solving the Schrödinger equation at the surface

The first stage in understanding the behaviour of electrons in solids and at surfaces is to separate their motion from that of the atomic nuclei. This is because the electrons (at $\mathbf{r}_i$) are much lighter, and consequently they satisfy the Schrödinger equation[1] in which the nuclei (at $\mathbf{r}_I$) are at rest:

$$\left\{ -\frac{1}{2}\sum_i \nabla_i^2 + \frac{1}{2}\sum_{i \neq j} \frac{1}{|\mathbf{r}_i - \mathbf{r}_j|} - \sum_{i,I} \frac{Z_I}{|\mathbf{r}_i - \mathbf{r}_I|} \right\} \Psi(\{\mathbf{r}_i\}) = E_0 \Psi(\{\mathbf{r}_i\}). \quad (1)$$

The sums run over all the electrons i and the nuclei I, Z_I is the atomic number of the Ith nucleus, and Ψ is the many-electron wavefunction. The motion of the nuclei is then governed by equations of motion (usually taken to be classical) in which the ground-state electronic energy E_0, which is a function of nuclear coordinates $\mathbf{r}_I$, behaves like a potential energy. $-\nabla_I E_0$ plus the Coulomb force from the other nuclei is then the force on the Ith nucleus, and by minimizing E_0 plus the nucleus–nucleus electrostatic energy with respect to the $\mathbf{r}_I$'s the equilibrium geometry can be found. This separation of nuclear and electronic motion is the adiabatic or Born–Oppenheimer approximation (Ziman, 1972). Corrections to it give the electron–phonon interaction, which looms large in transport theory but fortunately not here.

2.1. The many-electron problem and density-functional theory

The main problem in solving the electronic Schrödinger equation (1) is the electron–electron interaction, the repulsive Coulomb potential between the 10^{23} electrons in a reasonably sized piece of material. The most important development in electronic structure calculations in the last 30 years was the realization by Kohn, Hohenberg and Sham (Hohenberg and Kohn, 1964; Kohn and Sham, 1965) that this many-electron equation can be expressed in single-particle form, each electron moving in a potential field determined by the charge density of all the other electrons, in addition to the electrostatic potential due to the nuclei. Something like this picture is familiar to us from the Hartree–Fock method of theoretical chemists (Callaway, 1991; Szabo and Ostlund, 1982; Pisani et al., 1988), in which each electron moves in the

---

[1] Atomic units are used, with $e = \hbar = m_e = 1$. The unit of energy is the Hartree, 27.2 eV, and the unit of distance the hydrogen Bohr radius, 0.5292 Å.

electrostatic (Hartree) potential of the charge density of the other electrons, modified by a non-local exchange potential (non-local means that it depends on the wavefunction under consideration). This exchange potential comes from the hole in the electron distribution surrounding our electron due to the Pauli exclusion principle. The Hartree–Fock method is based on the variational method, starting from a trial many-electron wavefunction — a Slater determinant of one-electron wavefunctions. These are varied until the expectation value of the energy is minimized, and of course the accuracy of the method is limited by this form of wavefunction. By contrast, in the density-functional theory of Kohn, Hohenberg and Sham the aim is not to construct an accurate many-electron wavefunction, but simply (relatively speaking!) the ground-state charge density of the system. It turns out that this is enough to determine all other ground-state properties of the system, in particular for our purposes the ground-state energy.

Density-functional theory is based on the fact that there is a one-to-one correspondence between the ground-state charge density of interacting electrons, and the external potential (due to the nuclei) in which they are sitting (Hohenberg and Kohn, 1964; Kohn and Vashishta, 1983). In other words, knowing the charge density the potential can in principle be found, and *vice versa*. So all physical properties of the system are functionals of the ground-state charge density $\rho_0(\mathbf{r})$ (a functional is a function of a function), in particular the ground-state energy E_0:

$$E_0 = E[\rho_0(\mathbf{r})] \tag{2}$$

— if we know this functional we can find the ground-state energy and charge density by minimizing E_0.

To make progress in finding this functional dependence, and in fact to transform the problem into one-electron form, we consider *non-interacting* electrons moving in an effective potential v_{eff} so that the ground-state charge density equals $\rho_0(\mathbf{r})$. First, we define the exchange-correlation energy E_{xc} as the difference between E_0, and the kinetic energy T of this non-interacting system plus the electron–nucleus and Hartree potential energy:

$$E[\rho_0(\mathbf{r})] = T[\rho_0(\mathbf{r})] + \int d\mathbf{r}\rho_0(\mathbf{r})v_{\text{nuc}}(\mathbf{r})$$
$$+ \frac{1}{2} \int d\mathbf{r} \int d\mathbf{r}'\rho_0(\mathbf{r}) \frac{1}{|\mathbf{r} - \mathbf{r}'|}\rho_0(\mathbf{r}') + E_{\text{xc}}[\rho_0(\mathbf{r})]. \tag{3}$$

Now the kinetic energy can be written in terms of the ground-state energy of the non-interacting electron gas E', by subtracting the potential energy of the electrons in the field of v_{eff}:

$$T = E' - \int d\mathbf{r}\rho_0(\mathbf{r})v_{\text{eff}}(\mathbf{r}). \tag{4}$$

So we obtain:

$$E[\rho_0(\mathbf{r})] = E'[\rho_0(\mathbf{r})] - \int d\mathbf{r}\rho_0(\mathbf{r})v_{\text{eff}}(\mathbf{r}) + \int d\mathbf{r}\rho_0(\mathbf{r})v_{\text{nuc}}(\mathbf{r})$$

$$+ \frac{1}{2}\int d\mathbf{r} \int d\mathbf{r}'\rho_0(\mathbf{r})\frac{1}{|\mathbf{r}-\mathbf{r}'|}\rho_0(\mathbf{r}') + E_{\text{xc}}[\rho_0(\mathbf{r})]. \qquad (5)$$

Taking the functional derivative with respect to changes in $\rho_0(\mathbf{r})$ gives us:

$$\frac{\delta E}{\delta\rho_0(\mathbf{r})} = \frac{\delta E'}{\delta\rho_0(\mathbf{r})} - v_{\text{eff}}(\mathbf{r}) + v_{\text{nuc}}(\mathbf{r}) + \int d\mathbf{r}'\frac{1}{|\mathbf{r}-\mathbf{r}'|}\rho_0(\mathbf{r}') + \frac{\delta E_{\text{xc}}}{\delta\rho_0(\mathbf{r})}. \quad (6)$$

As $\delta E/\delta\rho_0$ and $\delta E'/\delta\rho_0$ are both zero — the energies are stationary with respect to changes in density — we see that the total effective potential, which gives the required charge density of non-interacting electrons, is given by:

$$v_{\text{eff}}(\mathbf{r}) = v_{\text{nuc}}(\mathbf{r}) + \int d\mathbf{r}'\frac{1}{|\mathbf{r}-\mathbf{r}'|}\rho_0(\mathbf{r}') + \frac{\delta E_{\text{xc}}}{\delta\rho_0(\mathbf{r})}. \qquad (7)$$

The second term in (7) is just the Hartree potential $V_{\text{H}}(\mathbf{r})$, and the third term is the exchange-correlation potential:

$$V_{\text{xc}}(\mathbf{r}) = \frac{\delta E_{\text{xc}}}{\delta\rho_0(\mathbf{r})}. \qquad (8)$$

The ground-state energy of the non-interacting electrons in the field of the effective potential can be found from the eigenvalues of the single-particle Schrödinger equation (Kohn and Sham, 1965):

$$-\frac{1}{2}\nabla^2\psi_i(\mathbf{r}) + v_{\text{nuc}}(\mathbf{r})\psi_i(\mathbf{r}) + V_{\text{H}}(\mathbf{r})\psi_i(\mathbf{r}) + V_{\text{xc}}(\mathbf{r})\psi_i(\mathbf{r}) = \epsilon_i\psi_i(\mathbf{r}). \quad (9)$$

We fill the N lowest states (each state is doubly degenerate, as the spin-up and spin-down states have the same energy) with the N electrons, giving:

$$E' = \sum_i \epsilon_i$$

$$\rho_0(\mathbf{r}) = \sum_i |\psi_i(\mathbf{r})|^2. \qquad (10)$$

By substituting this into (3) we finally obtain our required expression for the ground-state energy of the interacting system:

$$E_0 = \sum_i \epsilon_i - \frac{1}{2}\int d\mathbf{r}V_{\text{H}}(\mathbf{r})\rho_0(\mathbf{r}) - \int d\mathbf{r}V_{\text{xc}}(\mathbf{r})\rho_0(\mathbf{r}) + E_{\text{xc}}[\rho_0(\mathbf{r})]. \quad (11)$$

The exchange-correlation energy functional E_{xc}, and the exchange-correlation potential V_{xc} contain all the complexities of the electron–electron interaction, and we have reduced the many-body problem to single-particle form — at least we have a single-particle problem from which we can find the ground-state energy and charge density of the interacting system.

Have we eliminated the many-body problem? No — we have simply put all our ignorance about the many-body problem into E_{xc}, and we have to know its functional dependence on $\rho_0(\mathbf{r})$ in order to find $\rho_0(\mathbf{r})$ and E_0. The reason why the density-functional method is useful is that we can make use of the local density approximation (LDA) for E_{xc} (Kohn and Sham, 1965), writing:

$$E_{xc} \approx \int d\mathbf{r}\rho_0(\mathbf{r})\epsilon_{xc}(\rho_0(\mathbf{r})), \tag{12}$$

where $\epsilon_{xc}(\rho_0(\mathbf{r}))$ is the exchange-correlation energy, per electron, of an infinite, homogeneous electron gas with density equal to the local density $\rho_0(\mathbf{r})$. The exchange-correlation potential is then given by:

$$V_{xc}(\mathbf{r}) = \frac{d}{d\rho}\rho\epsilon_{xc}(\rho)|_{\rho=\rho_0}. \tag{13}$$

ϵ_{xc} is well-known in certain limits — at high electron densities where the random phase approximation is valid, and at low densities where the electrons crystallize into a Wigner lattice, and some sort of interpolation scheme or RPA generalization can be used in between (Pines, 1963). Currently the best form to use for $\epsilon_{xc}(\rho)$ comes from the numerical simulations of the electron gas by Ceperley and Alder (1980).

Density-functional theory is ideal for studying bonding, as the ground-state quantities which it is designed to calculate — principally ground-state charge density and total energy — are just those needed for understanding bonding. The individual eigenvalues and eigenfunctions in (9), ϵ_i, $\psi_i(\mathbf{r})$, have no meaning, except as constructs from which the ground-state energy and charge density can be found. However, in many cases the eigenvalues do correspond rather well to the energies needed to remove an electron from occupied states (as in photoemission) or to add an electron to the unoccupied states (as in inverse photoemission). There are many famous exceptions to this — the band gaps in semiconductors are invariably too small (Godby, 1992), and this means that care is needed in comparing calculated semiconductor surface state energies with experiment, for example.

2.2. The reduced symmetry at the surface

Having obtained an effective one-electron Schrödinger equation using density-functional theory, the next problem is solving it. In a bulk crystal this is relatively straightforward, as the three-dimensional periodicity of the crystal lattice makes it only necessary to solve the Schrödinger equation in one unit cell — Bloch's theorem tells us that the wavefunction in a cell displaced by lattice vector $\mathbf{r}_L$ is the same, apart from a phase factor $\exp(i\mathbf{k}.\mathbf{r}_L)$, containing the Bloch wavevector $\mathbf{k}$. At a surface the periodicity is broken in the perpendicular direction, and there remains only two-dimensional periodicity parallel to the surface. The wavefunctions are then labelled by a two-dimensional Bloch wavevector $\mathbf{K}$ (Inglesfield, 1982; Zangwill, 1988) such that:

$$\psi_{\mathbf{K}}(\mathbf{R} + \mathbf{R}_M, z) = \exp(i\mathbf{K} \cdot \mathbf{R}_M)\psi_{\mathbf{K}}(\mathbf{R}, z), \tag{14}$$

where $(\mathbf{R}, z)$ is the position vector with components parallel and perpendicular to the surface, and $\mathbf{R}_M$ is a vector of the surface mesh. So the Schrödinger equation needs to be solved in only one surface unit cell, but in the perpendicular direction it has to be solved from $-\infty$ deep in the vacuum to $+\infty$ inside the bulk.

There are two classes of eigenstate at the surface (Inglesfield, 1982). Bulk states hit the surface, and are reflected by the surface potential barrier; and at energies at which bulk states cannot propagate at the particular surface wavevector $\mathbf{K}$ under consideration, localized surface states may occur, decaying exponentially both into the bulk and into the vacuum.

Rather than consider individual states (rather meaningless in the continuum), it is convenient to consider the local density of states, given by:

$$\sigma_{\mathbf{K}}(\mathbf{r}, E) = \sum_i |\psi_{\mathbf{K},i}(\mathbf{r})|^2 \, \delta(E - \epsilon_{\mathbf{K},i}), \tag{15}$$

— the energy distribution of the charge density of states with wavevector $\mathbf{K}$. Figure 1 gives $\sigma_{\mathbf{K}}(\mathbf{r}, E)$ integrated through the surface and sub-surface atoms for Al(001) at $\mathbf{K} = 0$, and shows the continua of bulk states at the surface as well as a discrete surface state. Already by the sub-surface layer, the local density of states is looking more bulk-like. The results of fig. 1 were found using the embedding method (Inglesfield and Benesh, 1988), which is a way of including the scattering of the wavefunctions by the bulk in the Hamiltonian for a surface region of finite thickness. This is one technique amongst several for solving the Schrödinger equation for the surface of a true semi-infinite solid (Inglesfield, 1987). The peak in fig. 1 at $E = 0.21$ a.u. comes from a genuinely localized surface state (Inglesfield and Benesh, 1988), but here we should mention that for certain bulk band structures surface resonances can occur — broadened peaks in $\sigma_{\mathbf{K}}(E)$ coming from a surface state coupling weakly with a bulk continuum (Inglesfield, 1982; Zangwill, 1988).

The most widely used technique for finding surface electronic structure is, in fact, to treat a thin slab (typically 5 or so atomic layers thick) rather

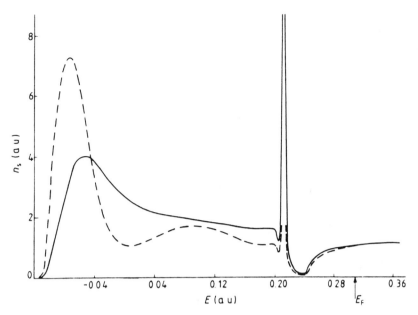

Fig. 1. Surface density of states with **K** = 0 on Al(001) (Inglesfield and Benesh, 1988). Full curve, top layer of atoms; broken curve, second layer. The densities of states are calculated with an imaginary part of the energy = 0.001 a.u., which broadens the discrete surface state by this amount.

than the semi-infinite solid, giving a finite problem in the z-direction. If the slab is repeated periodically, with vacuum in between, we obtain a "slab superlattice" with full three-dimensional periodicity which can be treated by crystal band-structure methods. The eigenstates of the slab are different from those of the actual surface problem, with discrete states at fixed wavevector **K** (there is negligible dispersion with the perpendicular component of the wavevector as the slabs barely interact with each other — see sect. 2.3.1). Figure 2 shows results from a 7-layer slab calculation for Al(001) (Benesh and Inglesfield, 1984), and although we can clearly see the relationship between this spectrum and the actual surface density of states (fig. 1), the discretization may be a problem for comparison with photoemission experiments, for example, in which the surface density of states (at fixed **K** in angle-resolved experiments) is probed (Van Hoof et al., 1992). However, quantities like charge density and total energy, which are our main concern in studies of surface bonding, depend on sums over states, and these are much more local properties than the individual eigenstates: local, that is, in the sense that the charge density for example at the surface is not much affected by the presence of the second surface of the slab, as long as it is more than a few screening lengths away. The

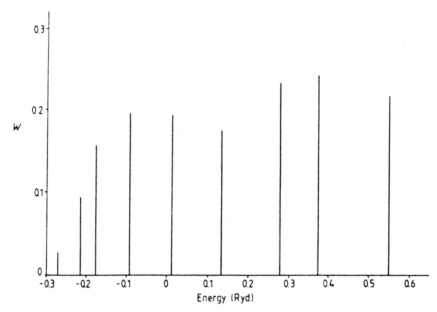

Fig. 2. Al(001) seven-layer slab calculation at **K** = 0: weight of states in the surface layer (Benesh and Inglesfield, 1984).

thickness of the vacuum between the slabs has to be fairly thick, say 20 a.u., so that the tails of the charge density outside each slab do not overlap significantly.

2.3. Basis functions for slab calculations

The advantage of slabs is that well-tried, conventional band-structure methods can be used to solve the Schrödinger equation. Here we shall discuss two. The most widely used basis set for expanding the wavefunctions consists of plane waves, with relatively weak pseudopotentials replacing the deep ionic potentials. Plane waves have the advantage of being relatively simple, the matrix elements of the full potential (including the rapid variation at the surface) can easily be found, and there is a well-developed method for finding forces so that atomic positions can be optimized (sect. 3). The disadvantage of plane waves is that even for semiconductors, with relatively simple pseudopotentials, a large number of plane waves is needed per atom, leading to a large matrix representing the Hamiltonian. A much more economical basis set is provided by LMTOs (Linearized Muffin Tin Orbitals) which we shall also describe.

2.3.1. Pseudopotentials and plane waves

A pseudopotential scatters the valence or conduction electrons in the same way as the actual potential (Bachelet et al., 1982; Pickett, 1989), and consequently a crystal built up out of pseudopotentials has the same band-structure as the original crystal potential. The advantage of the pseudopotential is that it is weaker than the actual potential, without any core states and consequently having smoother valence or conduction band wavefunctions inside the core. A plane wave basis can then be used:

$$\psi_{\mathbf{k}}(\mathbf{r}) = \sum_{\mathbf{g}} a_{\mathbf{g}} \exp i (\mathbf{k} + \mathbf{g}) \cdot \mathbf{r} \tag{16}$$

— the sum is over the reciprocal lattice vectors of the three-dimensional slab superlattice. Given the lack of interaction between the slabs we may take $\mathbf{k} = \mathbf{K}$, with the z-component of the three-dimensional Bloch wavevector equal to zero.

The pseudopotential is in principle an energy- and angular momentum-dependent operator. However it is possible to construct an energy-independent pseudopotential which has the same scattering properties as the actual atomic potential over a range of electron energies (Bachelet et al., 1982). Moreover a pseudopotential with this property has the additional property that the pseudo-wavefunction, which is different from the actual wavefunction (smoother) inside the atomic core, has the same integrated charge density in the atomic core as the actual wavefunction; the normalized pseudo-wavefunction is exactly the same in amplitude as well as form as the actual wavefunction outside the core. This means that the solid built up out of pseudopotentials will have just the same bonding properties and electronic energy spectrum as the actual solid. A complete set of norm-conserving/energy-independent pseudopotentials has been given by Bachelet et al. (1982) (fig. 3).

Plane wave basis sets of reasonable size (of the order of 100 plane waves per atom) can be used in surface calculations of semiconductors like Si, Ge, GaAs (Hebenstreit et al., 1991) and s–p bonded metals like Al (Needs and Godfrey, 1990). For transition and noble metals the d-electrons which participate in the bonding are relatively localized and correspondingly feel a stronger potential than the s- and p-electrons. A plane-wave basis for such systems is unwieldy, and localized Gaussian orbitals of the form:

$$\phi_{\alpha l m}(\mathbf{r}) = e^{-\alpha r^2} Y_{lm}(\Omega) \tag{17}$$

may be used either to supplement the plane waves (Louie et al., 1979), or by themselves in a linear combination of atomic orbitals approach. In this way pseudopotentials can be used to study the electronic structure of such

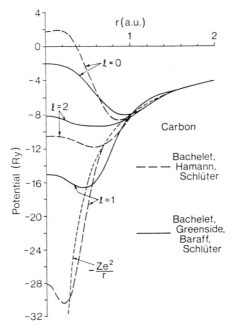

Fig. 3. The s, p and d pseudopotentials for the C atom, found with two different prescriptions (Bachelet et al., 1982; Pickett, 1989). Both scatter electrons in the same way as the actual potential.

systems as Rh(001) (Feibelman, 1991), Rh on Au(001) (Zhu et al., 1991), and Pd on Ag(001) (Zhu et al., 1990) in slab or slab superlattice geometry.

2.3.2. Linearized muffin tin orbitals

Linearized muffin tin orbitals (LMTOs) (Andersen, 1975; Skriver, 1984; Zeller, 1992) form a very economical basis set, with one basis function for each valence state angular momentum (l, m) on each atom. Each atom is put into a "muffin tin", within which the potential is almost spherically symmetric; between the muffin tins the potential is fairly flat, so the wavefunctions in this interstitial region satisfy the free-electron Schrödinger equation. The (l, m) basis function centred on the atom at the origin, say, is taken to be $r^{-(l+1)} Y_{lm}(\Omega)$ in the interstitial region, a spherical free-electron wave with zero kinetic energy. Within each muffin tin the solution of the Schrödinger equation is linearized around some fixed energy (an average energy in the range of interest), and the free-electron wave in the interstitial region is matched in amplitude and derivative over the surface of the muffin tin onto a linear combination of atomic solutions $u_{l'}$ and energy derivatives $\dot{u}_{l'}$ evaluated at this energy. Figure 4 shows the form of the LMTO, and the

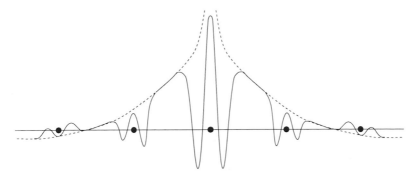

Fig. 4. Schematic LMTO, with $r^{-(l+1)}$ matched onto atomic solutions.

trial function for the whole system then consists of a linear combination of these atom-centred functions. LMTOs are usually used within the atomic sphere approximation (ASA), in which the muffin tins are expanded into space-filling atomic spheres, so that the contribution of the interstitial region to the Hamiltonian and overlap matrix elements can be dropped. Within each atomic sphere the potential is spherically averaged.

To apply LMTOs to slab superlattice geometry, the vacuum between the slabs is packed with empty spheres (Van Leuken et al., 1992). For many purposes standard LMTO methods can be used to solve the self-consistency part of the problem — only the monopole part of the potential produced by each atomic sphere is taken into account when constructing the potential — and this gives surprisingly good surface densities of states and magnetic moments for example. Good work functions and surface energies can be obtained when the dipole contribution to the potential outside each atomic sphere is taken into account as well as the monopole (Skriver and Rosengaard, 1992), still taking the spherical average of the total potential within each atomic sphere for solving the Schrödinger equation. Recently a full potential version of LMTO has been developed, using non-overlapping muffin-tin spheres and taking into account not only the non-spherical part of the potential within these spheres but also the full interstitial potential (Methfessel, 1988; Methfessel et al., 1992).

A linear combination of the LMTO basis functions can be taken to construct a new set of very short range basis functions, whose range extends only to nearest neighbours (Andersen et al., 1986); this gives an effectively tight-binding Hamiltonian. This is particularly useful if the surface problem is treated not within slab geometry, but using Dyson's equation to find the Green function for the solid with a surface from the bulk Green function (Skriver and Rosengaard, 1991, 1992). This is because the perturbation due to making the surface couples relatively few basis functions if these are short range. Tight binding LMTOs are also very useful in treating large systems, because of the simplification in evaluating the structure constants (essentially re-expanding the spherical free-electron wave about a distant atomic site for the purpose of matching the atomic solutions). A widely used basis set closely related to LMTOs is the Augmented Spherical Wave (ASW) basis (Williams et al., 1979), in which the basis function in the interstitial region is taken as a spherical free-electron wave with negative kinetic energy, in other words a spherical Hankel function rather than $r^{-(l+1)}$. This is matched on to atomic solutions $u_{l'}$ within each atomic sphere (no $\dot{u}_{l'}$), evaluated at energies such that amplitude and derivative are continuous across the surface of the muffin tin. A transformation of ASWs into short range Localized Spherical Waves (LSWs) can be made (Van Leuken et al., 1990), using ideas related to the LMTO transformation.

2.4.   Self-consistency in surface calculations

An important aspect of density-functional theory (like Hartree–Fock and other mean-field theories) is that the Schrödinger equation (9) has to be solved *self-consistently*, because the potentials V_H and V_{xc} depend on the wavefunctions themselves. In its simplest form this means making an initial guess at the effective potential (7), then using the output wavefunctions and charge density to construct a new potential and iterating till convergence.

In practice, straightforward iteration is wildly unstable in surface calculations, and one way of proceeding is to mix a small fraction α of the potential $v^{(o)}$ from the output charge density with the input potential to obtain an input potential for the next iteration (Pickett, 1989; Zeller, 1992):

$$v|_{n+1} = (1 - \alpha)v|_n + \alpha v^{(o)}|_n. \tag{18}$$

The origin of the instability is the long range of the Coulomb potential, and the problem is especially acute at surfaces where the surface dipole produced by shifting a small amount of electronic charge into the vacuum gives rise to a shift of the potential throughout the whole solid. Another way of looking at this is that in slab calculations, as described in sect. 2.2, the unit cell in the z-direction is very big, corresponding to small components of reciprocal lattice vectors. As Poisson's equation in reciprocal space is:

$$V_{\mathbf{g}} = \frac{4\pi}{g^2}\rho_{\mathbf{g}} \tag{19}$$

we see that small g's are associated with huge shifts in potential for small changes in charge density. Despite this sensitivity of the potential to the exact charge density, convergence can always be achieved using simple linear mixing (18), for sufficiently small mixing parameter α (Dederichs and Zeller,

1983). In surface calculations α is typically a few per cent but may be smaller, and then the problem is that a very large number of iterations is needed to achieve self-consistency.

A great improvement can be achieved by using Newton–Raphson methods (Bendt and Zunger, 1982; Srivastava, 1984) for solving our self-consistency problem, which consists of finding the value of input potential $v(\mathbf{r})$ for which the output potential $v^{(o)}(\mathbf{r})$, a functional of $v(\mathbf{r})$, satisfies:

$$v^{(o)}[v] - v = 0. \tag{20}$$

Let us consider an input potential $\tilde{v}$ away from self-consistency, which we vary by δv. Then (schematically):

$$v^{(o)}[\tilde{v} + \delta v] - (\tilde{v} + \delta v) \approx v^{(o)}[\tilde{v}] - \tilde{v} + J\delta v, \tag{21}$$

where:

$$J = \frac{\partial}{\partial v}(v^{(o)}[v] - v) \tag{22}$$

— the variation in the left hand side of (20) with input potential. So the approximate solution of (20) is:

$$v = \tilde{v} - J^{-1}(v^{(o)}[\tilde{v}] - \tilde{v}). \tag{23}$$

The Newton–Raphson method consists of using the potential given by (23) as the input for the next iteration. Of course v and $v^{(o)}$ are functions of $\mathbf{r}$, so J is a functional derivative; normally the potentials are expanded in terms of some basis functions, then J becomes a matrix relating the vectors of expansion coefficients. J is a linear response function of the system (related to the dielectric function), and in some implementations of the Newton–Raphson method this is calculated explicitly (Vanderbilt and Louie, 1984). In the Broyden method, one of the most widely used methods for accelerating convergence, a gradually improving approximation to J^{-1} is built up out of input and output potentials directly (Srivastava, 1984). The Broyden method can lead to spectacular improvements in convergence rate (Singh et al., 1986). We shall describe other approaches to the self-consistency problem in sect. 3.3, when we come to discuss structural optimization.

It is important in actual calculations to use a properly variational expression for the energy, so that errors in energy are second order in errors in charge density and effective potential (Pickett, 1989; Weinert et al., 1985). Suppose the effective potential $\tilde{v}_{\text{eff}}$ in the Schrödinger equation (9) gives an output charge density $\rho^{(o)}$, which is away from complete self-consistency. The kinetic energy corresponding to $\rho^{(o)}$ is:

$$T[\rho^{(o)}(\mathbf{r})] = \sum_i \epsilon_i - \int d\mathbf{r}\tilde{v}_{\text{eff}}(\mathbf{r})\rho^{(o)}(\mathbf{r}), \tag{24}$$

which adding on to the Hartree energy of $\rho^{(o)}$ and the other contributions gives a variational expression for the energy:

$$E = T[\rho^{(o)}(\mathbf{r})] + \int d\mathbf{r}\rho^{(o)}(\mathbf{r})v_{\text{nuc}}(\mathbf{r})$$
$$+ \frac{1}{2}\int d\mathbf{r}\int d\mathbf{r}'\rho^{(o)}(\mathbf{r})\frac{1}{|\mathbf{r}-\mathbf{r}'|}\rho^{(o)}(\mathbf{r}') + E_{\text{xc}}[\rho^{(o)}(\mathbf{r})]. \tag{25}$$

(25) is *not* identical to (11) except at exact self-consistency. This is the expression which is used in self-consistent calculations, providing an upper limit to the energy and converging quadratically with errors in ρ.

3. Forces and optimization

One of the aims of surface electronic structure calculations is to find the energy of the combined electron–atom system, and then minimize the energy with respect to atomic positions to determine the equilibrium structure. As density-functional theory is designed specifically to give the ground-state charge density and energy, this programme rests on firm ground, and indeed there are some spectacularly good results.

3.1.   Forces and the Hellmann–Feynman theorem

Assuming that we can solve the density-functional Schrödinger equation self-consistently for a particular set of nuclear coordinates $\mathbf{r}_I$, energy minimization is helped if we can find the forces on the atoms:

$$\mathbf{F}_I = -\frac{\partial E_0}{\partial \mathbf{r}_I} \tag{26}$$

(we include the nucleus–nucleus Coulomb repulsion in E_0 as well as the electron energy E_0). The Hellmann–Feynman theorem tells us that this is given by the electrostatic force on the nucleus due to the other nuclei and the electronic charge density (Ihm et al., 1979). To show that the force due to the electrons appears in this classical way we start from the many-electron Hamiltonian H with ground-state electron wavefunction Ψ and energy E_0:

$$H\Psi = \text{E}_0\Psi. \tag{27}$$

Varying the Hamiltonian, to first order:

$$\delta H\Psi + H\delta\Psi = \delta\text{E}_0\Psi + \text{E}_0\delta\Psi, \tag{28}$$

hence:

$$\langle\Psi|\delta H|\Psi\rangle + \langle\Psi|H|\delta\Psi\rangle = \delta\text{E}_0\langle\Psi|\Psi\rangle + \text{E}_0\langle\Psi|\delta\Psi\rangle, \tag{29}$$

and with normalized Ψ:

$$\delta E_0 = \langle \Psi | \delta H | \Psi \rangle. \tag{30}$$

From (1) the variation in H produced by moving atom I is:

$$\delta H = - \sum_i \frac{\partial}{\partial \mathbf{r}_I} \left(\frac{Z_I}{|\mathbf{r}_i - \mathbf{r}_I|} \right) \cdot \delta \mathbf{r}_I, \tag{31}$$

hence $\langle \Psi | \delta H | \Psi \rangle$ is just the electrostatic force on the nucleus due to the electron density.

The Hellmann–Feynman theorem holds for the density-functional expression for E_0, and it is instructive to work through this. Writing the one-electron eigenvalues in terms of the Hamiltonian h in (9), (11) becomes:

$$E_0 = \sum_i \langle \psi_i | h | \psi_i \rangle$$
$$- \frac{1}{2} \int d\mathbf{r} V_H(\mathbf{r}) \rho_0(\mathbf{r}) - \int d\mathbf{r} V_{xc}(\mathbf{r}) \rho_0(\mathbf{r}) + E_{xc}[\rho_0(\mathbf{r})], \tag{32}$$

so using the result that $\delta \langle \psi_i | h | \psi_i \rangle = \langle \psi_i | \delta h | \psi_i \rangle$:

$$\delta E_0 = \sum_i \langle \psi_i | \delta h | \psi_i \rangle - \frac{1}{2} \int d\mathbf{r} \delta V_H(\mathbf{r}) \rho_0(\mathbf{r}) - \frac{1}{2} \int d\mathbf{r} V_H(\mathbf{r}) \delta \rho_0(\mathbf{r})$$
$$- \int d\mathbf{r} \delta V_{xc}(\mathbf{r}) \rho_0(\mathbf{r}) - \int d\mathbf{r} V_{xc}(\mathbf{r}) \delta \rho_0(\mathbf{r}) + \delta E_{xc}. \tag{33}$$

But:

$$\sum_i \langle \psi_i | \delta h | \psi_i \rangle = \int d\mathbf{r} (\delta v_{nuc}(\mathbf{r}) + \delta V_H(\mathbf{r}) + \delta V_{xc}(\mathbf{r})) \rho_0(\mathbf{r}), \tag{34}$$

and $\int d\mathbf{r} \delta V_H \rho_0$ cancels with the second and third terms on the right hand side of (33). Moreover, from the definition of the exchange-correlation potential (8) we have:

$$\delta E_{xc} = \int d\mathbf{r} V_{xc}(\mathbf{r}) \delta \rho_0(\mathbf{r}). \tag{35}$$

So we are left with:

$$\delta E_0 = \int d\mathbf{r} \delta v_{nuc}(\mathbf{r}) \rho_0(\mathbf{r}) \tag{36}$$

— the same as (30) and (31), and the force on the atom is just the electrostatic force on the nucleus once again.

In pseudopotential calculations with plane wave basis sets, forces are determined using essentially this Hellmann–Feynman result (Ihm et al., 1979). If the pseudopotential is local (independent of angular momentum) we can take over (36) directly, with δv_{nuc} replaced by the shift in the ionic pseudopotential. With non-local pseudopotentials (the usual state of affairs), the integrand must be separated into the charge density of angular momentum components of the wavefunctions multiplied by the relevant angular momentum component of the pseudopotential. This can all be done most conveniently in terms of reciprocal lattice summations.

3.2. Pulay corrections to the force

The Hellmann–Feynman theorem gives the force on an atom in terms of the electric field produced by the exact charge density. Of course we never deal with the exact charge density, because a finite basis set is used in the wavefunction expansion like (16). With basis functions like plane waves which do not move with the displacement of the atom, (36) still holds even when a finite basis is used, as long as self-consistency is achieved within the basis. However, when the basis functions are themselves dependent on atomic positions, as with LMTOs (sect. 2.3.2) or plane waves augmented with local functions, an extra term has to be added on to the Hellmann–Feynman force — unless we are in the impossible situation of knowing the exact charge density.

The extra contribution to the force comes from the occurrence of the overlap matrix S in the matrix form of the Schrödinger equation with position-dependent basis functions (Yu et al., 1991). Expanding the wavefunctions in terms of basis functions χ_l:

$$\psi(\mathbf{r}) = \sum_l \psi_l \chi_l(\mathbf{r}), \tag{37}$$

the eigencoefficients are given by the matrix equation:

$$\sum_m H_{lm} \psi_m = \epsilon \sum_m S_{lm} \psi_m, \tag{38}$$

where H_{lm} and S_{lm} are the matrix elements:

$$H_{lm} = \int d\mathbf{r} \chi_l^* h \chi_m$$
$$S_{lm} = \int d\mathbf{r} \chi_l^* \chi_m. \tag{39}$$

Varying (38), exactly as in (27), we obtain (schematically):

$$\delta H \psi + H \delta \psi = \delta \epsilon S \psi + \epsilon \delta S \psi + \epsilon S \delta \psi, \tag{40}$$

hence:

$$\langle \psi | \delta H | \psi \rangle + \langle \psi | H | \delta \psi \rangle = \delta \epsilon \langle \psi | S | \psi \rangle + \epsilon \langle \psi | \delta S | \psi \rangle + \epsilon \langle \psi | S | \delta \psi \rangle. \tag{41}$$

Using $\langle\psi|H|\delta\psi\rangle = \epsilon\langle\psi|S|\delta\psi\rangle$ and the fact that the normalization of the wavefunction is $\langle\psi|S|\psi\rangle$ we obtain:

$$\delta\epsilon = \langle\psi|\delta H|\psi\rangle - \epsilon\langle\psi|\delta S|\psi\rangle. \tag{42}$$

The first term on the right hand side is what we had before. The second term is the new ingredient, and is the Pulay correction to the force. It clearly vanishes with basis functions independent of atomic position.

Expressions for the force with LAPW basis functions (linearized augmented plane waves — plane waves augmented by atomic solutions inside the muffin tins) have been given by Soler and Williams (1989), and Yu et al. (1991), including the Pulay contribution from the change in overlap matrix. Up to now there have been relatively few applications of their results, but as LAPWs provide an accurate, flexible and widely-used basis set we can expect them to be useful in the future. Methfessel and Van Schilfgaarde (1993) have used a quite different approach from what we have described so far to obtain a force theorem for full-potential LMTOs. Instead of starting from (11) or (25) for the total energy, they start from an expression derived by Harris (1985) and Foulkes and Haydock (1989) which gives the total energy in terms of the one-electron eigenvalue sum corrected by terms involving only the *input* charge density and potential. Knowing the actual self-consistent charge density for some configuration of atoms, an ansatz can then be made for its variation with atomic displacement which can be used in the Harris–Foulkes energy expression to find the corresponding variation in energy and force. This has been used to optimize the geometry of the large molecule Ti_8C_{12}, for example (Methfessel et al., 1993).

3.3.   Atomic structure and electronic structure optimization

Knowing the forces on the atoms at particular positions, the atoms can be moved in the direction of the forces, and the calculation repeated until the forces are zero and the total energy is minimum. This procedure is carried out nowadays in many ab initio studies of surface structure, and we shall see examples in subsequent sections of this chapter.

In order to tackle large and complex structures, new methods have recently been developed for solving the electronic part of the problem. These are mostly applied to the pseudopotential plane wave method, because of the simplicity of the Hamiltonian matrix elements with plane wave basis functions and the ease with which the Hellmann–Feynman forces can be found. Conventional methods of matrix diagonalization for finding the energy eigenvalues and eigenfunctions of the Kohn–Sham Hamiltonian in (9) can tackle matrices only up to about 1000×1000. As a basis set of about 100 plane waves per atom is needed, this restricts the size of problem to systems containing about 10 atoms per unit cell — not big, especially in a surface context where the supercell has to be quite big in the z-direction anyway. The fundamental problem with conventional matrix methods is that the time for finding eigenvalues and eigenvectors scales with the dimension M of the matrix as M^3. There is also the problem of storing the matrix elements in the first place. To handle big systems, the new methods calculate only the N wavefunctions which are actually occupied ($N \approx M/100$), using iterative techniques to relax the wavefunctions until they are solutions of the Kohn–Sham Schrödinger equation. The Hartree and exchange-correlation potentials can be determined from the ψ_i's at every iteration step, enabling self-consistency to be achieved simultaneously with finding the eigenvectors.

The first big breakthrough along these lines was the work of Car and Parrinello (1985), who derived an equation of motion for the electron wavefunctions which can be integrated using molecular dynamics techniques. By removing energy from the system (effectively adding a damping term to the equation of motion), the wavefunctions relax down to the solutions of the Kohn–Sham equation. At the same time, the forces on the atoms can be found from Hellmann–Feynman, and molecular dynamics applied to them. Removing energy from the atomic system (*simulated annealing*), the atoms relax down to the ground-state structure (there is always the possibility of a local minimum corresponding to a metastable structure, of course). The Car–Parrinello technique can be used for dynamical simulation as well as determining the equilibrium structure (Ancilotto et al., 1990).

Solving the self-consistent Schrödinger equation can be regarded as an optimization problem, minimizing E given by (2) as a functional of the N occupied wavefunctions (or a function of their expansion coefficients (16)) (Payne et al., 1992). Given a set of N trial ψ_i's, the most obvious way of improving the expectation value of the energy is the Steepest Descent method (Štich et al., 1989) — changing ψ_i in the direction (in function space) which gives the biggest decrease in E. This corresponds to changing ψ_i by:

$$\delta \psi_i(\mathbf{r}) \propto -\frac{\delta E}{\delta \psi_i^*(\mathbf{r})}$$
$$\propto -H\psi_i(\mathbf{r}), \tag{43}$$

subject to the normalization of ψ_i and its orthogonality to the other occupied wavefunctions. The process of following the line of steepest descent is continued, recalculating V_H and V_{xc} in H at every step, until the electron energy is minimized.

Steepest descents can in fact be beaten by the Conjugate Gradient method (Štich et al., 1989; Teter et al., 1989). Suppose the function to be optimized

Fig. 5. Steepest descents down a long narrow valley (Press et al., 1989). Copyright Cambridge University Press 1986, 1992; and reprinted with permission of Cambridge University Press.

(as a function of 2 variables) has the form of a long narrow valley (fig. 5). Then starting off from the right hand end of the arrow, steepest descents can take us down a very circuitous path to the valley floor. In conjugate gradients, we proceed after the first trajectory along the *conjugate* direction such that the minimum along this direction is the absolute minimum of the local quadratic form of the function. With a multi-dimensional function like the energy, the conjugate direction for $\delta\psi_i$ can be defined in an analogous way, and the conjugate gradient method is the best iterative procedure for reaching the ground-state electronic energy. As E expanded to second order in ψ_i contains information about the change in Hartree and exchange-correlation potentials, the way that these change with changes in wavefunction is taken into account in the direction and size of $\delta\psi_i$ at each step, and problems of instability in the self-consistency part of the problem (sect. 2.4) are reduced. Developments in the technique mean that the electronic structure of systems containing hundreds of atoms can now be determined (Štich et al., 1992).

When using the Conjugate Gradient method (or other iterative methods) to solve the electronic part of the problem, it has to be remembered that errors in the Hellmann–Feynman forces on the atoms are first order in errors in the wavefunctions, whereas the error in E is second order. This means that the electron system should be somewhat relaxed towards its ground state before calculating the Hellmann–Feynman forces and letting the atoms respond — especially near structural equilibrium where the forces are small (Payne et al., 1992).

4.   Clean metal surfaces

Metal surfaces usually show an inward relaxation of the top layer of atoms from the positions they would occupy in the bulk, and in some cases such as W(001) there is surface reconstruction. Experimental observations of these are described in ch. 1, sections 2.1 and 2.2. Here we shall discuss the physical origin of these effects.

4.1.   Surface energy

4.1.1.   Surface energy of simple metals

The surface energy — the work required to make unit area of surface — provides a measure of the change in bonding at the surface, so its understanding is fairly basic to the theme of this book.

The surface energy of the s–p bonded metals, the *simple* metals, was calculated 25 years ago by Lang and Kohn (1970) using a jellium model. In the s–p bonded metals, the nearly-free-electron band-structure shows that the pseudopotential with which we can replace the real ionic potential must be rather weak. Jellium provides a good starting point for the electronic structure, and in particular the surface of the simple metals can be modelled to zeroth order by electrons interacting with the positive background of jellium cut off abruptly at $z = 0$. The surface energy comes from the way that the electrons spill out of the surface (fig. 6): the increase in electrostatic energy (the electrons interact less favourably with the positively charged jellium) gives a positive contribution to the surface energy, whereas the fact that the electrons are more spread out lowers the kinetic energy and gives a negative contribution to the surface energy.

Unfortunately the nett surface energy of jellium goes negative for larger

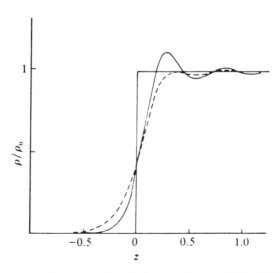

Fig. 6. Self-consistent surface charge density in the jellium model for K (solid line) and Al (dashed line) (Lang and Kohn, 1970). Distance is measured in Fermi wavelengths from the positive background edge; charge density is measured relative to the bulk density ρ_0.

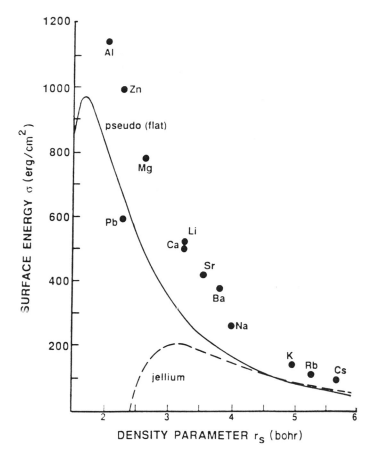

Fig. 7. Surface energy for jellium with structureless pseudopotential (solid line), compared with experiment (dots) (Perdew et al., 1990). The dashed line gives results for jellium without the pseudopotential.

electron densities ($r_s < 2.4$ a.u.[2]), and to obtain stability it is necessary to take the atomic pseudopotentials into account. This can be done by treating the pseudopotentials as a perturbation, and Lang and Kohn (1970) obtained good surface energies in this way. More recently Perdew et al. (1990) have treated the average pseudopotential in the bulk atomic cell as a constant to be added on to the potential in the jellium half-space; when this "structureless" pseudopotential is included, excellent results are obtained (fig. 7). Interestingly enough, Perdew et al. find the surface energy by minimizing the

[2] r_s is the radius of the sphere containing one electron.

Table 1
Surface energies calculated for jellium (Perdew et al.,
1990) and using the LMTO method (Skriver and Rosengaard, 1992), compared with experiment (Skriver and
Rosengaard, 1992). The bold letters indicate the stable
crystal structure.

Metal	Surface	Surface energy (erg/cm^2)		
		Jellium	LMTO	Experiment
Li	**bcc**(110)	326	458	525
	bcc(001)	371	436	
Na	**bcc**(110)	190	307	260
	bcc(001)	216	236	
K	**bcc**(110)	111	116	130
	bcc(001)	115	129	
	fcc(111)		112	
Rb	**bcc**(110)	86	92	110
	bcc(001)	98	107	
	fcc(111)		89	
Cs	**bcc**(110)	69	72	95
	bcc(001)	79	92	
	fcc(111)		70	
Be	**hcp**(001)		2122	2700
Mg	**hcp**(001)	554	642	760
Ca	**fcc**(111)	325	352	490
	bcc(110)		339	
Sr	**fcc**(111)	256	287	410
	bcc(110)		282	
Ba	**bcc**(110)	233	260	370
	fcc(111)		258	
Al	**fcc**(111)	921	1270	1160

Kohn–Sham functional written in terms of the density itself (3), rather than
writing the density in terms of one-electron wavefunctions (10).

Detailed calculations of the surface energy of some of the simple metals
have been carried out by Skriver and Rosengaard (1992), using tight-binding
LMTO basis functions in the atomic sphere approximation (sect. 2.3.2) and a
Green function technique to treat the semi-infinite solid (Skriver and Rosengaard, 1991). Their results are shown in table 1, together with the jellium
results of Perdew et al. (1990) and experimental values taken from surface
tension measurements. Agreement between the two theoretical studies is excellent, apart from the results for Li and Na, where the LMTO calculation is
in better agreement with experiment than the jellium model. The variation of
surface energy from surface to surface is important, particularly for the shape
of crystals: the more open surfaces (e.g. fcc(110), bcc(001) and (111)) tend
to have the higher surface energies, though the face-dependence is rather

small compared with the variation with electron density. Perdew et al. (1990) model the face-dependence of the surface energy by a factor which depends just on the degree of corrugation, varying from 1.15 for fcc(111) to 1.38 for fcc(110), 1.32 for bcc(001) and 1.55 for bcc(111). However it is clear from the detailed LMTO calculations that the variation is not as simple as this.

4.1.2.   Surface energy of transition metals

The surface energy of the 3d and 4d transition metals, calculated using the tight-binding LMTO Green function method (Skriver and Rosengaard, 1992), is shown in fig. 8 as a function of the number of valence electrons, together with experimentally derived surface energies. The most striking feature of these results is the roughly parabolic dependence of the surface energy on the valency, following in fact the behaviour of the cohesive energy. The origin of this behaviour is that increasing the number of electrons corresponds to filling up the tightly bound d band. Let us approximate the density of states in the d band by a constant, with bulk bandwidth W; at the surface the d band is narrowed by δW, and adding up the one-electron energies (corresponding to E' in (10)) the surface energy per surface atom is given by (Skriver and Rosengaard, 1992; Cyrot-Lackmann, 1969):

$$E_s = \frac{1}{2}n(1 - \frac{n}{10})\delta W, \tag{44}$$

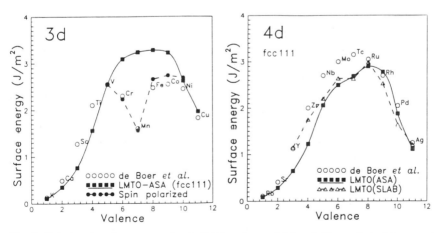

Fig. 8. Calculated surface energy for fcc(111) surfaces of 3d and 4d metals (solid squares), compared with experiment (open circles) (Skriver and Rosengaard, 1992; tight-binding LMTO-ASA, with Green function method). For the 3d metals, the dashed line connecting solid circles gives results from spin-polarized calculations. For the 4d metals, the dashed line connecting open triangles gives results from Methfessel et al. (1992; full potential LMTO, slab geometry).

where n is the number of d electrons. This is only part of the surface energy, of course, but this equation correctly describes the observed parabolic dependence of E_s on n.

There is a large anomaly in the surface energy of the 3d metals around Mn (fig. 8). The origin of this is the magnetism of the elements in this region of the 3d series, as we can see by comparing spin-polarized calculations (i.e. magnetized) with the non-magnetic results, which show the same overall behaviour as the 4d and 5d elements (Aldén et al., 1992). The anomaly is deepest for Cr, Mn and Fe, where calculations show that it is due to an increase in magnetism at the surface of these elements, which lowers the surface energy.

Table 2 shows the calculated surface energies for different surfaces of the 4d elements (Skriver and Rosengaard, 1992; Methfessel et al., 1992). The surface energies per atom increase with decreasing coordination of the surface atoms: fcc(111) < (001) < (110), bcc(110) < (001). Going from the surface energy per atom to the surface energy per unit area, this varies much less from surface to surface, though the open surfaces (fcc(110), bcc(001)) usually have the highest surface energy. The reason for the variation in surface energy per atom is that the bandwidth of the local density of states[3] varies with the square root of the number of neighbours, in tight binding, so δW in (44) is given by (Methfessel et al., 1992):

$$\delta W \propto \sqrt{C_b} - \sqrt{C_s}, \tag{45}$$

where C_b, C_s are the coordination numbers of the bulk and surface atoms. The reduction in bandwidth at the surface is shown very clearly in fig. 9 giving the density of states on surface, sub-surface and bulk-like atoms in a slab calculation for unreconstructed W(001) (Posternak et al., 1980). Bulk W is bcc, and the bulk density of states shows the bonding–antibonding shape characteristic of this structure. At the open (001) surface the number of nearest neighbours is reduced from 8 to 4, giving a comparatively large reduction in bandwidth in the surface density of states with a peak in the middle of the band. As we shall see shortly, this peak has dramatic consequences for the stability of W(001), and in fact is responsible for the surface reconstruction.

4.2. Surface relaxation

LEED experiments show that the outermost interlayer spacing tends to contract, especially on open surfaces like fcc(110) and bcc(001) — in both

---

[3] The local density of states here is (15) integrated over wavevector and projected onto an atomic d-orbital, i.e. the energy distribution of the occupancy of the d-orbital.

Table 2
Surface energies compared with experiment, and surface energy per atom.
LMTO1 results are from Methfessel et al. (1992; full potential LMTO, slab
geometry), and LMTO2 from Skriver and Rosengaard (1992; tight-binding
LMTO-ASA, with Green function method). The bold letters indicate the
stable crystal structure.

| Metal | Surface | Surface energy (erg/cm^2) | | | Per atom (eV) |
		LMTO1	LMTO2	Experiment	LMTO1
Y	**hcp**(001)		680	1130	
	fcc(111)	1150	650		0.73
	fcc(001)	1120			0.82
	fcc(110)	1180			1.21
Zr	**hcp**(001)		1530	2000	
	fcc(111)	1750	1220		0.91
	fcc(001)	1620			0.97
	fcc(110)	1850			1.56
Nb	**bcc**(110)	2360	1640	2700	1.08
	bcc(001)	2860			1.86
	fcc(111)	2200	2060		1.02
	fcc(001)	2110			1.13
	fcc(110)	2260			1.70
Mo	**bcc**(110)	3140	3180	3000	1.34
	bcc(001)	3520			2.13
	fcc(111)	2640	2500		1.11
	fcc(001)	2980			1.45
	fcc(110)	2770			1.90
Tc	**hcp**(001)		2800	3150	
	fcc(111)	2630	2690		1.04
	fcc(001)	3340			1.53
	fcc(110)	3000			1.94
Ru	**hcp**(001)		3320	3050	
	fcc(111)	2990	2900		1.16
	fcc(001)	3520			1.58
	fcc(110)	3450			2.17
Rh	**fcc**(111)	2530	2780	2700	0.99
	fcc(001)	2810	2900		1.27
	fcc(110)	2880			1.84
Pd	**fcc**(111)	1640	1880	2050	0.68
	fcc(001)	1860	1900		0.89
	fcc(110)	1970			1.33
Ag	**fcc**(111)	1210	1120	1250	0.55
	fcc(001)	1210	1200		0.63
	fcc(110)	1260	1290		0.93

these cases the decrease in top spacing Δd_{12} can be as much as 10% (ch. 1,
sect. 2.1, table 1). Surface contraction is not universal, and on Al(111) for
example there is a small outward expansion ($\sim$1%), but it is the general

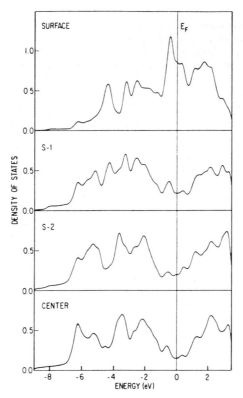

Fig. 9. Local density of states for different layers in a 7-layer slab calculation for ideal W(001)(1×1) (Posternak et al., 1980).

rule. The sub-surface interlayer spacings change, in many cases with an alternation of sign. In the case of Al(110) LEED analysis gives (Noonan and Davis, 1984; ch. 1, table 1):

Δd_{12}	Δd_{23}	Δd_{34}	Δd_{45}
-8.5%	$+5.6\%$	$+2.4\%$	$+1.7\%$.

First principles calculations, minimizing the energy with respect to atomic positions as in sect. 3, give results for surface relaxations in reasonable agreement with experiment, for example a pseudopotential study of Al(110) gives (Ho and Bohnen, 1985):

Δd_{12}	Δd_{23}	Δd_{34}	Δd_{45}
-6.8%	$+3.5\%$	-2.0%	$+1.6\%$.

The tendency for surface contraction has been explained by Finnis and Heine (1974) using the Hellmann–Feynman theorem (sect. 3.1). If the

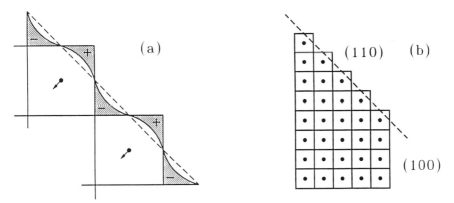

Fig. 10. Smoothing of surface charge (schematic), showing (a) the resulting inward electro-static force on the surface ions; and (b) surface dependence of smoothing, hence inward relaxation (Methfessel et al., 1992).

charge density in the surface atomic cells were undistorted, exactly the same as in the bulk, each ion would remain at the centre of its own atomic cell, feeling no nett electric field from the charge in its own cell, nor from the other cells, because they are nearly spherical: there would be no surface relaxation. However the surface charge (at least on s–p bonded metals) tends to be somewhat smoothed — this smoothing, first invoked by Smoluchowski (1941) to account for work function variations from surface to surface — lowers the kinetic energy of the electrons. The effect of this redistribution of charge at the surface produces a Hellmann–Feynman force on the surface ions in their ideal positions (fig. 10). If the surface charge is cut off on a planar surface, corresponding to complete smoothing, the electrostatic "centre of gravity" at which an ion experiences no nett field corresponds to contractions on fcc(111) of −1.6%, fcc(001) −4.6%, and fcc(110) −16%. These are the right trends, though are overestimates because the surface smoothing is nothing like as dramatic as shown in fig. 10.

We should use a different argument for a qualitative understanding of the inward relaxation of transition metal surfaces (ch. 1, table 1). Not that the Hellmann–Feynman force is inapplicable — it is just that it involves a more subtle redistribution of charge than surface smoothing. In bulk transition metals the equilibrium volume per atom is determined by competition between the d and s–p electrons: the s–p electrons exert an outward pressure, counteracting the effects of d–d bond formation which tends to decrease the interatomic spacing (Pettifor, 1978). At the surface the s–p electrons spill out into the vacuum to lower their kinetic energy, hence the d electrons can now pull the surface atoms inwards to increase their interaction with the substrate (Fu et al., 1984). This competition between s–p

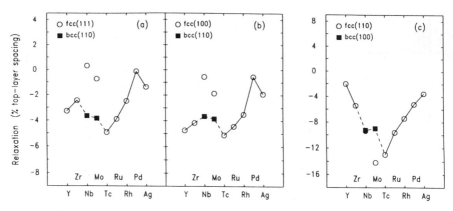

Fig. 11. Calculated top layer relaxation for 4d metals, relative to bulk interlayer spacing (Methfessel et al., 1992). (a) fcc(111) and bcc(110) surfaces; (b) fcc(100) and bcc(110); (c) fcc(110) and bcc(100).

and d electrons gives a roughly parabolic variation with d band filling for the (calculated) surface relaxations of the open fcc(110) and bcc(001) surfaces in the 4d elements (fig. 11) (Methfessel et al., 1992). Calculation reproduces the observed tendency for open surfaces to show greater inward relaxation than the more close-packed surfaces, for the transition metals just as for the simple metals.

The oscillatory behaviour of surface relaxation — inward for Δd_{12}, outward for Δd_{23} — seems to be fairly universal (Fu et al., 1984; Landman et al., 1980; Jiang et al., 1986). It is found not only experimentally and in fully self-consistent calculations, but also in simplified calculations à la Heine–Finnis. If a frozen charge density is used, for example a step density or the Lang–Kohn jellium surface profile, and the ions are relaxed to positions of zero force, oscillatory relaxations are found (Landman et al., 1980). This shows that it is not a consequence of the Friedel oscillations in the surface charge density.

4.3. Surface reconstruction

4.3.1. W(001) and Mo(001)

The W(001) surface reconstruction below room temperature consists of lateral zig-zag displacements of the surface atoms, giving the $(\sqrt{2} \times \sqrt{2})R45°$ structure shown in fig. 12 (Debe and King, 1979; ch. 1, sect. 2.2). A similar reconstruction occurs on Mo(001), but in this case the structure is modulated along the direction of the displacements ($\langle 11 \rangle$ with respect to the cubic x and y axes) to give a unit cell 7 times longer than the $(\sqrt{2} \times \sqrt{2})R45°$

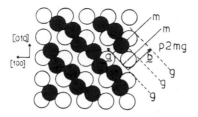

Fig. 12. W(001)($\sqrt{2}\times\sqrt{2}$)R45° reconstruction (Debe and King, 1979).

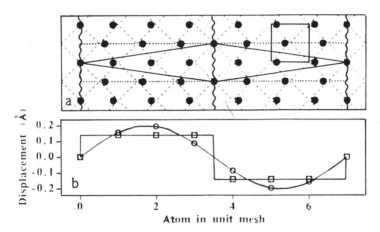

Fig. 13. Reconstruction of Mo(001) (Daley et al., 1993). (a) Antiphase domain structure for T below 125 K; (b) displacements in the antiphase domain structure (squares), and the periodic lattice displacement structure for T above 125 K.

(fig. 13) (Daley et al., 1993). (For a long time it was thought that the Mo(001) reconstruction was incommensurate with the underlying lattice (Felter et al., 1977).) The modulation consists of a sinusoidal modulation of the displacements, which sharpens up as the temperature is lowered into something like domains of ($\sqrt{2}\times\sqrt{2}$)R45° reconstruction separated by antiphase boundaries.

The origin of these reconstructions — or perhaps we should say the instability of the ideal (1×1) surface — lies in the peak of the density of states on the (1×1) surface at the Fermi energy (fig. 9) (Singh and Krakauer, 1988). A peak in the density of states at E_F frequently leads to one sort of instability or another, either structural or magnetic, because a change in structure can split the peak and lower the energy of the occupied states. There has been a long discussion, however, about the role of a surface state Fermi surface in this.

The fact that both W(001) and Mo(001) reconstruct, with a slightly

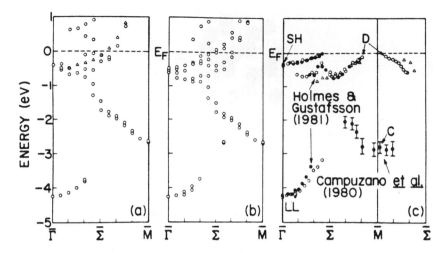

Fig. 14. Dispersion of surface states on W(001) along $\bar{\Sigma}$ (Mattheiss and Hamann, 1984). (a) Relativistic calculation without spin-orbit coupling; (b) fully relativistic calculation; and (c) experimental photoemission results.

different periodicity, is reminiscent of the structural phase transitions which occur in many layer compounds, driven by flattened pieces of the 2D Fermi surface (Felter et al., 1977): the phase transition produces new Brillouin zone boundaries which touch the Fermi surface, lowering the energy of the occupied states, and the precise periodicity of the new unit cell depends on the size of the Fermi surface. This is a sort of Peierls transition, and is sometimes called a charge density wave (CDW) transition, because it is associated with long range oscillations in the screening charge around the displaced atoms. For this mechanism to be effective, the 2D Fermi surface has to be fairly flat, so that the new Brillouin zone boundary makes contact with it over a reasonable length to lower the energy of an appreciable number of electrons. In the surface context, these ideas have been applied to the 2D Fermi surface of the surface states and surface resonances (sect. 2.2) on W(001) and Mo(001) (Inglesfield, 1978; Tosatti, 1978). On these surfaces, a surface state disperses up through E_F about half-way along the $\bar{\Sigma}$ symmetry line where the new Brillouin zone boundary appears in the $(\sqrt{2} \times \sqrt{2})R45°$ reconstruction (fig. 14) (Mattheiss and Hamann, 1984). This has been confirmed by photoemission experiments (Holmes and Gustafsson, 1981), which also show evidence of flattening favourable to reconstruction (Smith et al., 1990).

The Brillouin zone/Fermi surface mechanism is local in reciprocal space, corresponding to long range effects in real space; it clearly provides a nice explanation for the long periodicity modulation of the Mo(001) reconstruc-

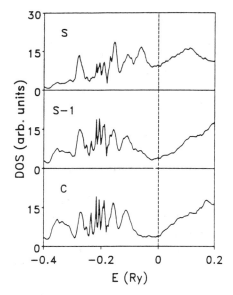

Fig. 15. Local density of states on W(001)($\sqrt{2} \times \sqrt{2}$)R45° (Singh and Krakauer, 1988). (s) Surface layer; (s − 1) sub-surface layer; and (c) bulk-like layer.

tion in terms of the dimensions of the Fermi surface. However there is computational evidence that local (in real space) bonding effects actually dominate. A calculation by Singh and Krakauer (1988) for W(001) gives the minimum energy for the observed reconstruction with displacements of 0.27 Å, in excellent agreement with the measured displacements. The effect of the reconstruction is to split the peak in the surface density of states at E_F (fig. 15), thereby lowering the energy (Singh and Krakauer, 1988). The fact that the calculation is not very sensitive to the number of K-points at which the Schrödinger equation is solved (sect. 2.2) shows that the energy gain is not associated with a limited region of K-space, as in the CDW mechanism. Rather, the peak in the (1×1) surface density of states comes from surface states and surface resonances over a large region of the surface Brillouin zone, corresponding to short range forces in real space. Singh and Krakauer (1988) found an instability of the ideal W(001) surface to many different atomic displacements, confirming the idea that this is quite a general instability.

The Singh–Krakauer calculation does not rule out Fermi surface effects as an additional effect, and these still offer the most plausible explanation for the long periodic reconstruction of Mo(001). The peak in the surface density of states — the main driving force — does come from a large region of the Brillouin zone, but the Fermi surface may be in there somewhere.

4.3.2. *Missing row reconstructions on late 5d metals*

The (110) surfaces of Ir, Pt and Au at the end of the 5d series show a
(1×2) reconstruction, in which alternate ⟨1̄10⟩ rows of atoms are removed,
giving close-packed facets (fig. 16) (Moritz and Wolf, 1985; Fery et al., 1988;
Copel and Gustafsson, 1986; ch. 1, sect. 2.2, table 2). (1×3) reconstructions
also occur, similar to (1×2) but with larger facets (Fery et al., 1988).
A small coverage (∼0.1 monolayer) of adsorbed alkali induces the same
reconstruction on Ag(110), Pd(110) and Cu(110) (Barnes et al., 1985; Hu et
al., 1990).

This reconstruction is connected with the balance between s–p and d
electron contributions to the energy (Heine and Marks, 1986). A pseudopotential calculation on Au(110) by Ho and Bohnen (1987) shows that the
(1×2) reconstruction is stabilized over the unreconstructed surface by a
reduction in the kinetic energy of the s–p electrons — these electrons can
spread out into the missing rows. The reason why the 5d elements undergo
the reconstruction but not the 3d or 4d may be that the 5d orbitals are
more extended, giving a stronger bond which puts the s–p electrons under
more compression in the bulk and on the unreconstructed surface (Ho and
Bohnen, 1987). The reconstruction then gives a greater release of energy for
the s–p electrons. In fact the energy balance between the unreconstructed
and reconstructed surface is quite fine on Ag(110) for example, and it
is likely that the increase in the number of s–p electrons, due to charge
transfer, drives the reconstruction on alkali adsorption. Calculations show
that the (1×2) reconstruction of Ag(110) can also be driven by an external
electric field (Fu and Ho, 1989), whose effect is to induce extra s–p electrons

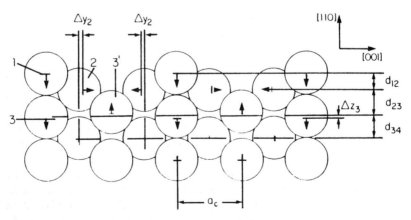

Fig. 16. Missing row (1×2) reconstruction on Ir, Pt and Au (110) surfaces (Ho and Bohnen,
1987).

on top of the surface atoms just as in alkali adsorption. In fact the reversible reconstruction of Au(110) in an electrolytic cell by varying the voltage across the cell has been measured (Magnussen et al., 1993), though it is unclear whether it is the strong electric field at the Au(110) electrode or adsorption of ions from the electrolyte which is responsible.

5. Adsorbates on metals

Adsorption of atoms and molecules on surfaces plays a fundamental role in catalysis; a distinction can be made between physisorption, in which weak Van der Waals forces bind the atom/molecule to the surface, and chemisorption in which chemical bonds dominate. Much experimental and theoretical work is devoted to studying energy changes as a molecule approaches the surface and dissociates (or doesn't) into separate atoms on the surface. Here we concentrate on the relation between structure and bonding for chemisorbed atoms in their equilibrium sites on the surface (ch. 1, sect. 2.4).

5.1. Hydrogen adsorption on metals

Hydrogen adsorption plays a part in several catalytic processes, and the way that it incorporates into the bulk is important for understanding hydrogen embrittlement (Nordlander et al., 1984). Of course as the simplest atom, H has been one of the first to be studied in chemisorption calculations: Lang and Williams (1978) and Gunnarsson et al. (1976) have studied the adsorption of a single H atom on the surface of jellium. This is a good model for adsorption on s–p bonded metals like Na, Mg or Al. Solving the Schrödinger equation for a single adsorbate atom even on jellium is difficult because it destroys the translational invariance of the clean surface; however scattering theory can be used to do this, relying on the fact that the changes in charge density around the adatom are rather localized (Lang and Williams, 1978; Gunnarsson et al., 1976). It is found that the H 1s level becomes a resonance due to its interaction with the metal electrons. The chemisorption energy as a function of H-jellium distance is shown in fig. 17 for substrates corresponding to Na, Mg and Al (Hjelmberg, 1979), and in all cases the binding energy turns out to be 1.5–2.0 eV. The equilibrium distance is determined by the balance between attractive bonding forces, and the repulsion due to the increase in electron kinetic energy when the atom overlaps significantly with the substrate electron density.

A simple picture of chemisorption comes from effective medium theory (Nørskov and Lang, 1980), based on the idea (related to density-functional theory, sect. 2.1) that to a first approximation the adsorption energy at point

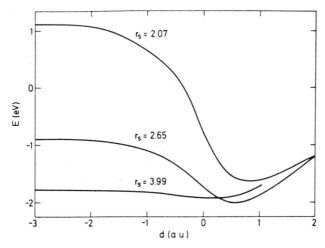

Fig. 17. Energy for H adsorbed on jellium as a function of metal–adatom distance, for $r_s = 2.07$ a.u. (Al), $r_s = 2.65$ a.u. (Mg), and $r_s = 3.99$ a.u. (Na) (Hjelmberg, 1979).

r is the same as the heat of solution of the atom in a uniform electron gas with the electron density of this point on the clean surface:

$$\Delta E = \Delta E^{\text{uniform}}(\rho_0(\mathbf{r})). \tag{46}$$

The heat of solution of H in a uniform electron gas has a minimum at an electron density of 0.002 e/(Bohr radius)^3, with $\Delta E^{\text{uniform}}$ given by -1.7 eV, in just the range of chemisorption energy found in fig. 17. At a surface, the picture is then that the H atom seeks out the optimum electron density, giving this rather universal (for s–p metals) binding energy.

To build the atomic structure of the surface into the jellium picture of chemisorption, the interaction of the self-consistent induced charge density of the H-jellium system with the atomic pseudopotentials of the substrate can be treated by first-order perturbation theory. In this way Hjelmberg (1979) found that bridge sites are favoured for H on Al(001) and Al(110), with an atop or bridge site favoured for Al(111). The energy barrier between different adsorption sites on the surface is 0.1–0.2 eV. Electron energy loss experiments, which probe vibrations of adsorbed atoms, are consistent with the bridge site (Paul, 1988). On Mg(0001) on the other hand Hjelmberg (1979) found that the three-fold coordinated site is most favoured.

A monolayer of H adsorbed on Be(0001) has been studied in a slab calculation, using Linearized Augmented Plane Wave (LAPW) basis functions (Feibelman, 1993), as well as pseudopotentials (Yu and Lam, 1989). The electronic energy levels as a function of wavevector **K** parallel to the surface are shown in fig. 18, and we see that a surface state, localized on the H

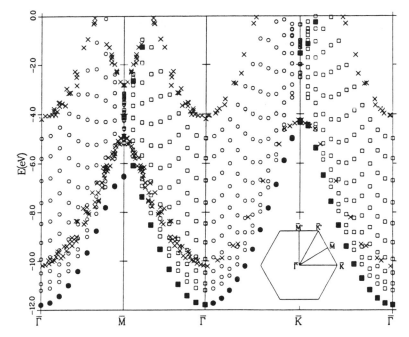

Fig. 18. Electron energy levels (circles and squares) as a function of **K** in a 7-layer slab calculation for Be(0001) with 1 ml H adsorbed in bridge sites (Feibelman, 1993). The solid circles and squares represent states heavily weighted on the surface Be and H atoms. The crosses are experimental results.

and the outermost Be layers, is pulled off the bottom of the conduction band. Photoemission from H/Be(0001) shows a H-induced surface resonance rather than this surface state (surface states do not overlap with bulk states, resonances do, sect. 2.2), but this is for a lower coverage and there is uncertainty about the adsorbate structure. Energetically the bridge sites are preferred, with a binding energy of 1.38 eV (Feibelman, 1993). The energy difference between this and three-fold sites is about 0.2 eV, similar to the sort of energies found by Hjelmberg (1979). The three-fold site corresponding to fcc stacking is preferred over the hcp, which seems to be a fairly general rule. It is interesting that in these studies of H on simple metals, an adsorption site with less than optimum coordination is often preferred. In the case of adsorption on say Al, this must be related to the way that the H-induced charge density interacts with the substrate pseudopotential which has (effectively) a repulsive core. For H on Be(0001) Feibelman (1993) has suggested that the bridge site is favoured by the requirements of Be–Be bonding.

The characteristic feature of the electronic structure of H adsorbed on transition metals is a H-induced bonding state pulled off the bottom

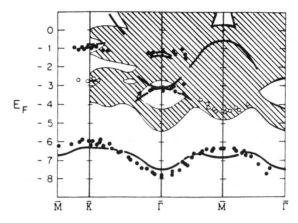

Fig. 19. Calculated (solid lines) and measured (circles) surface states for Pd(111) with 1 ml H adsorbed (Eberhardt et al., 1983). Shaded area represents Pd bulk states.

of the conduction band. Figure 19 shows the calculated surface state on Pd(111) with a monolayer of H, compared with photoemission results — the bonding state is pulled off by about 2 eV and is made up of the H 1s orbitals mixing with 5s and 4d valence orbitals from the surface metal atoms (Eberhardt et al., 1983). It is the interaction with the transition metal d states which pushes the resonance found in jellium to below the bottom of the band. This bonding state has been calculated and measured on many surfaces including Ti(0001) (Feibelman et al., 1980), Ni(111) (Eberhardt et al., 1981; Greuter et al., 1988), Pt(111) (Eberhardt et al., 1981), Ru(0001) (Hofmann and Menzel, 1985), with both theory (Chubb and Davenport, 1985) and experiment (Greuter et al., 1988) suggesting that at lower coverage the state moves closer to the bottom of the band. Fully self-consistent electronic structure calculations, usually for monolayer coverage, show that on low index transition metal surfaces the H atoms prefer to sit in high coordination sites (ch. 1, sect. 2.4.1, table 5), three-fold hollow sites on Ru(0001) (Feibelman and Hamann, 1987a), Cu(111) (Feibelman and Hamann, 1986) and Pt(111) (Feibelman and Hamann, 1987b) for example, and four-fold hollow sites on Rh(001) (Hamann and Feibelman, 1988; Feibelman, 1991), Ni(001) (Weinert and Davenport, 1985; Umrigar and Wilkins, 1985) and Pd(001) (Tománek et al., 1986). In all these cases the H atoms are well embedded into the surface, and on Rh(001) for example, energy minimization puts the H layer 1.23 a.u. above the top Rh layer (Feibelman, 1991). An exception to the high coordination sites is H on W(001), where the bridge sites are energetically favoured: at low coverages this stabilizes a reconstruction in which the W atoms are displaced in the ⟨10⟩ direction rather than ⟨11⟩ as on the clean surface (sect. 4.3.1) (Biswas and Hamann, 1986; Weinert et al., 1986), and at saturation coverage (2 H's per surface W) the W atoms return to their bulk-like positions with the H's occupying all bridge sites. The bridge sites are energetically favoured because of the interaction of the H atoms with W surface states made up of $d_{x^2-y^2}$ orbitals, some of the surface states which contribute to the peak in the surface density of states on the unreconstructed clean surface (fig. 9) (Weinert et al., 1986).

Effective medium theory (Nordlander et al., 1984) can give a simple description of H adsorption on transition metal surfaces, provided that corrections are made for the hybridization of the adatom wavefunctions with the substrate d orbitals. Again the H atom seeks out the energetically favoured charge density, but the favoured site (highly coordinated) is determined by the hybridization. A trend for the binding energy of H to decrease with substrate d-band filling is also due to hybridization, with the d-electrons filling antibonding H–transition metal states.

5.2. Alkali adsorption on metals

At low coverages, the electropositive alkali metal atoms transfer charge to the substrate, giving a large ionic contribution to the bonding (Lang and Williams, 1978; Scheffler et al., 1991). The charge distribution of positively charged adatoms plus screening charge on the surface of the substrate sets up a surface dipole, leading to a large reduction in work-function (Muscat and Newns, 1979). The reduction in work-function is linear with coverage at low coverages, reaching a minimum at a coverage of typically 0.1–0.2 monolayer, and then rising to the metallic alkali work-function at saturation coverage (fig. 20) (Bonzel, 1987; Kiskinova et al., 1983). The work-function minimum and the subsequent increase is associated with a change in bonding, towards a metallic overlayer with a less ionic form of bonding to the substrate (Lamble et al., 1988). This transition can be understood classically in terms of depolarization effects due to the dipoles interacting with one another (Neugebauer and Scheffler, 1992). The charge transfer at low coverages is responsible for alkali adsorbates driving the (1×2) reconstruction on Ag(110), as we have seen in sect. 4.3.2; the electrostatic field produced by the dipole and the charge transfer are involved in the role of alkali adsorbates as catalyst promoters; and the reduction in work-function with alkali adsorption is important for producing low work-function electron emitters (Bonzel, 1987).

The ionic bonding is reflected in the density of states of an alkali adsorbate. Figure 21 shows the change in density of states calculated by Lang and Williams (1978), for adsorbates on jellium with an electron density appropriate to Al, with the 2s state on the adsorbed Li broadened into a resonance centred above E_F. If the 2s state just broadened into a half-filled Lorentzian,

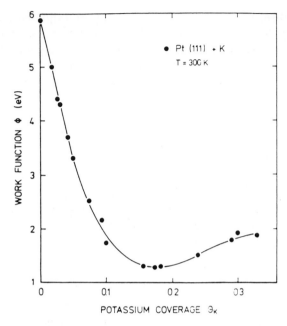

Fig. 20. Work-function versus coverage of K on Pt(111) (Bonzel, 1987).

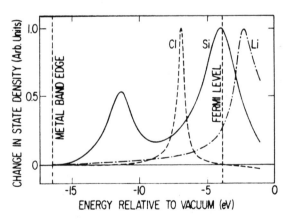

Fig. 21. Change in density of states for adsorbates on jellium, with electron density appropriate to Al (Lang and Williams, 1978).

we would describe the bonding as covalent or metallic; in fact the peak lies above E_F, so this 2s level is depopulated, with a redistribution of charge into metallic states spread out below E_F. Ishida (1990) has calculated the density of states and the electron density of Na overlayers on Al jellium at

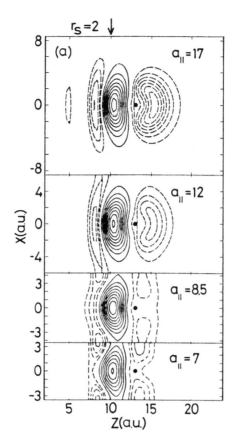

Fig. 22. Contours of electron density difference, for Na adsorbed on jellium (Al), compared with clean jellium plus isolated Na (Ishida, 1990). The different plots are for different Na spacings ($a_\parallel$). Solid and dashed contours correspond to positive and negative changes in density, respectively.

several coverages, and obtains results at low coverages similar to those of Lang and Williams (1978). The change in electron density of the adsorbed alkali, compared with a superposition of clean jellium plus isolated alkali monolayer charge densities reflects the ionic bonding (fig. 22) (Ishida, 1990), with a depletion of charge on the vacuum side of the alkali, and an increase of charge between the alkali and the substrate which we can think of as the screening charge on the substrate screening out the electric field due to the alkali ions. Photoemission experiments by Horn et al. (1988) from K on Al(111) support the ionic picture of bonding, with the 4s state on the K appearing just below E_F only at coverages of about 1/3 monolayer. At lower coverages the 4s resonance is apparently completely unoccupied, the electron transferred from the K 4s level presumably being distributed through the metal density of states.

Although this picture of ionic bonding at low coverages seems straightforward enough, it became highly controversial partly because of calculations by Wimmer et al. (1983) for c(2×2) (50% coverage) Cs on W(001). They interpreted the calculated reduction in work-function from the value for clean W(001) of 4.77 eV to about 2.5 eV in terms of polarized Cs valence electrons rather than charge transfer from Cs to W. This is largely semantics, for as Benesh and King (1992) point out, a charge redistribution *outside* the surface W atoms is just what we expect in the classical ionic picture — an external field, as provided by the alkali ions, is screened by charge on top of the substrate atoms. In any case, depolarization effects — leading to a decrease in ionicity — are certainly important at the coverage considered by Wimmer et al. (1983). Experimental evidence apparently refuting the ionic picture came from photoemission experiments on W(001) by Riffe et al. (1990), who found only very small shifts in the binding energy of the surface W 4f core level on adsorption of alkalis. Transfer of electronic charge from the adsorbate to the surface W atoms might be expected to lead to a decrease in binding energy, but this argument is over-simplified: core-level shifts are affected by atomic coordination, an increase of which (as on adsorption) tends to increase binding energy; screening of the core hole is an uncertain contribution. In any case, the charge transferred to the W atoms in the process of ionic bonding is sitting — as we have just emphasised — on the top of the atoms, and its effect is purely one of screening. A quantitative measure of ionicity is difficult, because of the arbitrariness in assigning charge to individual atoms. However the dynamic effective charge, which gives the derivative of dipole moment with atomic position, can be determined uniquely. In the case of Li on jellium, Lang and Williams (1978) found an effective charge on the Li of $+0.4|e|$; for Cs on W(001) at 50% coverage the results of Wimmer et al. (1983) give an effective charge on the Cs of $+0.2|e|$, so there is clear evidence here of metallization at higher coverage.

The dipole moment associated with an adsorbed alkali ion plus the screening charge leads to an electrostatic repulsion between the adatoms varying like $1/r^3$ — especially important at low coverages where the dipoles are largest. A consequence of this is that the alkali atoms tend to spread out fairly uniformly over the surface (Bonzel, 1987). In many systems (e.g. Na and K on Ni(001) (Gerlach and Rhodin, 1969; Fisher and Diehl, 1992), K on Cu(001) (Aruga et al., 1986), Cs on Ru(0001) (Over et al., 1992)) the adatom–adatom repulsion leads at low coverage to an isotropic fluid phase, with a quite well-defined nearest neighbour distance which decreases uniformly with increasing coverage. This gives rise to a characteristic LEED pattern of rings around the integer order spots. The fluid phase can go over at higher coverage (for a coverage $\theta > 0.08$ monolayer in the case of K on Cu(111) (Fan and Ignatiev, 1988)) to a phase in which the alkalis are arranged in a hexagonal array, incommensurate with the substrate, with an adatom spacing varying continuously with coverage. This phase melts to an "orientationally ordered phase" (characteristic of 2D systems), which has been studied in detail by Chandavarkar and Diehl (1989) for K on Ni(111). For this behaviour the adsorbate–adsorbate interaction must dominate the adsorbate–substrate interaction. However in general these interactions compete with one another, and in the case of K on Ir(001) with coverages up to 0.5 monolayer five different coincidence structures have been measured at $T = 100$ K — distorted hexagons with dipole repulsion trying to keep atoms as well separated as possible, but adsorbate–substrate interactions forcing the K atoms to sit in high-symmetry adsorption sites (Heinz et al., 1985).

In many systems, alkali–substrate interactions favour adatoms sitting in highly-coordinated hollow sites (ch. 1, sect. 2.4.2, table 6). However in the p(2×2) structure of Cs, Rb and K adsorbed on several close-packed surfaces, LEED and SEXAFS analysis suggests that the alkali atoms are in on-top positions [4] (Adler et al., 1993; Kaukasoina et al., 1993; ch. 1, table 6). The Cs/Ru(0001) system is quite complicated (Over et al., 1992): up to a coverage of about 0.15 monolayer the system shows the characteristic ring LEED pattern, then with increasing coverage there follows a (2×2) phase, a series of structures with rotated unit cells, and around $\theta = 1/3$ a $(\sqrt{3} \times \sqrt{3})R30°$ phase. What is remarkable is that in the (2×2) phase the Cs atoms occupy on-top sites, and in the $(\sqrt{3} \times \sqrt{3})R30°$ structure three-fold hollow sites. The change in adsorption site may reflect the decrease in ionicity with increasing coverage — the work-function minimum coincides more or less with the (2×2) phase, and in the $(\sqrt{3} \times \sqrt{3})R30°$ phase the dipole moment at each adsorbate is about 30% smaller. The decrease in ionicity probably accounts for the change in Cs–Ru bond length, for which LEED analysis gives 3.25 Å in the (2×2) phase, and 3.52 Å in the $(\sqrt{3} \times \sqrt{3})R30°$ phase. A similar change in the apparent size of the Cs atom was discovered by Lamble et al. (1988) in a SEXAFS experiment on Cs/Ag(111) (ch. 1, sect. 2.4.2, table 6). At a coverage of 0.15 monolayer the Cs–Ag distance was found to be 3.20 Å, increasing to 3.50 Å at 0.3 monolayer, again indicative of a change to a less ionic type of bonding with increasing coverage.

Total energy calculations for Na on Al(111) have shown that in the $(\sqrt{3} \times \sqrt{3})R30°$ structure which is observed at $\theta = 0.33$, the minimum energy corresponds to Na atoms *substituting* for top-layer Al atoms (Neugebauer and Scheffler, 1992, 1993). These slab-pseudopotential calculations (sect. 2.3.1), in which the geometrical structure is optimized à la sect. 3.1,

---

[4] Cs/Cu(111), Cs/Ru(0001), Rb/Al(111), K/Ni(111), K/Cu(111), K/Al(111).

suggest that the Na atoms, substituting for Al atoms which presumably diffuse away to surface steps, stick out of the surface rather than lying completely within the Al surface plane. Substitution is energetically more favourable than the three-fold hollow site by 0.16 eV per adatom, but in the lower coverage (2×2) structure the energy gain is much smaller, only 0.04 eV. A comparison of binding energies in the two coverages shows a repulsive interaction between alkali adsorbates in on-top sites, but in the stable substitutional sites the interaction is actually attractive, leading to the formation of islands of $(\sqrt{3}\times\sqrt{3})$R30° structure for all coverages less than 1/3. Experiments have confirmed this theoretical work of Neugebauer and Scheffler (1992, 1993) (ch. 1, table 6). SEXAFS experiments (Schmalz et al., 1991) show that for $0.16 < \theta < 0.33$ the alkalis substitute for Al with a Na–Al nearest neighbour distance of 3.31 Å compared with the prediction of 3.13 Å and X-ray diffraction experiments (Kerkar et al., 1992) are also compatible with this. The preference for a substitutional site seems to be connected with the very effective screening of the dipole repulsion for atoms in these sites; this outweighs the energy cost of creating an Al surface vacancy, which in any case is quite low.

5.3. Oxygen adsorption on metals

The interaction of O with surfaces is very important because of oxide formation, which in some cases may be preceded by adsorption of O atoms on the surface (Brundle and Broughton, 1990). O is electronegative, and its chemisorption leads to an increase in work-function due to electronic charge being transferred from the substrate to the adsorbate.

There are several surface reconstructions induced by O chemisorption (ch. 1, sect. 2.4.5.1, table 9). On Cu(110) a (2×1) reconstruction occurs at an O coverage of around 0.5 monolayer, with missing rows in the ⟨001⟩ direction (*perpendicular* to the missing rows in the (1×2) reconstructions described in sect. 4.3.2) (fig. 23) (Feidenhans'l et al., 1990; Parkin et al., 1990; Coulman et al., 1990; Jensen et al., 1990a). The O adatoms are located in the long-bridge sites along the ⟨001⟩ rows. At lower coverages, islands of reconstruction form, and STM studies suggest that the reconstruction proceeds via the formation of "added" O–Cu rows, with Cu atoms diffusing to the reconstruction from terraces (Coulman et al., 1990; Jensen et al., 1990a). A similar reconstruction is found on Ni(110) (Kleinle et al., 1990), and on Ag(110) several p(n×1)–O phases occur, which also most likely involve missing row reconstructions (Bracco et al., 1990). The driving force for the reconstructions seems to be the formation of O–Cu bonds along the ⟨001⟩ chains (DiDio et al., 1984; Courths et al., 1987; Weimert et al., 1992). Angle-resolved photoemission experiments show three bonding bands derived from O 2p orbitals, one of which disperses very strongly in the k_z

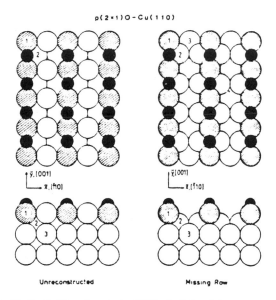

p(2×1)O–Cu(110)

Unreconstructed   Missing Row

Fig. 23. Cu(110)p(2×1)–O (Courths et al., 1987). Left hand figure: unreconstructed; right hand figure: missing row reconstruction.

direction, and which is made up of O $2p_z$ and Cu $3d_{z^2}$ orbitals (DiDio et al., 1984; Courths et al., 1987). The antibonding band corresponding to this state is unoccupied — hence the O–Cu interaction leads to a nett gain in energy — and is seen in inverse photoemission (Jacob et al., 1980).

The tendency to form O–Cu chains shows up also in the O-induced $(2\sqrt{2}\times\sqrt{2})$R45° reconstruction of Cu(001), which also forms at a local coverage of 0.5 monolayer (Asensio et al., 1990; Zeng and Mitchell, 1990). Several surface crystallographic techniques show that this involves missing rows of Cu atoms, with near-coplanar O atoms occupying former hollow sites adjacent to the missing rows (fig. 24) (Asensio et al., 1990; Zeng and Mitchell, 1990; Wuttig et al., 1989; Robinson et al., 1990). STM results suggest that in this case 25% of the Cu atoms are squeezed out, forming islands of Cu elsewhere on the surface (Jensen et al., 1990b). Again the energy gain is due to O–Cu bonding, and this has been studied theoretically by Jacobsen and Nørskov (1990) — they find an energy gain of 1.1 eV per O atom on reconstruction. What happens is that the surface reconstruction reduces the coordination number of the surface Cu atoms, raising the energy of their 3d levels and thereby decreasing the occupancy of the O–Cu antibonding levels.

This $(2\sqrt{2}\times\sqrt{2})$R45° reconstruction is the stable structure of O on Cu(001), but there has been much recent controversy about the occur-

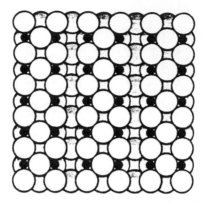

Fig. 24. Cu(001)$(2\sqrt{2}\times\sqrt{2})$R45°–O missing row reconstruction (Zeng and Mitchell, 1990).

rence of an unreconstructed chemisorption state (Arvanitis et al., 1993; Lederer et al., 1993). It seems that samples can be prepared in which the O atoms are sitting in a precursor state, probably in four-fold hollow sites in a roughly c(2×2) structure but with some disorder, resulting in a fuzzy LEED pattern (Arvanitis et al., 1993). O on Ni(001) forms a stable, well-ordered c(2×2) structure of this sort (Chubb et al., 1990). In the O/Cu(001) system this precursor state seems to go over to the stable $(2\sqrt{2}\times\sqrt{2})$R45° structure as the coverage is increased (Arvanitis et al., 1993). An argument based on effective charges (sect. 5.2) can explain why the four-fold hollow structure becomes unstable at higher coverage (Colbourn and Inglesfield, 1991). The effective charge turns out to be rather large, about −0.9 |e| for the O atom adsorbed in the four-fold hollow site of Cu(001). Now a large effective charge can cause a surface instability if the coverage is large enough, because a displacement of an adsorbate atom produces a long range dipole field which acts on the effective charge on other adsorbate atoms, tending to displace them — the surface tends to buckle. In the case of O on Cu(001) theory suggests that the surface becomes unstable above a critical coverage of about 0.3 monolayer. Evidence from electron energy loss spectroscopy (EELS), which measures the vibrational frequencies of the atoms at the surface, suggests that the $(2\sqrt{2}\times\sqrt{2})$R45° structure begins to appear at just about this coverage (Wuttig et al., 1989). The effective charge argument is related to more general arguments about surface stress and stability, but it cannot predict to which structure the unstable phase will transform.

6.   Semiconductor surfaces

Semiconductors are held together by covalent bonds, and creating a surface chops the bonds in two. These dangling bonds at the surface are unstable, and semiconductor surfaces invariably reconstruct (or relax if there is no change in the two-dimensional unit cell from the ideal surface) to eliminate the dangling bonds as far as possible.

6.1. Elemental semiconductors

6.1.1. Si and Ge(001) surfaces

The (2×1) reconstructions on these surfaces involve the formation of dimers between atoms in the top layer which are tilted out of the surface plane (asymmetric dimers) (fig. 25) (Roberts and Needs, 1990; ch. 1, sect. 3.1.1.1, table 15). This was proposed by Chadi (1979) for Si(001) on the basis of a semi-empirical tight-binding calculation of the total energy (Chadi, 1978). In this approach, which is quite widely used for getting insight into the physics of surface reconstruction, the one-electron energies ϵ_i (11) are found from a tight-binding Hamiltonian containing parametrized hopping integrals, and then the structural energy is written as a sum over the occupied one-electron energies corrected by an empirical repulsive two-body interaction:

$$E = \sum_i \epsilon_i + \sum_{I,J} U_{I,J}. \tag{47}$$

The repulsive interaction $U_{I,J}$ is assumed to be short-range, normally taken over nearest neighbour atoms I and J — it replaces all the terms in (11) which are added on to the one-electron energy. Termination of the

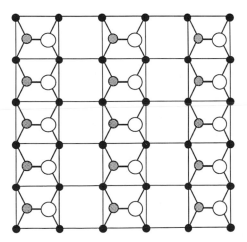

Fig. 25. Si(001)(2×1) reconstruction (Roberts and Needs, 1990). Plan view, white and grey circles representing atoms in the top layer (white circles: atoms moved outwards; grey circles: atoms moved inwards); and black circles: atoms in the second layer.

bulk structure leaves two dangling bonds per surface atom on the (001) surface, and dimerization satisfies one of these bonds. In the language of surface state bands, the remaining dangling bonds correspond to a metallic, partially filled surface state band. Chadi (1979) suggested that the tilting of the dimer, which results in electron transfer from the "down" atom to the "up" atom, increases the gap between the surface state bands, resulting in a semiconducting surface.

Calculations of the electronic structure and total energy using pseudopotentials in supercell geometry, with structural optimization, have largely confirmed this picture (Roberts and Needs, 1990; Zhu et al., 1989). In the case of Si(001)(2×1) the asymmetric dimers are found to occur, but the (2×1) arrangement shown in fig. 25 results in fact in a metallic surface state band, in disagreement with photoemission experiments. However a variety of periodicities have been seen experimentally, which probably correspond to a different arrangement of buckled dimers. The c(4×2) reconstruction of Si(001) (fig. 26) turns out to be slightly favoured energetically over the (2×1), by an energy of 0.07 eV per dimer (Zhu et al., 1989) — other arrangements of dimers such as p(2×2) have very similar energy. In these structures there is indeed a gap between occupied and unoccupied surface states. The interaction energy *between* the tilted dimers is very much smaller than the energy gained by the dimerization and tilting, which amounts to about 2 eV per dimer (Roberts and Needs, 1990).

Experimentally the situation was unclear because of STM experiments showing apparently symmetric dimers on an almost defect-free terrace of Si(001)(2×1), with tilting only near steps (Wiesendanger et al., 1990). However, recent temperature-dependent STM work has shown that on cooling to 120 K, the number of buckled dimers increases (Wolkow, 1992). It seems likely that the bistability of the asymmetric dimer results in flipping between

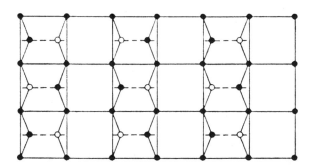

Fig. 26. Si(001)c(4×2) reconstruction (Kevan, 1985). Plan view, with white and black circles connected by dashed lines representing atoms in the top layer (white circles in top layer: atoms moved outwards, black circles in top layer: atoms moved inwards).

the two stable tilts at higher temperatures, resulting in an apparently symmetric dimer. A comparison between optical measurements and calculated optical properties of Si(001)(2×1) based on a tight-binding calculation of electronic structure gives strong evidence for dimer tilting (Shkrebtii and Del Sole, 1993), and it is also supported by grazing incidence X-ray diffraction experiments (Jedrecy et al., 1990).

In the case of Ge(001)(2×1), STM investigations show asymmetric dimers at room temperature (Kubby et al., 1987), and this has also been confirmed by grazing incidence X-ray diffraction (Rossmann et al., 1992). In the STM experiments regions of (2×1) and c(4×2) reconstructions are found (Kubby et al., 1987), and LEED suggests that there is an order–disorder transition at about 200 K between the low temperature c(4×2) structure and a (2×1) structure in which the dimer tilts are disordered (Kevan, 1985). All this is evidence for a small energy of interaction between the tilted dimers. Angle-resolved photoemission suggests that the (2×1) phase might have a metallic surface state, whereas the c(4×2) surface is semiconducting (Kevan, 1985) — the same picture as we discussed above. Pseudopotential calculations with molecular dynamics structure optimization show that the c(4×2) structure is most stable, favoured over the (2×1) buckled dimer by about 0.05 eV per dimer (Needels et al., 1987). This energy gain is apparently associated with subsurface atomic relaxations.

6.1.2.   Si(111) surface

On the clean Si(111) surface at low temperature a (2×1) reconstruction occurs, but this is metastable and on annealing the stable (7×7) structure develops (fig. 27) (Haneman, 1987; ch. 1, sect. 3.1.1.2). Several experimental techniques have confirmed the Takayanagi "dimer–adatom–stacking fault" model, characterized — as the name suggests — by dimerization of the second-layer atoms, the presence of adatoms, and stacking faults between the first and second layers (Takayanagi et al., 1985). This structure has been found theoretically in structure-optimization calculations, with pseudopotentials and a plane wave basis set (Štich et al., 1992; Brommer et al., 1992). These are huge calculations, equivalent to treating 700 atoms in the slab geometry which was used, and were carried out on massively parallel computers. There is generally very good agreement with experimental values for the structural parameters. Again the energy gain is due to removing dangling bonds, this time via adatoms — as shown in earlier semi-empirical tight-binding calculations (Qian and Chadi, 1987).

In the metastable (2×1) structure the dangling bonds are removed by the formation of zig-zag chains of atoms in the top two atomic layers — π-bonded chains (fig. 28). This was first suggested by Pandey (1981, 1982) on the basis of a comparison of surface state dispersion with photoemission

(a)

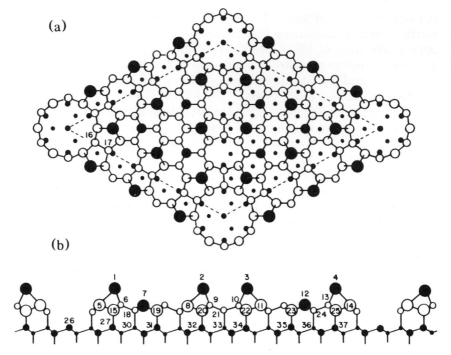

(b)

Fig. 27. Si(111)(7×7) reconstruction (Brommer et al., 1992). (a) plan view; (b) side view.

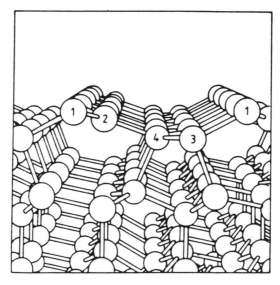

Fig. 28. Si(111)(2×1) reconstruction (Haneman, 1987).

experiments. A first-principles structure optimization calculation confirms the π-bonded chain structure, with buckling once again (Ancilotto et al., 1990). There is excellent agreement between the calculated surface states found for this structure, and photoemission experiments.

6.2. Compound semiconductors

6.2.1. III–V (110) surfaces

The (110) cleavage surfaces of the III–V semiconductors relax from the ideal termination of the bulk structure — *relax* rather than *reconstruct* because the atomic displacements maintain the ideal two-dimensional unit cells (ch. 1, sect. 3.2.1). The relaxation consists of a rotation of the pairs of atoms in the surface layer by about 30°, maintaining the bond length (fig. 29) (Alves et al., 1991; ch. 1, table 18). In the case of GaAs(110) the relaxation lowers the surface energy by about 0.3 eV per surface unit cell compared with a surface energy of 1.2 eV, so the relaxation has an appreciable effect on the energy (Alves et al., 1991).

Pseudopotential slab calculations have been carried out by Alves et

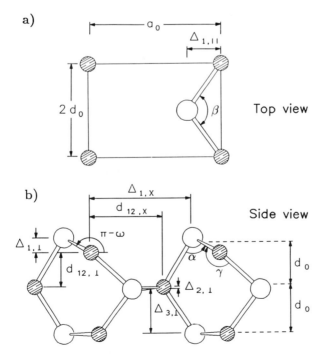

Fig. 29. Relaxed III–V (110) surface (Alves et al., 1991). (a) Plan view; (b) side view.

Table 3

Structural parameters for III–V semiconductor (110) surfaces (Alves et al., 1991). $\Delta_{1,\perp}$ and ω are shown in fig. 29.

Compound	$\Delta_{1,\perp}$ (Å)		ω (°)	
	Theory	Experiment	Theory	Experiment
GaP	0.61	0.63	29.2	27.5
InP	0.67	0.73	30.1	29.9
GaAs	0.67	0.69	30.2	31.1
InAs	0.75	0.78	32.0	36.5

al. (1991) for GaP, InP, GaAs and InAs (110) surfaces, with structural optimization à la Car and Parrinello (1985). Basis sets of 5500 plane waves were used. They obtain rotation angles and atomic displacements in excellent agreement with experiment (table 3). The driving force for the relaxation seems to be rehybridization at the surface, the group III atom preferring the more planar sp^2 bonding, and the group V, p bonding to its neighbours.

The surface state bands are shown in fig. 30 for GaAs(110), band A_5 corresponding to an occupied dangling bond localized mainly on the surface As, and C_3 to an unoccupied dangling bond on the surface Ga. These bands are pushed out of the fundamental gap (except at the C_3 band minimum at $\bar{X}$) by the relaxation (Alves et al., 1991; Schmeits et al., 1983).

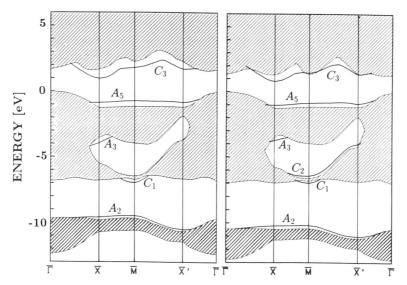

Fig. 30. Surface states on relaxed GaAs(110) (Alves et al., 1991). Shaded area represents bulk GaAs states. Left hand figure uses smaller plane wave basis set than right hand.

7. Conclusions

Especially for cases like the (7×7) reconstruction of Si(111) (sect. 6.1.2), first-principles calculations have been very successful in describing and explaining the structure of surfaces. The explanation and understanding — *why* the huge computer solution of the Schrödinger equation predicts this or that structure — have also benefitted from simpler, sometimes semi-empirical approaches, like the early work of Chadi (1979) on Si(001). It isn't surprising that calculations have been most spectacularly successful for semiconductor surfaces, as these are the systems for which pseudopotentials with plane wave basis functions can be most successfully applied (sect. 2.3.1), and semiconductors with their covalent bonding show the most impressive surface reconstructions. Studies of metal surfaces have also been very successful — the work by Singh and Krakauer (1988) on the W(001) reconstruction (sect. 4.3.1) has helped in the understanding of the phase transition; and the work by Neugebauer and Scheffler (1992, 1993) on alkali adsorption on Al(111) has real predictive power.

There is still much to do. Do we understand why the Mo(001) surface reconstruction involves a long period modulation (sect. 4.3.1)? The dynamics of reconstruction is a vast field for the future, especially as this can be explored in the STM. There are many problems to study in adsorption, the role of surface defects in adsorption, the effect of adsorption on surface structure, and the processes of surface chemistry so important in catalysis (see, for example, De Vita et al., 1993). And anyone who has ever done a surface electronic structure calculation knows that we are a long way from having the proverbial black box which can give us all the answers about a surface without a lot of hard work.

Acknowledgements

I thank my friends Simon Crampin and John Michiels for their comments.

References

Adler, D.L., I.R. Collins, X. Liang, S.J. Murray, G.S. Leatherman, K.D. Tsuei, E.E. Chaban, S. Chandavarkar, R. McGrath, R.D. Diehl and P.H. Citrin, 1993, Phys. Rev. B **48**, 17445.

Aldén, M., H.L. Skriver, S. Mirbt and B. Johansson, 1992, Phys. Rev. Lett. **69**, 2296.

Alves, J.L.A., J. Hebenstreit and M. Scheffler, 1991, Phys. Rev. B **44**, 6188.

Ancilotto, F., W. Andreoni, A. Selloni, R. Car and M. Parrinello, 1990, Phys. Rev. Lett. **65**, 3148.

Andersen, O.K., 1975, Phys. Rev. B **12**, 3060.

Andersen, O.K., Z. Pawlowska and O. Jepsen, 1986, Phys. Rev. B **34**, 5253.

Andrews, P.T., I.R. Collins and J.E. Inglesfield, 1992, in: Unoccupied Electronic States, eds. J.C. Fuggle and J.E. Inglesfield (Springer, Berlin), pp. 243–276.

Aruga, T., H. Tochihara and Y. Murata, 1986, Surface Sci. **175**, L725.

Arvanitis, D., G. Cornelli, T. Lederer, H. Rabus and K. Babeschke, 1993, Chem Phys. Lett. **211**, 53.

Asensio, M.C., M.J. Ashwin, A.L.D. Kilcoyne, D.P. Woodruff, A.W. Robinson, T. Lindner, J.S. Somers, D.E. Ricken and A.M. Bradshaw, 1990, Surface Sci. **236**, 1.

Bachelet, G.B., D.R. Hamann and M. Schlüter, 1982, Phys. Rev. B **26**, 4199.

Barnes, C.J., M.Q. Ding, M. Lindroos, R.D. Diehl and D.A. King, 1985, Surface Sci. **162**, 59.

Bendt, P., and A. Zunger, 1982, Phys. Rev. B **26**, 3114.

Benesh, G.A., and J.E. Inglesfield, 1984, J. Phys. C: Solid State Phys. **17**, 1595.

Benesh, G.A., and D.A. King, 1992, Chem. Phys. Lett. **191**, 315.

Biswas, R., and D.R. Hamann, 1986, Phys. Rev. Lett. **56**, 2291.

Bonzel, H.P., 1987, Surface Sci. Reports **8**, 43.

Bracco, G., R. Tatarek and G. Vandoni, 1990, Phys. Rev. B **42**, 1852.

Brommer, K.D., M. Needels, B.E. Larsen and J.D. Joannopoulos, 1992, Phys. Rev. Lett. **68**, 1355.

Brundle, C.R., and J.Q. Broughton, 1990, in: The Chemical Physics of Solid Surfaces and Heterogeneous Catalysis, Vol. 3A, eds. D.A. King and D.P. Woodruff (Elsevier, Amsterdam), pp. 131–388.

Callaway, J., 1991, Quantum Theory of the Solid State (Academic Press, San Diego, CA).

Car, R., and M. Parrinello, 1985, Phys. Rev. Lett. **55**, 2471.

Ceperley, D.M., and B.J. Alder, 1980, Phys. Rev. Lett. **45**, 566.

Chadi, D.J., 1978, Phys. Rev. Lett. **41**, 1062.

Chadi, D.J., 1979, Phys. Rev. Lett. **43**, 43.

Chandavarkar, S., and R.D. Diehl, 1989, Phys. Rev. B **40**, 4651.

Chubb, S.R., and J.W. Davenport, 1985, Phys. Rev. B **31**, 3278.

Chubb, S.R., P.M. Marcus, K. Heinz and K. Müller, 1990, Phys. Rev. B **41**, 5417.

Colbourn, E.A., and J.E. Inglesfield, 1991, Phys. Rev. Lett. **66**, 2006.

Copel, M., and T. Gustafsson, 1986, Phys. Rev. Lett. **57**, 723.

Coulman, D.J., J. Wintterlin, R.J. Behm and G. Ertl, 1990, Phys. Rev. Lett. **64**, 1761.

Courths, R., B. Cord, H. Wern, H. Saalfeld and S. Hüfner, 1987, Solid State Commun. **63**, 619.

Cyrot-Lackmann, F., 1969, Surface Sci. **15**, 535.

Daley, R.S., T.E. Felter, M.L. Hildner and P.J. Estrup, 1993, Phys. Rev. Lett. **70**, 1295.

Debe, M.K., and D.A. King, 1979, Surface Sci. **81**, 193.

Dederichs, P.H., and R. Zeller, 1983, Phys. Rev. B **28**, 5462.

De Vita, A., I. Štich, M.J. Gillan, M.C. Payne and L.J. Clarke, 1993, Phys. Rev. Lett. **71**, 1276.

DiDio, R.A., D.M. Zehner and E.W. Plummer, 1984, J. Vac. Sci. Technol. A **2**, 852.

Eberhardt, W., F. Greuter and E.W. Plummer, 1981, Phys. Rev. Lett. **46**, 1085.

Eberhardt, W., S.G. Louie and E.W. Plummer, 1983, Phys. Rev. B **28**, 465.

Fan, W.C., and A. Ignatiev, 1988, Phys. Rev. B **37**, 5274.

Feibelman, P.J., 1991, Phys. Rev. B **43**, 9452.

Feibelman, P.J., 1993, Phys. Rev. B **48**, 11270.

Feibelman, P.J., and D.R. Hamann, 1986, Surface Sci. **173**, L582.

Feibelman, P.J., and D.R. Hamann, 1987a, Surface Sci. **179**, 153.

Feibelman, P.J., and D.R. Hamann, 1987b, Surface Sci. **182**, 411.

Feibelman, P.J., D.R. Hamann and F.J. Himpsel, 1980, Phys. Rev. B **22**, 1734.

Feidenhans'l, R., F. Grey, R.L. Johnson, S.G.J. Mochrie, J. Bohr and M. Nielsen, 1990, Phys. Rev. B **41**, 5420.

Felter, T.E., R.A. Barker and P.J. Estrup, 1977, Phys. Rev. Lett. **38**, 1138.

Fery, P., W. Moritz and D. Wolf, 1988, Phys. Rev. B **38**, 7275.

Finnis, M.V., and V. Heine, 1974, J. Phys. F: Metal Phys. **4**, L37.

Fisher, D., and R.D. Diehl, 1992, Phys. Rev. B **46**, 2512.

Foulkes, W.M.C., and R. Haydock, 1989, Phys. Rev. B **39**, 12520.

Fu, C.L., and K.M. Ho, 1989, Phys. Rev. Lett. **63**, 1617.

Fu, C.L., S. Ohnishi, E. Wimmer and A.J. Freeman, 1984, Phys. Rev. Lett. **53**, 675.

Gerlach, R.L., and T.N. Rhodin, 1969, Surface Sci. **17**, 32.

Godby, R.W., 1992, in: Unoccupied Electronic States, eds. J.C. Fuggle and J.E. Inglesfield (Springer, Berlin), pp. 51-88.

Greuter, F., I. Strathy, E.W. Plummer and W. Eberhardt, 1988, Phys. Rev. B **33**, 736.

Gunnarsson, O., H. Hjelmberg and B.I. Lundqvist, 1976, Phys. Rev. Lett. **37**, 292.

Hamann, D.R., and P.J. Feibelman, 1988, Phys. Rev. B **37**, 3847.

Haneman, D., 1987, Rep. Prog. Phys. **50**, 1045.

Harris, J., 1985, Phys. Rev. B **31**, 1770.

Hebenstreit, J., M. Heinemann and M. Scheffler, 1991, Phys. Rev. Lett. **67**, 1031.

Heine, V., and L.D. Marks, 1986, Surface Sci. **165**, 65.

Heinz, K., H. Hertrich, L. Hammer and K. Müller, 1985, Surface Sci. **152/153**, 303.

Hjelmberg, H., 1979, Surface Sci. **81**, 539.

Ho, K.M., and K.P. Bohnen, 1985, Phys. Rev. B **32**, 3446.

Ho, K.M., and K.P. Bohnen, 1987, Phys. Rev. Lett. **59**, 1833.

Hofmann, P., and D. Menzel, 1985, Surface Sci. **152/153**, 382.

Hohenberg, P.C., and W. Kohn, 1964, Phys. Rev. **136**, B864.

Holmes, M.I., and T. Gustafsson, 1981, Phys. Rev. Lett. **47**, 443.

Hörmandinger, G., 1994, Phys. Rev. B **49**, 13897.

Horn, K., A. Hohlfeld, J. Somers, T. Lindner, P. Hollins and A.M. Bradshaw, 1988, Phys. Rev. Lett. **61**, 2488.

Hu, Z.P., B.C. Pan, W.C. Fan and A. Ignatiev, 1990, Phys. Rev. B **41**, 9692.

Ihm, J., A. Zunger and M.L. Cohen, 1979, J. Phys. C: Solid State Phys. **12**, 4409.

Inglesfield, J.E., 1978, J. Phys. C: Solid State Phys. **11**, L69.

Inglesfield, J.E., 1982, Rep. Prog. Phys. **45**, 223.

Inglesfield, J.E., 1987, Prog. Surface. Sci. **25**, 57.

Inglesfield, J.E., and G.A. Benesh, 1988, Surface Sci. **200**, 135.

Ishida, H., 1990, Phys. Rev. B **42**, 10899.

Jacob, W., V. Dose and A. Goldmann, 1980, Applied Phys. A **41**, 145.

Jacobsen, K.W., and J.K. Nørskov, 1990, Phys. Rev. Lett. **65**, 1788.

Jedrecy, N., M. Sauvage-Simkin, R. Pinchaux, J. Massies, N. Greiser and V.H. Etgens, 1990, Surface Sci. **230**, 197.

Jensen, F., F. Besenbacher, E. Læsgaard and I. Stensgaard, 1990a, Phys. Rev. B **41**, 10233.

Jensen, F., F. Besenbacher, E. Læsgaard and I. Stensgaard, 1990b, Phys. Rev. B **42**, 9206.

Jiang, P., P.M. Marcus and F. Jona, 1986, Solid State Commun. **59**, 275.

Kaukasoina, P., M. Lindroos, R.D. Diehl, D. Fisher, S. Chandavarkar and I.R. Collins, 1993, J. Phys.: Condensed Matter **5**, 2875.

Kerkar, M., D. Fisher, D.P. Woodruff, R.G. Jones, R.D. Diehl and B. Cowie, 1992, Surface Sci. **278**, 246.

Kevan, S.D., 1985, Phys. Rev. B **32**, 2344.

Kiskinova, M., G. Pirug and H.P. Bonzel, 1983, Surface Sci. **133**, 321.

Kleinle, G., J. Wintterlin, G. Ertl, R.J. Behm, F. Jona and W. Moritz, 1990, Surface Sci. **225**, 171.

Kohn, W., and L.J. Sham, 1965, Phys. Rev. **140**, A1133.

Kohn, W., and P. Vashishta, 1983, in: Theory of the Inhomogeneous Electron Gas, eds. S. Lundqvist and N.H. March (Plenum, New York, NY), pp. 79-147.

Kubby, J.A., J.E. Griffith, R.S. Becker and J.S. Vickers, 1987, Phys. Rev. B **36**, 6079.

Lamble, G.M., R.S. Brooks, D.A. King and D. Norman, 1988, Phys. Rev. Lett. **61**, 1112.

Landman, U., R.N. Hill and M. Mostoller, 1980, Phys. Rev. B **21**, 448.

Lang, N.D., and W. Kohn, 1970, Phys. Rev. B **1**, 4555.

Lang, N.D., and A.R. Williams, 1978, Phys. Rev. B **18**, 616.

Lederer, T., D. Arvanitis, G. Cornelli, L. Tröger and K. Babeschke, 1993, Phys. Rev. B **48**, 15390.

Louie, S.G., K.M. Ho and M.L. Cohen, 1979, Phys. Rev. B **19**, 1774.

Magnussen, O.M., J. Wiechers and R.J. Behm, 1993, Surface Sci. **289**, 139.

Mattheiss, L.F., and D.R. Hamann, 1984, Phys. Rev. B **29**, 5372.

Methfessel, M., 1988, Phys. Rev. B **38**, 1537.

Methfessel, M., D. Hennig and M. Scheffler, 1992, Phys. Rev. B **46**, 4816.

Methfessel, M., and M. van Schilfgaarde, 1993, Phys. Rev. B **48**, 4937.

Methfessel, M., M. van Schilfgaarde and M. Scheffler, 1993, Phys. Rev. Lett. **70** 29; **71**, 209.

Moritz, W., and D. Wolf, 1985, Surface Sci. **163**, L655.

Muscat, J.P., and D.M. Newns, 1979, Surface Sci. **84**, 262.

Needels, M., M.C. Payne and J.D. Joannopoulos, 1987, Phys. Rev. Lett. **58**, 1765.

Needs, R.J., and M.J. Godfrey, 1990, Phys. Rev. B **42**, 10933.

Neugebauer, J., and M. Scheffler, 1992, Phys. Rev. B **46**, 16067.

Neugebauer, J., and M. Scheffler, 1993, Phys. Rev. Lett. **71**, 577.

Noonan, J.R., and H.L. Davis, 1984, Phys. Rev. B **29**, 4349.

Nordlander, P., S. Holloway and J.K. Nørskov, 1984, Surface Sci. **136**, 59.

Nørskov, J.K., and N.D. Lang, 1980, Phys. Rev. B **21**, 2131.

Over, H., H. Bludau, M. Skottke-Klein, G. Ertl, W. Moritz and C.T. Campbell, 1992, Phys. Rev. B **45**, 8638.

Pandey, K.C., 1981, Phys. Rev. Lett. **47**, 1913.

Pandey, K.C., 1982, Phys. Rev. Lett. **49**, 223.

Parkin, S.R., H.C. Zeng, M.Y. Zhou and K.A.R. Mitchell, 1990, Phys. Rev. B **41**, 5432.

Paul, J., 1988, Phys. Rev. B **37**, 6164.

Payne, M.C., M.P. Teter, D.C. Allan, T.A. Arias and J.D. Joannopoulos, 1992, Rev. Mod. Phys. **64**, 1045.

Perdew, J.P., H.Q. Tran and E.D. Smith, 1990, Phys. Rev. B **42**, 11627.

Pettifor, D.G., 1978, J. Phys. F: Metal Phys. **8**, 219.

Pickett, W.E., 1989, Comp. Phys. Reports **9**, 115.

Pines, D., 1963, Elementary Excitations in Solids (Benjamin, New York, NY).

Pisani, C., R. Dovesi and C. Roetti, 1988, Hartree-Fock ab initio Treatment of Crystalline Systems (Lecture Notes in Chemistry, Vol. 48) (Springer, Berlin).

Posternak, M., H. Krakauer, A.J. Freeman and D.D. Koelling, 1980, Phys. Rev. B **21**, 5601.

Press, W.H., B.P. Flannery, S.A. Teukolsky and W.T. Vetterling, 1989, Numerical Recipes (Cambridge University Press, Cambridge).

Qian, G.X., and D.J. Chadi, 1987, Phys. Rev. B **35**, 1288.

Riffe, D.M., G.K. Wertheim and P.H. Citrin, 1990, Phys. Rev. Lett. **64**, 571.

Roberts, N., and R.J. Needs, 1990, Surface Sci. **236**, 112.

Robinson, I.K., E. Vlieg and S. Ferrer, 1990, Phys. Rev. B **42**, 6954.

Rossmann, R., H.L. Meyerheim, V. Jahns, J. Wever, W. Moritz, D. Wolf, D. Dornisch and H. Schulz, 1992, Surface Sci. **279**, 199.

Scheffler, M., C. Droste, A. Fleszar, F. Máca, G. Wachutka and G. Barzel, 1991, Physica B **172**, 143.

Schmalz, A., S. Aminipirooz, L. Becker, J. Haase, J. Neugebauer, M. Scheffler, D.R. Batchelor, D.L. Adams and E. Bøgh, 1991, Phys. Rev. Lett. **67**, 2163.

Schmeits, M., A. Mazur and J. Pollmann, 1983, Phys. Rev. B **27**, 5012.

Shkrebtii, A.I., and R. Del Sole, 1993, Phys. Rev. Lett. **70**, 2645.

Singh, D., and H. Krakauer, 1988, Phys. Rev. B **37**, 3999.

Singh, D., H. Krakauer and C.S. Wang, 1986, Phys. Rev. B **34**, 8931.

Skriver, H.L., 1984, The LMTO Method (Springer-Verlag, Berlin).

Skriver, H.L., and N.M. Rosengaard, 1991, Phys. Rev. B **43**, 9538.

Skriver, H.L., and N.M. Rosengaard, 1992, Phys. Rev. B **46**, 7157.

Smith, K.E., G.S. Elliott and S.D. Kevan, 1990, Phys. Rev. B **42**, 5385.

Smoluchowski, R., 1941, Phys. Rev. **60**, 661.

Soler, J.M., and A.R. Williams, 1989, Phys. Rev. B **40**, 1560.

Srivastava, G.P., 1984, J. Phys. A: Math. Gen. **17**, L317.

Štich, I., R. Car, M. Parrinello and S. Baroni, 1989, Phys. Rev. B **39**, 4997.

Štich, I., M.C. Payne, R.D. King-Smith, J.S. Lin and L.J. Clarke, 1992, Phys. Rev. Lett. **68**, 1351.

Szabo, A., and N.S. Ostlund, 1982, Modern Quantum Chemistry (Macmillan, New York, NY).

Takayanagi, K., Y. Tanishiro, S. Takahashi and M. Takahashi, 1985, Surface Sci. **164**, 367.

Teter, M.P., M.C. Payne and D.C. Allan, 1989, Phys. Rev. B **40**, 12255.

Tománek, D., S.G. Louie and C.T. Chan, 1986, Phys. Rev. Lett. **57**, 2594.

Tosatti, E., 1978, Solid State Commun. **25**, 637.

Umrigar, C., and J.W. Wilkins, 1985, Phys. Rev. Lett. **54**, 1551.

Vanderbilt, D., and S.G. Louie, 1984, Phys. Rev. B **30**, 6118.

Van Hoof, J.B.A.N., S. Crampin and J.E. Inglesfield, 1992, J. Phys.: Condensed Matter **4**, 8477.

Van Hove, M.A., and S.Y. Tong, 1979, Surface Crystallography by LEED (Springer, Berlin).

Van Leuken, H., A. Lodder, M.T. Czyżyk, F. Springelkamp and R.A. de Groot, 1990, Phys. Rev. B **41**, 5613.

Van Leuken, H., A. Lodder and R.A. de Groot, 1992, Phys. Rev. B **45**, 4469.

Weimert, B., J. Noffke and L. Fritsche, 1992, Surface Sci. **264**, 365.

Weinert, M., and J.W. Davenport, 1985, Phys. Rev. Lett. **54**, 1547.

Weinert, M., A.J. Freeman and S. Ohnishi, 1986, Phys. Rev. Lett. **56**, 2295.

Weinert, M., R.E. Watson and J.W. Davenport, 1985, Phys. Rev. B **32**, 2115.

Wiesendanger, R., D. Bürgler, G. Tarrach and H.J. Güntherodt, 1990, Surface Sci. **232**, 1.

Williams, A.R., J. Kübler and C.D. Gelatt, 1979, Phys. Rev. B **19**, 6094.

Wimmer, E., A.J. Freeman, J.R. Hiskes and A.M. Karo, 1983, Phys. Rev. B **28**, 3074.

Wolkow, R.A., 1992, Phys. Rev. Lett. **68**, 2636.

Woodruff, D.P., and T.A. Delcher, 1986, Modern Techniques of Surface Science (Cambridge University Press, Cambridge).

Wuttig, M., R. Franchy and H. Ibach, 1989, Surface Sci. **213**, 103.

Yu, R., and P.K. Lam, 1989, Phys. Rev. B **39**, 5035.

Yu, R., D. Singh and H. Krakauer, 1991, Phys. Rev. B **43**, 6411.

Zangwill, A., 1988, Physics at Surfaces (Cambridge University Press, Cambridge).

Zeller, R., 1992, in: Unoccupied Electronic States, eds. J.C. Fuggle and J.E. Inglesfield (Springer, Berlin), pp. 25–49.

Zeng, H.C., and K.A.R. Mitchell, 1990, Surface Sci. **239**, L571.

Zhu, M.J., D.M. Bylander and L. Kleinman, 1990, Phys. Rev. B **42**, 2874.

Zhu, M.J., D.M. Bylander and L. Kleinman, 1991, Phys. Rev. B **43**, 4007.

Zhu, Z., N. Shima and M. Tsukuda, 1989, Phys. Rev. B **40**, 11868.

Ziman, J.M., 1972, Principles of the Theory of Solids (Cambridge University Press, Cambridge).

Chapter III

PHASE TRANSITIONS AT SURFACES

KURT BINDER

Institut für Physik, Johannes Gutenberg-Universität Mainz, Staudinger Weg 7,
D-55099 Mainz, Germany

Contents

Cohesion and Structure of Surfaces
Edited by D.G. Pettifor
© *1995 Elsevier Science B.V. All rights reserved.*

121

Abstract

The statistical mechanics of phase transitions is briefly reviewed, with an emphasis on surfaces. Flat surfaces of crystals may act as a substrate for adsorption of two-dimensional ($d = 2$) monolayers and multilayers, offering thus the possibility to study phase transitions in restricted dimensionality. Critical phenomena for special universality classes can thus be investigated which have no counterpart in $d = 3$. Also phase transitions can occur that are in a sense "in between" different dimensionalities (e.g., multilayer adsorption and wetting phenomena are transitions in between two and three dimensions, while adsorption of monolayers on stepped surfaces allows phenomena in between one and two dimensions to be observed).

Related phenomena concern transitions of surface layers of semi-infinite bulk systems: such singularities of a surface excess free energy may be related to a bulk transition (e.g., surface-induced ordering or disordering, surface melting, etc.) or may be a purely interfacial phenomenon (e.g. the roughening and facetting transitions of crystal surfaces). This article gives an introductory survey of these phenomena, discussing also illustrative model calculations employing computer simulation techniques.

1. Introduction: surface phase transitions versus transitions in the bulk

This chapter gives a tutorial introduction to the theory of phase transitions, emphasizing aspects which are particularly relevant in surface science.

If a surface of a crystal at low enough temperatures is used as a substrate for the adsorption of layers of atoms or molecules, one often — though not always — may treat the substrate as perfectly rigid and describe its effect simply by a potential $V(x, z)$ (fig. 1), z being the distance perpendicular to the substrate surface. Often this potential possesses a rather deep minimum at a preferred distance z_o, and then one can adsorb a two-dimensional monolayer, ideally of infinite extent in two space directions, if the substrate surface is perfect. In this way it is conceivable to have two-dimensional counterparts of all the phases which are familiar states of matter in three space dimensions: gas, fluid, and various solid phases (fig. 2). If these adsorbed species have internal degrees of freedom (e.g. electric or magnetic dipole moments, electric quadrupole moments, etc.), these internal degrees of freedom may exhibit two-dimensional order–disorder phenomena (e.g., oxygen molecules adsorbed on grafoil exhibit antiferromagnetic order (McTague and Nielsen, 1976); nitrogen molecules adsorbed on grafoil display a quadrupolar ordering of the molecule's orientations in the herring-

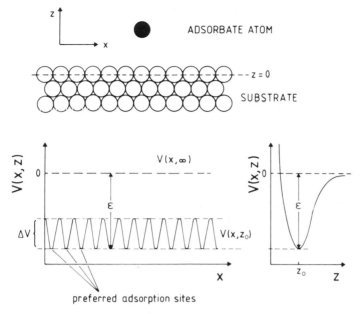

Fig. 1. Schematic description of adsorption on regular crystal surfaces. While circles show the atoms of the three topmost layers of the substrate, one adsorbate atom near the surface is shown as a black circle. The corrugation potential $V(x, z)$ has periodically arranged minima of depth ϵ separated by barriers of height ΔV from each other. These minima occur in a two-dimensional plane at a distance z_0 from the surface plane $z = 0$.

bone structure (Eckert et al., 1979); etc.) Depending whether the periodic variation of the "corrugation" potential $V(x, z)$ is weak or strong (in comparison to the thermal energy $k_B T$ at the temperatures T of interest), it may be appropriate to consider this formation of an adsorbed layer as a problem of statistical mechanics in two-dimensional continuous space or on a two-dimensional lattice (the lattice sites for this "lattice gas"-problem are given by the minima of the corrugation potential). Since the lattice spacing preferred by the (pairwise) interactions of the adatoms need not be commensurate with the lattice spacing offered by the substrate, this misfit of lattice spacings may give rise to commensurate–incommensurate phase transitions, and the occurrence of "striped phases" [characterized by regular arrangements of "domain walls" or misfit seams, respectively (Bak, 1984; Selke, 1992)]; these phenomena do not have obvious counterparts in the three-dimensional bulk.

Understanding the phase transitions and ordering phenomena in two-dimensional adsorbed monolayers is not only important in order to characterize the surface properties of various materials, but also is of fundamental

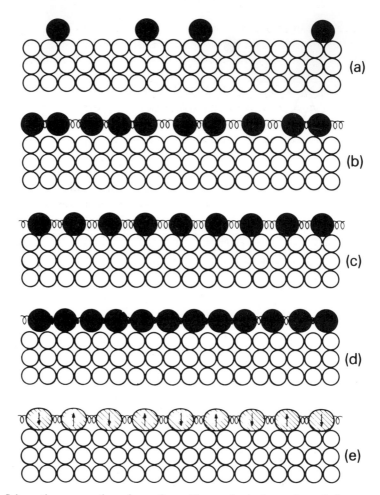

Fig. 2. Schematic cross section of a surface with an adsorbed monolayer (substrate atoms are denoted as open circles, adsorbate atoms as full circles). Springs indicate adsorbate-adsorbate interactions. Lattice gas (a), fluid (b), commensurate (c) and incommensurate (d) solid phases are shown, while case (e) indicates ordering of internal degrees of freedom of non-dissociated physisorbed molecules, such as the antiferromagnetic structure of O_2 on graphite (McTague and Nielsen, 1976). From Binder (1979a).

interest for the theory of phase transitions and critical phenomena (Fisher, 1974). As is well known, effects of statistical fluctuations are much stronger in $d = 2$ dimensions than in $d = 3$. As a consequence, certain ordering phenomena are destroyed by fluctuations: in $d = 2$ there is no long range order for isotropic magnets (Mermin and Wagner, 1966) and for other systems with isotropic n-component order parameters with $n \geq 2$ (Hohenberg,

1967). A related destruction of long range order also occurs for crystals in $d = 2$ (Mermin, 1968), their corresponding Bragg peaks no longer being delta functions at $T > 0$ but rather power-law singularities occur reflecting also a power law decay of spatial positional correlations. Under these circumstances, the nature of the melting transition may change from first-order to a sequence of two continuous transitions, controlled by topological defects (Nelson and Halperin, 1979): dislocation pair unbinding transforms the two dimensional "crystal" into the so-called hexatic phase, characterized by a power-law decay of orientational correlations. Disclination pair unbinding in a second continuous transition (of Kosterlitz–Thouless (1973) type) transforms the hexatic solid into a true liquid, where all correlations are short ranged. This Kosterlitz–Thouless (1973) transition was first proposed for planar magnets in $d = 2$ (XY-ferromagnets, that have $n = 2$), the topological objects that unbind at the transition are vortex–antivortex pairs. The most important application of this picture is the superfluid–normal fluid transition of He^4 layers in $d = 2$ (Dash, 1978).

Also for other orderings where long range order exist fluctuations are important — e.g. for the Ising model critical exponents differ much more from the Landau mean field values in $d = 2$ than in $d = 3$ (Baxter, 1982). For the Potts model (Potts, 1952; Wu, 1982) with $q = 3$ or $q = 4$ states, mean field theory even fails in predicting the order of the transition correctly: Landau theory (Landau and Lifshitz, 1958) symmetry arguments imply a discontinuous vanishing of the order parameter (first-order transition), while in reality in $d = 2$ the transition is of second order, as known from exact solutions. Also for the prediction of phase diagrams molecular field theory is a bad guide, often the ordering temperatures being overestimated by a factor of two or more, and sometimes even the topology of phase diagrams being predicted incorrectly (Binder et al., 1982). Thus the phase transitions of two-dimensional monolayers are a welcome laboratory, where sophisticated methods of statistical mechanics can be put to work, and various approximations can be tested, as well as new concepts. A concept specifically useful in $d - 2$ is the "conformal invariance", which allows one to predict exactly all the critical exponents (Cardy, 1987).

Of course, the situation is not always as simple as in the idealized case sketched in figs. 1 and 2. First of all, the substrate often is not ideal over distances of infinite extent, and so one often has to consider finite size effects due to the limited size of the linear dimensions over which the substrate is homogeneous. An interesting and important case are vicinal (high-indexed) crystal surfaces, produced by cutting a crystal under a small angle to a close-packed crystal plane, such that a staircase-like structure is formed (fig. 3) with terraces of width L separated by steps. (Albano et al., 1989a). If the linear dimension M in the direction parallel to the steps is much larger than L, the system behaves in many aspects quasi-one-dimensionally.

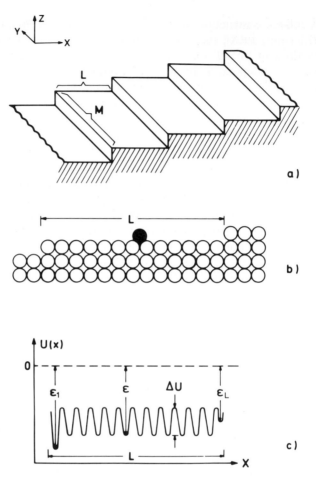

Fig. 3. (a) Schematic view of regularly stepped surface, where steps a distance L apart in the x-direction run parallel to each other a distance M in the y-direction, to form a "staircase" of $L \times M$ terraces, on which adsorption can take place. (b) Cross section through one terrace of width L. Open circles represent substrate atoms, full circle represents an adsorbate atom. (c) Corrugation potential corresponding to the geometry of case (b). We assume that the substrate creates a lattice of preferred sites, at which adatoms can be bound to the surface with an energy ϵ (cf. fig. 1). In the rows adjacent to the terrace boundaries, however, one assumes in general different binding energies ϵ_1, ϵ_L which correspond in the Ising magnet terminology to the "boundary magnetic fields" $H_1 = J - (\epsilon_1 - \epsilon)/2$, $H_L = J - (\epsilon_L - \epsilon)/2$ for a case of nearest-neighbor interaction J. The energy barrier ΔU separates neighboring preferred sites. From Albano et al. (1989a).

At the terrace edges, the corrugation potential may differ from the potential in the terrace interior, and hence many different cases may need consideration.

Another complication is that the substrate often does not possess a rigid structure but responds to the forces exerted by the adsorbate by some local deformation. Such an adsorbate-induced relaxation of the topmost substrate layer (with a corresponding change of lattice parameters but no change of lattice symmetry) or adsorbate-induced reconstruction (with a corresponding change of lattice symmetry of the topmost substrate layer) will not be discussed further here.

Rather we focus on the possibility that a second layer of adsorbate atoms may condense on top of the first one, a third layer on top of the second, etc. (fig. 4; Patrykiejew et al., 1990). By this "multilayer adsorption" (de Oliveira and Griffiths, 1978; Pandit et al., 1982; Dietrich, 1988) one may proceed from two-dimensional to three-dimensional geometry of the adsorbed layer. An alternative — and useful — view of this phenomenon is to consider it as a surface effect to the bulk three-dimensional gas exposed to forces at the walls of the container confining it. Then the gas–liquid condensation (or gas–solid crystallization) can occur at the wall already at a smaller pressure than the pressure where gas and liquid (or gas and crystal, respectively) could coexist in thermal equilibrium in the bulk. Thus while in a thin film geometry (such as a fluid confined in slit like pores) the surfaces affect the phase transitions of the bulk layer ("capillary condensation", see e.g. Binder and Landau, 1992a), in a semi-infinite geometry it is the surface excess free energy which exhibits singularities reflecting surface phase transitions such as "layering transitions" (multilayer adsorption) or "wetting transitions" (Dietrich, 1988). In a wetting transition, a macroscopically thick fluid layer condenses at the surface, while in the bulk one still has a saturated gas (with a density according to the gas liquid coexistence curve). Alternatively, one may interpret it as an interface unbinding transition: in the non-wet state of the gas surface, a gas–liquid interface is very tightly bound to the wall, and thus there is a density enhancement only over a few atomic diameters near the surface (fig. 5). In the wet state, at the wall the density enhancement even exceeds the density of the liquid branch of the gas–liquid coexistence curve, and then the density profile decays in two steps, one first reaches the liquid coexistence density and at a large distance the density falls off to the gas density in a liquid–gas interfacial profile.

The density profiles of fig. 5 treat the atomic mass density as a variable in continuous space, so that unlike multilayer adsorption (fig. 4) one does not identify discrete atomic layers adsorbed at the surface. This also means that the gas–liquid interface is a smooth, delocalized object (fig. 6), while in the case of multilayer adsorption the interface to the gas is sharp on the scale of atomic diameters. Although these two pictures of an interface between coexisting phases are mutually exclusive, they can occur in the same system: the interface of a solid crystal which is rough and localized at

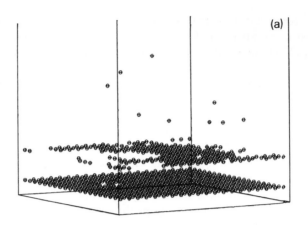

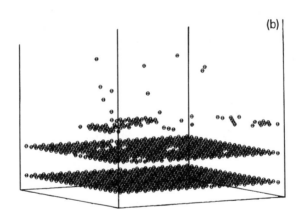

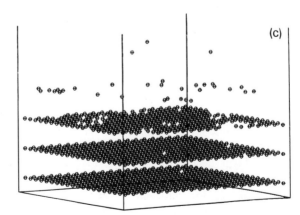

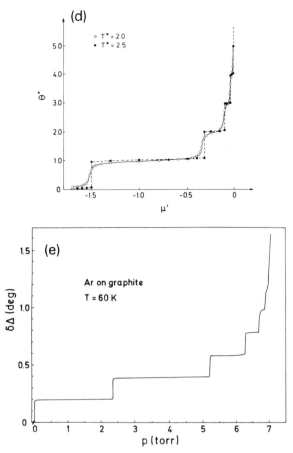

Fig. 4. Snapshot pictures of a Monte Carlo simulation of multilayer adsorption (a)–(c). A simple cubic lattice gas with $L \times L \times D$ thin film geometry is treated, where in two directions periodic boundary conditions are applied, while the two hard walls (of surface area $L \times L$) which are a distance D apart exert an attractive potential $V(z) = -A/z^3$ on the gas. Measuring all lengths in units of the lattice spacing and temperature ($T^\star$) in units of the exchange constant J of the resulting Ising model, parameters of the simulation shown are $L = 30$, $D = 40$, $T^\star = 2.32$, $A = 2.5$. Three choices of the chemical potential difference $\mu' = (\mu - \mu_o)/J$ relative to the chemical potential μ_o, where gas–liquid condensation occurs in the bulk, are shown here: $\mu' = -0.34$ (a), $\mu' = -0.11$ (b), and $\mu' = -0.095$ (c). For clarity, the distances in the z-direction are displayed on a strongly expanded scale relative to the scale of the x- and y-directions. One can see that in case (a) the adsorption of the first layer is nearly completed, and the second layer is about half full, while in (b) also the second layer and in (c) the third layer are completed. In (d) the corresponding adsorption isotherms (coverage $\theta^\star$ vs. μ') are plotted for a temperature below the layering critical temperatures $T_c^\star(N)$, $T^\star = 2.0$, and a temperature $T^\star > T_c^\star(N)$, $T^\star = 2.5$. Corresponding experimental adsorption isotherms (Argon on graphite at $T = 60$ K observed by ellipsometry) are shown in (e) taken from Volkmann and Knorr (1989), while the simulation results are taken from Patrykiejew et al. (1990).

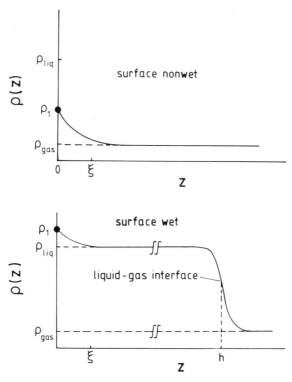

Fig. 5. Density profile $\rho(z)$ for a fluid at a surface near a wetting transition. In the non-wet state of the surface the local density ρ_1 at the surface is less than the density ρ_{liq} at the liquid branch of the coexistence curve describing gas–liquid condensation (upper part). Then the density profile $\rho(z)$ decays to the gas density ρ_{gas} in the bulk at a microscopic distance (which is of the order of the correlation length ξ). In the wet state of the surface (lower part), the bulk gas is saturated (ρ_{gas} must have the value of the gas branch of the coexistence curve) and $\rho_1 > \rho_{\text{liq}}$, and a (macroscopic) liquid layer condenses at the surface, separated at large distances from the gas by a liquid–gas interface centered at $z = h(x, y)$.

low temperatures can undergo an interfacial roughening transition (Weeks, 1980; van Beijeren and Nolden, 1987), where the interface gets delocalized because a localized interface would then be unstable against the formation of long wavelength capillary waves. While interfaces of true off-lattice fluids are rough for all temperatures $T > 0$, this roughening transition can occur for surfaces of lattice fluids, crystal surfaces against vacuum, antiphase domain boundaries in ordered alloys, etc.; the quantitative characteristics of this roughening transition are again closely related to the Kosterlitz–Thouless transition of XY-ferromagnets mentioned above. In surface physics, the roughening of crystal surfaces (fig. 7) is also related to "facetting transitions" of macroscopic equilibrium crystal shapes (Rottmann and Wortis, 1984), as

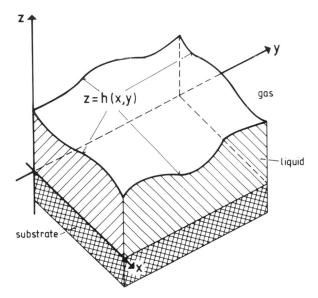

Fig. 6. Coarse-grained description of a liquid-gas interface, where the "intrinsic" profile and local structure of the interface is disregarded, and one rather treats the interface as an "elastic membrane" at position $z = h(x, y)$ ("sharp kink"-approximation for the interfacial profile).

well as the spontaneous formation of surface steps since the step free energy vanishes (Mon et al., 1989).

Now assuming rough interfaces (fig. 6), one can describe wetting phenomena by introducing an effective potential $V_{\text{eff}}(h)$ for the local interface position $z = h(x, y)$, assuming a "sharp kink"-picture for this local position (Dietrich, 1988). On this coarse-grained level of theoretical description, fig. 6 not only describes wetting in fluids, but many related interfacial unbinding phenomena. If ordered crystals undergo a first-order order–disorder transition at some temperature T_c, already at $T < T_c$ a disordered layer may intrude at their surface. As $T \to T_c$ from below, the thickness of this disordered layer at the surface diverges, as the interface separating this disordered surface layer from the ordered bulk unbinds from the surface and wanders into the bulk (Lipowsky, 1984). "Surface melting" (Van der Veen and Frenken, 1986; Van der Veen et al., 1990) of crystals can be viewed as the analogue of surface-induced disordering for the solid–fluid transition: then the gas phase in fig. 6 is to be replaced by the crystal. Although in these transitions there is no diverging correlation length in the bulk, the thickness of the disordered layer at the surface exhibits a critical divergence, and at the same time there is a critical vanishing of the local order parameter at the surface. Also the opposite phenomenon, where the surface orders at a higher

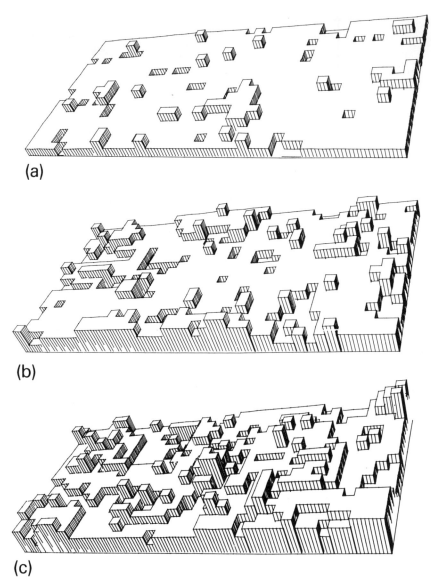

Fig. 7. Snapshot pictures of a Monte Carlo simulation of the crystal–vacuum interface in the framework of a solid-on-solid (SOS) model, where bubbles and overhangs are forbidden. Each lattice site i is characterized by a height variable h_i and the Hamiltonian then is $\mathcal{H} = -\phi \sum_{\langle i,j \rangle} |h_i - h_j|$. Three temperatures are shown: $kT/\phi = 0.545$ (a), 0.600 (b) and 0.667 (c). The roughening transition temperature T_R roughly coincides with case (b). From Weeks et al. (1973).

temperature than the bulk, is conceivable: in this "surface induced ordering" the order propagates more and more into the bulk as $T \rightarrow T_c$, and again the transition can be interpreted as an interface unbinding transition (Lipowsky, 1984, 1987).

For a theoretical discussion of these various phase transitions associated with surfaces and interfaces, the proper thermodynamic functions need to be characterized. In order to be specific, we consider an Ising lattice model (Binder, 1983; fig. 8) in which each site i of a d-dimensional cubic lattice carries an Ising spin $S_i = \pm 1$. Let us assume a thin film-geometry, where a film of thickness $L = (N_1 - 1)a$ (i.e., we have N_1 atomic layers and a is the lattice spacing) has two free surfaces. The Hamiltonian of such a model could be (in the case of nearest-neighbor interaction)

$$\mathcal{H} = - \sum_{\substack{\langle i,j \rangle \\ \text{interior layers}}} J S_i S_j - \sum_{\substack{\langle i,j \rangle \\ i \in 1 \\ j \in 2}} {}_{\text{or}} {}_{\substack{i \in N_1 - 1 \\ j \in N_1}} J_\perp S_i S_j - \sum_{\substack{\langle i,j \rangle \\ i,j \in 1 \text{ or } i,j \in N_1}} J_\| S_i S_j$$

$$- H \sum_i S_i - H_1 \sum_{\substack{i \\ i \in 1 \text{ or } i \in N_1}} S_i, \qquad (1)$$

where J is the exchange interaction in the bulk, $J_\|$ the exchange interaction in the two free surface planes in which a "surface magnetic field" H_1 also acts, in addition to the bulk field acting on all the spins, and $J_\perp$ is the coupling between spins in the surface planes and spins in the adjacent layers. This model is not only useful for describing surface magnetism, but can also describe surface properties of binary alloys as well as adsorption of fluids. For the binary alloy (AB) application, we use $S_i = +1$ if site i is taken by an A-atom and $S_i = -1$ if site i is taken by a B-atom. The "field" H then translates to the bulk chemical potential difference $\Delta\mu$ between A and B and is eliminated if one fixes the thermodynamically conjugate variable, the relative concentration c_A of A in the bulk, $c_A = (1 + \langle S_i \rangle)/2$. The field H_1 on the other hand, is related to the difference between the pairwise interactions, i.e. $v_{AA} - v_{BB}$, and hence in general is non-zero. Remember that the exchange interaction J is proportional to the combination $v_{AB} - (v_{AA} + v_{BB})/2$ only. As a consequence, surface enrichment of one species in alloys must be generally expected to occur, and there may then be an interplay of surface enrichment and order–disorder phenomena at the surfaces of alloys (Kroll and Gompper, 1987; Helbing et al., 1990; Schmid, 1993).

If one wishes to use eq. (1) to model adsorption of fluids, one refers to the lattice gas interpretation of eq. (1), where $S_i = -1$ if a lattice site is filled while $S_i = +1$ if it is empty, $\rho_i = (1 - S_i)/2$ being the local density. Then again H is related to the chemical potential in the bulk, while H_1 relates to the binding energy to the walls (in the framework of a model where the range

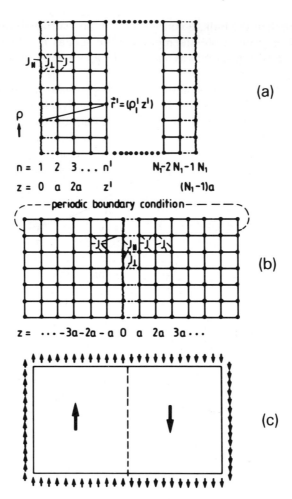

Fig. 8. (a) d-Dimensional Ising film of N_1 layers shown schematically in cross-section. Each vertical line represents a $(d-1)$ dimensional layer, with coordinate ρ. The layers are indexed by n, which goes from 1 to N_1, or by z, going from 0 to $(N_1-1)a$. An arbitrary point is denoted by the vector $r' = (\rho', z')$. For $N_1 \to \infty$ the system is a halfspace with a free surface at $z = 0$. While all nearest neighbor interactions are taken to be the same (J) in the bulk, the interactions within the surface plane are $J_\parallel$, and the coupling between spins in the surface plane and in the adjacent plane is $J_\perp$. In the direction parallel to the film, where one assumes a linear dimension $L_\parallel \to \infty$ and periodic boundary conditions, the system is translationally invariant. (b) d-Dimensional Ising system with a $(d-1)$ dimensional "defect plane" at $z = 0$ but periodic boundary conditions in all lattice directions. (c) d-Dimensional Ising system with homogeneous interactions but fixed spin boundary conditions as symbolized by the arrows, such that the ground state of the system contains one interface separating a domain with positive magnetization from a domain with negative magnetization.

of this binding potential is so short that it is felt in only the one layer right adjacent to the walls). Thus this model Hamiltonian is well suited to study wetting and multilayer adsorption phenomena (Binder and Landau, 1988).

The thermodynamic functions which are then derived from this model (or suitable other models) via statistical mechanics, such as the free energy F (T,H,H_1,L), can conveniently be discussed by splitting them into bulk and surface terms (see e.g. Binder (1983) for more details),

$$\frac{F(T, H, H_1, L)}{AL} = f_b(T, H) - \frac{2}{L} f_s(T, H, H_1), \qquad L \to \infty, \qquad (2)$$

where A is the area of the $(d - 1)$ dimensional wall, $f_b(T, H)$ is the bulk free energy per lattice site, and f_s is the surface free energy (or boundary free energy, respectively) per lattice site at the surface (remembering that the geometry of fig. 8 implies two surfaces). At the layering transitions, as well as at the wetting transitions, the singular behavior of $f_s(T, H, H_1)$ is sought, while $f_b(T, H)$ remains non-singular there. However, one may also study the singular behavior of $f_s(T, H, H_1)$ induced by the singular behavior due to phase transitions in the bulk: for a second-order bulk transition, the critical behavior of the local quantities at the surface differs from the bulk (this is termed the "ordinary" surface critical behavior); for a first-order bulk transition, $f_s(T, H, H_1)$ may reflect surface-induced disordering, for instance.

Thus, while it is rather straightforward to define excess free energies due to external boundaries, it is more subtle to obtain the excess free energies due to interfaces between coexisting phases (Widom, 1972; Jasnow, 1984). In an Ising magnet with periodic boundary conditions in all directions we may generate an interface by choosing a "defect plane" in between $z = 0$, $z = a$ in the system (fig. 8b), such that all bonds $J_\perp$ crossing this plane have $J_\perp = -J$, all other bonds in the system being equal to $+J$. Taking the ratio between partition functions Z_- (containing such a "defect plane" with negative bonds) and Z_+ with homogeneous bonds $+J$ throughout the system, one obtains $f_{int}(T, H)$,

$$f_{int}(T, H) = \frac{1}{k_B T A} \ln \left[\frac{Z_-(T, H, L)}{Z_+(T, H, L)} \right]. \qquad (3)$$

An alternative to this geometry is the "fixed spin" boundary condition, fig. 8c. If all spins adjacent to the boundary are up, the system is homogeneous, and does not contain boundaries. Its partition function is Z_{++}. If half the spins adjacent to the boundary are minus and only the other half are plus, we stabilize an interface in the system, and from the partition function Z_{+-} we hence obtain (Privman, 1992)

$$f_{int}(T, H) = \frac{1}{k_B T A} \ln \left[\frac{Z_{+-}(T, H)}{Z_{++}(T, H)} \right]. \qquad (4)$$

This "surface tension" $f_{int}(T, H)$ is singular at the roughening transition temperature T_R, as well as at the bulk critical temperature T_c of the Ising model.

2. Phenomenological theory of phase transitions: a brief review

This section summarizes the main facts of the theory of phase transitions, with an emphasis on aspects relevant for surface physics. It also serves to introduce the necessary terminology and notation. For more details, see Stanley (1971), Fisher (1974), Schick (1981), and Yeomans (1992).

2.1. Order parameters, second-order versus first-order transitions

We consider systems that can exist in several thermodynamic phases, depending on external thermodynamic variables which we take as intensive variables here (independent of the volume), such as temperature T, pressure p, external fields, etc. Assuming that an extensive thermodynamic variable (i.e., one which is proportional to the volume) can be identified that distinguishes between these phases, namely the "order parameter" ϕ, we introduce the conjugate thermodynamic variable, the "ordering field" H, such that

$$\phi = -\left(\frac{\partial F}{\partial H}\right)_T, \qquad S = -\left(\frac{\partial F}{\partial T}\right)_H, \tag{5}$$

where S is the entropy of the system. For a ferromagnet, ϕ is the magnetization and H is a magnetic field: for a monolayer of oxygen chemisorbed on Ru(001) surfaces in the p(2×2) structure (Piercy and Pfnür, 1987), ϕ is the amplitude of a mass density wave with a wavevector k characterizing the periodicity of that structure. These examples already indicate that in eq. (1) we have simplified matters — ϕ and H in general are not scalars but typically the order parameter has several components, e.g. the magnetization is a vector that has three components. Also H is not always physically realizable in the laboratory — for the case where ϕ relates to a mass density wave, H relates to the corresponding Fourier component of the chemical potential, $\mu(k)$. Nevertheless, the notion of the field conjugate to the order parameter is useful.

It is clear that thermodynamic relations as written in eq. (5) apply to any material: ϕ qualifies as an order parameter when a particular value of the ordering field exists where the order parameter exhibits a jump singularity between two distinct values (fig. 9). This means that for these values of the ordering field a first-order phase transition occurs, where a first derivative of the thermodynamic potential F exhibits a singularity. At this transition, two phases can coexist; i.e. at the liquid–gas transition

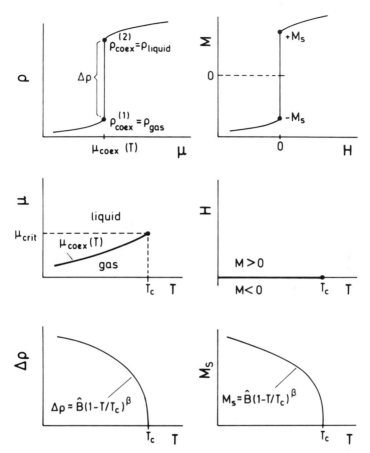

Fig. 9. The fluid–magnet analogy. On varying the chemical potential μ, at $\mu_{\text{coex}}(T)$ the density ρ jumps from the value at the gas branch of the gas–liquid coexistence curve ($\rho_{\text{gas}} = \rho_{\text{coex}}^{(1)}$) to the value at the liquid branch ($\rho_{\text{liquid}} = \rho_{\text{coex}}^{(2)}$); top left. Similarly, on varying the (internal) magnetic field H, the magnetization M jumps from the negative value of the spontaneous magnetization ($-M_s$) to its positive value (top right). While this first-order liquid–gas transition occurs at a curve $\mu_{\text{coex}}(T)$ in the μ–T-plane ending in a critical point (μ_c, T_c) where the transition then is of second order, the curve where phases with positive and negative spontaneous magnetization can coexist simply is $H = 0(T < T_c)$; middle part. The order parameter (density difference $\Delta\rho$, or spontaneous magnetization M_s) vanishes according to a power law at T_c (bottom part).

for a chemical potential $\mu = \mu_{\text{coex}}(T)$, two phases with different density coexist; and in a ferromagnet at zero magnetic field, phases with opposite sign of the spontaneous magnetization can coexist. Although the fluid–magnet analogy (fig. 9) goes further, since the first-order lines in the (μ,T) or (H,T) plane in both cases end in critical points which may even be characterized by the same critical exponents, there is also an important distinction: in the magnetic problem the Hamiltionian [e.g. eq. (1)] possesses a symmetry with respect to the change of sign of the magnetic field; reversing this sign and also reversing the sign of the magnetization leaves the Hamiltonian invariant. Owing to this symmetry, the transition line must occur at $H = 0$. Conversely, if the system at $H = 0$ is in a monodomain state with either positive or negative spontaneous magnetization, this symmetry is violated: "spontaneous symmetry breaking". No such obvious symmetry exists for the liquid–gas transition, and thus the curve $\mu = \mu_{coex}(T)$ is a non-trivial function in the $\mu - T$ plane (no simple symmetry operation acting on the gas phase atoms is known that would transform the gas into a liquid, or vice versa). The order parameter in an adsorbed monolayer in a square lattice geometry of adsorption sites, e.g. the $c(2\times2)$ structure, fig. 10, may be taken as the density difference of the two sublattices, $\psi = (\rho^{II} - \rho^{I})/2$. However, the two sublattices physically are completely equivalent; therefore the Hamiltonian possesses a symmetry against the interchange of the two sublattices, which implies that ψ changes sign, just as the (idealized!) ferromagnet does for $H = 0$ in fig. 9. Again in this example of a monolayer which may undergo an order–disorder transition where the permutation symmetry between the two sublattices is spontaneously broken, the "ordering field" conjugate to the order parameter is a chemical potential difference between the two sublattices, and hence this ordering field is not directly obtainable in the laboratory. The situation is comparable to the case of simple antiferromagnets, the order parameter being the "staggered magnetization" (= magnetization difference between the sublattices), and the conjugate ordering field would change sign from one sublattice to the other ("staggered field"). Although the action of such fields usually cannot be measured directly, they nevertheless provide a useful conceptual framework.

Another problem which obscures the analogy between different phase transitions is the fact that one does not always wish to work with the corresponding statistical ensembles. Consider, for example, a first-order transition where from a disordered lattice gas islands of ordered $c(2\times2)$ structure form. If we consider a physisorbed layer in full thermal equilibrium with the surrounding gas, then the chemical potential of the gas and the temperature would be the independent control variables. In equilibrium, of course, the chemical potential μ of subsystems is the same, and so the chemical potential of the lattice gas and that of the ordered islands would be the same, while the surface density (or "coverage" θ) in the islands will differ from that of the lattice gas. The three-dimensional gas acts as a reservoir which supplies adsorbate atoms to maintain the equilibrium value of the coverage in the ordered islands when one cools the adsorbed layer through the order–disorder transition. However, one often considers such a transition at

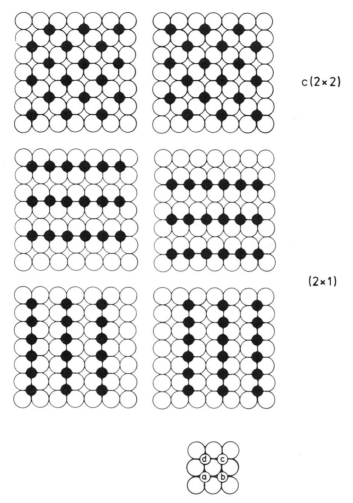

c(2×2)

(2×1)

Fig. 10. Adsorbate superstructures on (100) surfaces of cubic crystals. Atoms in the top-layer of substrate are shown as white circles, while adsorbate atoms are shown as full black circles. Upper part shows the two possible domains of the c(2×2) structure, obtained by dividing the square lattice of preferred adsorption sites into two sublattices following a checkerboard pattern: either the white sublattice or the black sublattice is occupied with adatoms. The (2×1) structure also is a 2-sublattice structure, where full and empty rows alternate. These rows can be interchanged and they also can run either in x-direction (middle part) or y-direction (lower part), so four possible domains result and one has a two-component order parameter.

fixed monolayer coverage (e.g. in studies of chemisorbed monolayers under ultrahigh vacuum there is no gas that could act as a reservoir). Consequently, the temperature T and coverage θ are the independent variables. Now, similarly as in a canonical ensemble description of fluids (fig. 9), the first-order transition shows up as a two-phase coexistence region of disordered lattice gas (at low coverage $\theta_{coex}^{(1)}$) with ordered c(2×2) islands (at higher coverage $\theta_{coex}^{(2)}$). For a given coverage θ with $\theta_{coex}^{(1)} < \theta < \theta_{coex}^{(2)}$, the system is an inhomogeneous mixture of both coexisting phases; the relative amounts of these two phases is given by the lever rule, if interfacial contributions to the thermodynamic potential can be neglected. These different statistical ensembles (grand-canonical ensemble, if full equilibrium with surrounding gas in $d = 3$ is established, or canonical ensemble at constant coverage) also have pronounced consequences on the dynamic properties of the considered systems: in the layers at constant coverage, the conservation law for the surface density means that density fluctuations can relax only by surface diffusion, while for a layer at constant chemical potential in equilibrium with the surrounding gas, density fluctuations relax by condensation/evaporation processes whereby atoms are exchanged between the layer and the gas.

An important characteristic to which we turn next is the order of a phase transition. In the examples shown in the upper part of fig. 9, a first derivative of the appropriate thermodynamical potential has a jump singularity and therefore such transitions are called *first-order transitions*. However, if we cool a ferromagnet down from the paramagnetic phase in zero magnetic field, the spontaneous magnetization sets in continuously at the critical temperature T_c (lower part of fig. 9). Similarly, on cooling hydrogen on Pd(100) at $\theta = 0.5$ down from high temperatures, where the adsorbed layer is in a state of a disordered lattice gas, on the square lattice, one observes at the critical temperature $T_c \approx 260$ K a continuous onset of ordering in the c(2×2) structure (Behm et al., 1980). Whereas the first derivatives of the thermodynamic potential at these continuous phase transitions are smooth, the second derivatives are singular, and therefore these transitions are called *second-order transitions*. For example, in a ferromagnet typically the isothermal susceptibility χ_T and the specific heat have power law singularities (fig. 11)

$$\chi_T \equiv -\left(\frac{\partial^2 F}{\partial H^2}\right)_T = \hat{C}^\pm \left|\frac{T}{T_c} - 1\right|^{-\gamma}, \qquad T \to T_c, \tag{6}$$

$$C_H \equiv -T\left(\frac{\partial^2 F}{\partial T^2}\right)_H\bigg|_{H=0} = \hat{A}^\pm \left|\frac{T}{T_c} - 1\right|^{-\alpha}, \qquad T \to T_c, \tag{7}$$

where α, γ are critical exponents, $\hat{A}^\pm$, $\hat{C}^\pm$ critical amplitudes (the $\pm$ signs refer to the sign of $T/T_c - 1$, and we have anticipated that there is no need to distinguish critical exponents γ, γ' or α, α' above or below T_c). Note that $\hat{B}$ and β refer to the order parameter (spontaneous magnetization of a ferromagnet, for example, see fig. 9).

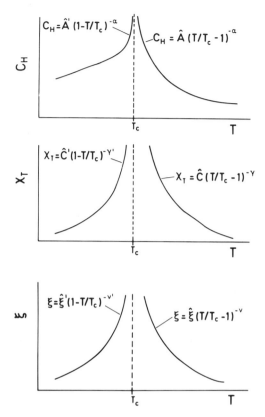

Fig. 11. Schematic variation with temperature T plotted for several quantities near a critical point T_c: specific heat C_H (top), ordering "susceptibility" χ_T (middle part), and correlation length ξ of order parameter fluctuations (bottom). The power laws which hold asymptotically in the close vicinity of T_c are indicated.

$$M_s = \hat{B}\left(1 - \frac{T}{T_c}\right)^{\beta}, \qquad T \to T_c. \tag{8}$$

A behavior of the specific heat as described by eq. (7) immediately carries over to systems other than ferromagnets, such as antiferromagnets, the liquid–gas system near its critical point, and order–disorder transitions of physisorbed layers such as the $(\sqrt{3}\times\sqrt{3})$R30° structure of He^4 adsorbed on grafoil (Bretz, 1977); one must remember, however, that H then means the appropriate ordering field. In fact, this is also true for eq. (6), but then the physical significance of χ_T changes. For a two-sublattice antiferromagnet, the ordering field is a "staggered field", which changes sign between the two sublattices, and hence is thermodynamically conjugate to the order parameter of the antiferromagnet. Although such a field normally cannot be applied in the laboratory, the second derivative, χ_T (in this case it is called "staggered susceptibility") is experimentally accessible via diffuse magnetic neutron scattering, as will be discussed below.

Similarly, for the ordering monolayer hydrogen on Pd(100), the ordering field stands for a chemical potential difference between the two sublattices; the response function χ_T is again physically meaningful, it measures the peak intensity of the diffuse LEED spots. These scattering peaks occur at the superlattice Bragg positions characteristic for the considered sublattice ordering (Behm et al., 1980).

As will be discussed in more detail in sect. 2.2, the divergences of second derivatives of the thermodynamic potential at a critical point [eqs. (6), (7); fig. 11] are linked to a diverging correlation length ξ of order parameter fluctuations (fig. 11). Hence any discussion of phase transitions must start with an identification of the order parameter. Expanding the thermodynamic potential in powers of the order parameter, in the spirit of Landau's theory, also gives some information on the "universality class" to which a transition belongs (Schick, 1981). We have already mentioned that the order parameter is not always a scalar quantity, as was assumed in eq. (5), although this is correct for the gas–liquid transition, for uniaxial ferromagnets, for the order–disorder transition of the $c(2\times2)$ structure (fig. 10) and other order–disorder transitions where only two sublattices need to be considered. But there are also other cases where the order parameter must have vector or tensor character: for an isotropic ferromagnet the order parameter in three-dimensional space is a three-component vector. In systems with a planar anisotropy the magnetization must lie, for instance, in the XY-plane, and hence a two-component order parameter applies. However, for describing order–disorder transitions with many sublattices multi-component order parameters are also needed, and the number of components of the order parameter, the so-called "order parameter dimensionality", is dictated by the complexity of the structure, and has nothing to do with the spatial dimension.

This is best understood by considering specific examples. Consider, for example, the ordering of the (2×1)-structure on the square lattice (Binder and Landau, 1980): whereas in the disordered phase the adatoms are distributed at random over the available lattice sites, consistent with the considered coverage $\theta = 1/2$ (although there may be some short-range order), in the ordered (2×1) phase the square lattice is split into four interpenetrating sublattices a, b, c, d of twice the lattice spacing (see assignment of sublattices in the lower part of fig. 10). In the Ising spin representation where $\rho_i = (1-S_i)/2$ a spin $S_i = -1$ corresponds to an adsorbed atom at site i, and hence it is convenient to characterize the ordering by the "magnetizations" m_a, m_b, m_c, m_d of the sublattices in this pseudospin representation, with $m_\mu = (1/N) \sum_{i\in\mu} S_i = (1/N) \sum_{i\in\mu} (1 - 2\rho_i)$. Then the order parameter of the c(2×2) structure is given by (N is the total number of sites and thus ψ is normalized to unity)

$$\psi^{c(2\times2)} = m_a + m_c - (m_b + m_d).\tag{9}$$

One easily recognizes that the two types of domains shown on the top of fig. 10 simply correspond to $\psi = \pm 1$. Since for this structure the sublattices (a,c) and (b,d) each can be combined to a single sublattice, the c(2×2) structure has a single order parameter component. But the situation differs for the (2×1) structure, where two components are needed:

$$\psi_{\mathrm{I}}^{(2\times1)} = m_a + m_b - (m_c + m_d), \quad \psi_{\mathrm{II}}^{(2\times1)} = m_a + m_d - (m_b + m_c)\tag{10}$$

The domains shown in the middle part of fig. 10 correspond to $\psi_{\mathrm{I}}^{(2\times1)} = \pm 1$, $\psi_{\mathrm{II}}^{(2\times1)} = 0$, while the domains shown in the lower part correspond to $\psi_{\mathrm{I}}^{(2\times1)} = 0$, $\psi_{\mathrm{II}}^{(2\times1)} = \pm 1$. Thus the (2×1) structure belongs (Krinsky and Mukamel, 1977) to the universality class of the XY-model with cubic anisotropy (obviously in the order parameter space spanned by $(\psi_{\mathrm{I}},\psi_{\mathrm{II}})$ the coordinate axes are singled out). If we consider the (2×2) structure with $\theta = 1/4$ on the square lattice, we would also have four kinds of domains but now with three components of the order parameter, which can be written in terms of the four sublattice densities ρ_a, ρ_b, ρ_c, ρ_d as

$$\psi_{\mathrm{I}}^{(2\times2)} = (\rho_a - \tfrac{1}{4}), \quad \psi_{\mathrm{II}}^{(2\times2)} = (\rho_b - \tfrac{1}{4}), \quad \psi_{\mathrm{III}}^{(2\times2)} = (\rho_c - \tfrac{1}{4}).\tag{11}$$

Note that due to the constraint $\rho_a + \rho_b + \rho_c + \rho_d = 1/4$, there is no fourth independent component. The order parameter components defined in eq. (11) are not orthogonal with each other, and do not bring out the symmetry properties of the structure in a natural way; thus in practice one proceeds differently, by considering the expansion of the ordering in terms of mass density waves, as will be discussed below.

Apart from this n-vector model allowing for a n-component order parameter, there is also the need to consider order parameters of tensorial character. This happens, for example, when we consider the adsorption of molecules such as N_2 on grafoil. For describing the orientational ordering of these dumbbell-shaped molecules, the relevant molecular degree of freedom which matters is their electric quadrupole moment tensor,

$$f_{\mu\nu} = \int d\boldsymbol{x}\, \rho_{\mathrm{el}}(\boldsymbol{x}) \left(x_\mu x_\nu - \frac{1}{3} \sum_{\lambda=1}^{3} x_\lambda^2 \delta_{\mu\nu} \right),\tag{12}$$

where $\rho_{\mathrm{el}}(\boldsymbol{x})$ is the charge density distribution function of a molecule, $\boldsymbol{x} = (x_1, x_2, x_3)$ are cartesian coordinates in its center of mass system, and $\delta_{\mu\nu}$ is the Kronecker-symbol. While eq. (12) considers molecular orientations that exist in the three-dimensional space, there may again occur anisotropies that restrict the molecular orientation to certain planes (e.g. in the herringbone phase of N_2 adsorbed on grafoil the quadrupole moments of the N_2 molecules lie in a plane parallel to the substrate surface).

Proper identification of the order parameter of a particular system often needs detailed physical insight, and sometimes is complicated because different degrees of freedom are coupled. For example, there are many reports in the literature that an order–disorder transition of adsorbates on loose-packed substrates causes an adsorbate-induced reconstruction of the substrate surface. In such a situation, the order parameter of the adsorbate order–disorder transition is the "primary order parameter" whereas the lattice distortion of the substrate surface is a "secondary order parameter". However, for pure surface reconstruction transitions (i.e. structural phase transitions of the surface of crystals where no adsorbates are involved) all considered degrees of freedom are atomic displacements relative to positions of higher symmetry. The proper distinction between primary and secondary order parameters is then much more subtle.

Since the identification of universality classes for surface layer transitions needs the Landau expansion as a basic step, we first formulate Landau's theory (Tolédano and Tolédano, 1987) for the simplest case, a scalar order parameter density $\phi(x)$. This density is assumed to be small near the phase transition and slowly varying in space. It can be obtained by averaging a microscopic variable over a suitable coarse-graining cell L^d (in d-dimensional space). For example, for the c(2×2) structure in fig. 10 the microscopic variable is the difference in density between the two sublattices I (a and c in fig. 10) or II (b and d in fig. 10), $\phi_i = \rho_i^{II} - \rho_i^{I}$. The index i now labels the elementary cells (which contain one site from each sublattice I, II). Then

$$\phi(x) = \frac{\displaystyle\sum_{i \in L^d} \phi_i}{L^d}, \tag{13}$$

x being the center of gravity of the cell. The linear dimension L of the coarse-grained cell must be much larger than the lattice spacing, in order for the continuum description to make sense. Then a free energy functional $F\{\phi(x)\}$ is assumed,

$$\frac{1}{k_B T} F\{\phi(x)\} = \frac{F_o}{k_B T} + \int dx \left\{ \frac{1}{2} r \phi^2(x) + \frac{1}{4} u \phi^4(x) \right.$$
$$\left. - \frac{H}{k_B T} \phi(x) + \frac{1}{2d} [R \bigtriangledown \phi(x)]^2 \right\}, \tag{14}$$

where F_o is the background free energy of the disordered phase, r, u and R being phenomenological constants (R can be interpreted as the effective range of interaction between the atomic degrees of freedom ϕ_i). Equation (14) is a Taylor series expansion of a free energy density $f(\phi, \nabla\phi)$ where just the lowest order terms are kept. This makes sense if both the coefficients u and R^2 are positive constants at T_c whereas the essential assumption which defines ϕ as playing the role of an order parameter of a second-order phase transition is that r changes sign at the transition, as the variable of interest (the temperature in the present case) is varied,

$$k_B T r = r'(T - T_c) \tag{15}$$

In eq. (14) we have *assumed* a symmetry in the problem against the change of sign of the order parameter for $H = 0$, and thus odd powers of ϕ such as $\phi^3(x)$ do not occur; this is true for magnets (no direction of the magnetization is preferred without a magnetic field in a ferromagnet) and for sublattice ordering of adsorbate layers such as hydrogen on Pd(100) in the $c(2\times2)$ structure [since whether the hydrogen atoms preferentially occupy sublattices a,c in fig. 10 or sublattices b,d is equivalent, and this changes the sign of the order parameter, eq. (9)]. But this assumption is not true in general, e.g. in the (2×2) structure at $\theta = 1/4$ the permutation of sublattices does not lead to sign changes of the order parameter components [eq. (11)], since there is no symmetry between $\psi^{(2\times2)}$ and $-\psi^{(2\times2)}$, and hence third-order terms can occur. The same also holds for the ordering of rare gas monolayers adsorbed on graphite at $\theta = 1/3$ in the $(\sqrt{3}\times\sqrt{3})R30°$ structure, fig. 12, as will be discussed below.

We first consider the fully homogeneous case in eq. (14), $\nabla\phi(x) \equiv 0$, $\phi(x) \equiv \phi_o$; then $F[\phi]$ is the standard (Helmholtz) free energy function of thermodynamics, which needs to be minimized with respect to ϕ in order to determine the thermal equilibrium state. With $V = \int dx$ being the total volume of the system, we have

$$\frac{1}{k_B T V}\left(\frac{\partial F}{\partial \phi_o}\right)_T\bigg|_{H=0} = r\phi_o + u\phi_o^3 = 0,$$

$$\phi_o = \pm\left(-\frac{r}{u}\right)^{1/2} = \pm\left(\frac{r'}{k_B u}\right)^{1/2}\left(\frac{T_c}{T} - 1\right)^{1/2}, \qquad T < T_c \tag{16}$$

while $\phi_o \equiv 0$ for $T > T_c$. Hence eqs. (14)–(16) indeed yield a second-order transition as T is lowered through T_c at $H = 0$. For $T < T_c$, a first-order transition as function of H occurs, since ϕ_o jumps from $(-r/u)^{1/2}$ to $-(-r/u)^{1/2}$ as H changes sign. This behavior is exactly that shown in fig. 9, with $\beta = 1/2$, $\hat{B} = (r'/k_B u)^{1/2}$ and $\phi_o(H = 0) = M_s$.

If $u < 0$ in eq. (14), however, one must not stop the expansion at fourth order but rather must include a term $\frac{1}{6}v\phi^6(x)$ (assuming now $v > 0$). Whereas in the second-order case $F(\phi)$ has two minima for $T < T_c$ which

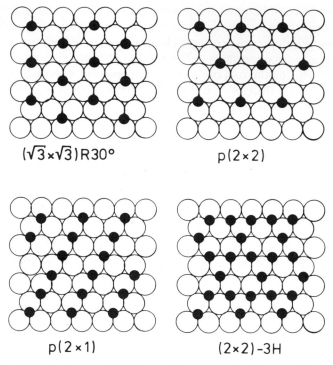

($\sqrt{3} \times \sqrt{3}$)R30°   p(2×2)

p(2×1)   (2×2)-3H

Fig. 12. Adsorbate structures on (111) faces of face-centered cubic crystals [e.g. Ni(111)] or (100) faces of hexagonal close packed crystals [e.g. Ru(100)]. The adsorption sites form a regular triangular lattice. Ordered structures that are discussed are ($\sqrt{3} \times \sqrt{3}$)R30° (coverage $\theta = 1/3$), p(2×2) ($\theta = 1/4$), p(2×1) ($\theta = 1/2$) and (2×2)–3H ($\theta = 3/4$), respectively.

continuously merge as $T \to T_c$ and only one minimum at $\phi = 0$ remains for $T > T_c$ (fig. 13a), $F(\phi)$ now has three minima for $T_o < T < T_c$, and the temperature T_o where r changes sign ($r = r'(T - T_o)$ now) differs from the phase transition temperature T_c where the order parameter jumps discontinuously from zero for $T > T_c$ to $\phi_o|_{T_c} = \pm(-u/4v)^{1/2}$, see fig. 13b. These results are found by analogy with eq. (16) from

$$\frac{1}{k_B T V} \left(\frac{\partial F}{\partial \phi_o} \right)_T \bigg|_{H=0} = \phi_o (r + u\phi_o^2 + v\phi_o^4) = 0,$$

$$\phi_o^2 = -\frac{u}{2v} + \left[\left(\frac{u}{2v} \right)^2 - \frac{r}{v} \right]^{1/2}. \tag{17}$$

Choosing the minus sign of the square root would yield the maxima rather than the minima in fig. 13b. On the other hand, $F = F(0)$ in the disordered

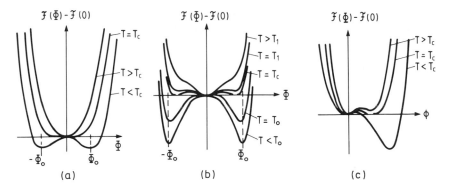

Fig. 13. Schematic variation of the Landau free energy at transitions of (a) second order and (b), (c) first order as a function of the (scalar) order parameter ϕ. Cases (a) and (b) assume a symmetry around $\phi = 0$, whereas case (c) allows a cubic term.

phase, and hence T_c can be found by equating the free energy of the ordered phase to this value, fig. 13b, i.e.

$$\frac{F(\phi_o) - F(0)}{V k_B T_c} = \phi_o^2 \left(\frac{1}{2} r + \frac{1}{4} u \phi_o^2 + \frac{1}{6} v \phi_o^4 \right)_{T=T_c} = 0. \tag{18}$$

With some simple algebra, eqs. (17) and (18) yield T_c and the "stability limit" T_1, where the minimum describing the metastable ordered phase in the disordered phase above T_c disappears,

$$T_c = T_o + \frac{3u^2}{16r'v}, \qquad T_1 = T_o + \frac{u^2}{4r'v}. \tag{19}$$

The alternative mechanism by which a first-order transition arises in the Landau theory with a scalar order parameter is the lack of symmetry of F against a sign change of ϕ. Then we may add a term $\frac{1}{3} w \phi^3$ to eq. (14), with another phenomenological coefficient w. For $u > 0$, $F(\phi)$ may have two minima (fig. 13c); again the transition occurs when the minima are equally deep. For $r = r'(T - T_o)$ this happens when

$$T_c = T_o + \frac{2w^2}{9ur'}, \qquad \phi_o|_{T_c} = -\frac{9r}{w}. \tag{20}$$

Again a stability limit of the ordered state in the disordered phase occurs, i.e.

$$T_1 = T_o + \frac{w^2}{4ur'} \tag{21}$$

At this point, we emphasize a caveat: free energy curves involving several minima and maxima as sketched in fig. 13 are so commonly used that many researchers believe these concepts to be essentially rigorous. However, general principles of thermodynamics require that in thermal equilibrium the thermodynamic potentials are convex functions of their variables. Thus, in fact, $F(\phi_o)$ should be convex as a function of ϕ_o, which exclude multiple minima! For fig. 13a, b this means that for $T < T_c$ in states with $-\phi_o < \phi < \phi_o$ (ϕ_o being the non-zero solution of eqs. (16) or (17), respectively) the thermal equilibrium state is not a pure homogeneous phase: rather the minimum free energy state is given by the double-tangent construction to $F(\phi)$ and this corresponds to a mixed phase state (the relative amounts of the coexisting phases are given by the lever rule). Now it is common "folklore" to interpret that part of $F(\phi)$ in fig. 13 which lies above the $F(\phi)$ found from the double-tangent construction as metastable states, provided $\chi_T = (\partial^2 F/\partial\phi^2)_T > 0$ is satisfied, while states with $\chi_T < 0$ are considered as intrinsically unstable states. Unfortunately, this notion is intrinsically a concept valid only in mean-field theory, but lacks any fundamental justification in statistical mechanics. Expansions such as eq. (14) make only general sense for a local "coarse-grained free energy function" which depends on the length scale L introduced in eq. (13), but not for the global free energy.

   How does one obtain the Landau expansion in particular cases? If one wishes to consider specific models, a straightforward approach uses a molecular field approximation (MFA), where one then obtains the free energy explicitly and expands it directly. We illustrate this approach here with the q-state Potts model (Potts, 1952). The Hamiltonian is

$$\mathcal{H}_{\text{Potts}} = -\sum_{\langle i,j \rangle} J\delta_{S_i,S_j}, \qquad S_i = 1, 2, \ldots, q \qquad (22)$$

Each lattice site i can be in one out of q states (labelled by S_i), and an energy J is won if two neighboring sites are in the same state. For example, imagine for $q = 3$ an uniaxial molecule that can be oriented along the x-axis, y-axis, or z-axis: the energy depends on the relative orientation of molecules and thus has one value for parallel orientation and another one for perpendicular orientation. In the MFA, we construct the free energy $F = U - TS$ simply by expressing both enthalpy U and entropy S in terms of the fractions n_α of lattice sites in states α. The entropy is simply the entropy of randomly mixing these species,

$$S = -V \sum_{\alpha=1}^{q} n_\alpha \ln n_\alpha \qquad (23)$$

In the enthalpy term, MFA neglects correlations in the occupation probability of neighboring sites. Hence the probability of finding a nearest neighbor pair in the state α is simply n_α^2. In a lattice with coordination number z there are $z/2$ pairs per site, and hence $U = -(zJV/2)\sum_{\alpha=1}^{q} n_\alpha^2$. Thus

$$\frac{F}{V k_B T} = -\left(\frac{zJ}{2k_B T}\right) \sum_{\alpha=1}^{q} n_\alpha^2 + \sum_{\alpha=1}^{q} n_\alpha \ln n_\alpha \qquad (24)$$

One can directly minimize F with respect to the n_α, subject to the constraint $\sum_{\alpha=1}^{q} n_\alpha = 1$.

In order to make contact with the Landau expansion, however, we consider now the special case $q = 3$ and expand F in terms of the two order parameter components $\phi_1 = n_1 - 1/3$ and $\phi_2 = n_2 - 1/3$ (note that all $n_i = 1/q$ in the disordered phase). One recognizes that the model for $q = 3$ has a two-component order parameter and there is no symmetry between ϕ_i and $-\phi_i$. So cubic terms in the expansion of F are expected and do occur, whereas for a properly defined order parameter, there cannot be any linear term in the expansion:

$$\frac{F}{V k_B T} = -\frac{zJ}{6k_B T} - \ln 3 + 3\left(1 - \frac{zJ}{3k_B T}\right)(\phi_1^2 + \phi_2^2 + \phi_1 \phi_2)$$
$$+ \frac{9}{2}(\phi_1^2 \phi_2 + \phi_1 \phi_2^2) + \dots \qquad (25)$$

As expected, there is a temperature $T_o(= zJ/3k_B)$ where the coefficient of the quadratic term changes sign.

Of course, for many phase transitions a specific model description is not available, and even if a description in terms of a model Hamiltonian is possible, for complicated models the approach analogous to eqs. (22)–(25) requires tedious calculation. Thus the elegant but abstract Landau approach based on symmetry principles (Landau and Lifshitz, 1958; Tolédano and Tolédano, 1987) is preferable for construction of the Landau expansion. One starts from the observation that usually the disordered phase at high temperatures is more "symmetric" than the ordered phase(s) occurring at lower temperature. Recalling the example of the (2×1) structure in fig. 10, we note that in the high temperature phase all four sublattices a,b,c,d are completely equivalent. This permutation symmetry among the sublattices is broken in the (2×1) structure, where the concentrations on the different sublattices are no longer equivalent.

In such cases the appropriate structure of the Landau expansion for F in terms of the order parameter $\boldsymbol{\phi} = (\phi_1, \dots, \phi_n)$ [in our case $n = 2$, $\phi_1 = \psi_I^{(2\times1)}$, $\phi_2 = \psi_{II}^{(2\times1)}$, eq. (10)] is found from the principle that F must be invariant against all symmetry operations of the symmetry group G_o describing the disordered phase. In the ordered phase G, some symmetry elements of G_o fall away ("spontaneously broken symmetry"); the remaining symmetry elements form a subgroup G of G_o. Now the invariance of F must hold separately for terms of ϕ^k of any order k and this fixes the character of the terms that may be present.

Rather than formulating this approach systematically, which would require a lengthy and very mathematical exposition (Tolédano and Tolédano, 1987), we rather illustrate it with the simple example of the (2×1) structure, with $\boldsymbol{\phi} = (\phi_1, \phi_2)$. F is then given as follows (Krinsky and Mukamel, 1977), for $H = 0$,

$$
\frac{F\{\boldsymbol{\phi}(x)\}}{k_B T} = \frac{F_o}{k_B T} + \int dx \left\{ \frac{1}{2} r \boldsymbol{\phi}^2 + \frac{1}{4} \left[u(\phi_1^4 + \phi_2^4) + u' \phi_1^2 \phi_2^2 \right] \right.
$$
$$
\left. + \frac{R^2}{2d} \left[(\nabla \phi_1)^2 + (\nabla \phi_2)^2 \right] \right\} \tag{26}
$$

In this case, there is a symmetry against the change of sign of $\boldsymbol{\phi}$ (fig. 10 shows that this simply corresponds to an interchange of sublattices) and hence a term $\boldsymbol{\phi}^3$ cannot occur. The fourth order term, however, now contains two "cubic invariants" rather than a single term $[(\boldsymbol{\phi}^2)^2 = (\phi_1^2 + \phi_2^2)^2]$ which would occur in the isotropic XY model, where there is also a rotational invariance in the order parameter space (ϕ_1, ϕ_2), since all directions in the (ϕ_1, ϕ_2) plane are equivalent. No such rotational symmetry applies to the (2×1) structure, of course. So the expansion eq. (26) results, which defines the universality class of the "XY model with cubic anisotropy". Of course, in this approach not much can be said on the phenomenological coefficients r, u, u', R in eq. (26).

Rather than visualizing the ordered structure for an order–disorder transition in real space (figs. 10, 12) and considering the symmetry of the structure by applying suitable operations of the point group, it is often more convenient to carry out a corresponding discussion of the ordering in reciprocal space rather than in real space. Remember that the ordering shows up in superlattice Bragg spots appearing in the reciprocal lattice in addition to the Bragg spots of the disordered phase, fig. 14. The superlattice Bragg spots of the $c(2 \times 2)$ and (2×1) superstructures on the square lattice occur at special points at the boundary of the first Brillouin zone, e.g. the $c(2 \times 2)$ structure is characterized by the point $\boldsymbol{q}_o = (\pi/a)(1, 1)$. Of course, other Bragg spots appear at additional positions such as $\pi/a(-1, -1)$, $\pi/a(1, -1)$, etc., but they need not be considered explicitly since they can be obtained from $\boldsymbol{q}_o$ by adding a suitable vector of the reciprocal lattice. On the other hand, for the (2×1) structure two vectors $\boldsymbol{q}_1 = \pi/a(1, 0)$ and $\boldsymbol{q}_2 = \pi/a(0, 1)$ are required, they are not related by a reciprocal lattice vector of the original square lattice. One can find all ℓ independent members $\boldsymbol{q}_i$ (the so-called "star" of $\boldsymbol{q}_1$) by applying the point–group operations of the lattice of adsorption sites, and keeps only those that are not related by a reciprocal lattice vector $\boldsymbol{g}$. For the $(\sqrt{3} \times \sqrt{3}) R 30°$ structure on the triangular lattice, one finds the two vectors $\boldsymbol{q}_1$ and $\boldsymbol{q}_2 = -\boldsymbol{q}_1$, with $\boldsymbol{q}_1 = 4\pi/3a(1, 0)$ and for the (2×2) structure on the triangular lattice one finds the vectors $\boldsymbol{q}_1'$, $\boldsymbol{q}_2'$, $\boldsymbol{q}_3'$, which are related to each other via rotations by 120°, see Schick (1981) or Einstein (1982),

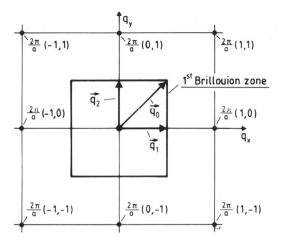

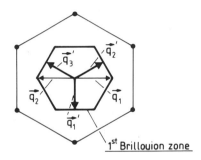

Fig. 14. Reciprocal lattice and 1st Brillouin zone for the square lattice (upper part) and triangular lattice (lower part). The c(2×2) structure is described by the single wavevector q_o in reciprocal space, while the (2×1) structure on the square lattice is described by a star (q_1, q_2), as well as the $\sqrt{3} \times \sqrt{3}$R30° structure on the triangular lattice. The star of the (2×2) structure on the triangular lattice contains three members q'_1, q'_2 and q'_3.

and $q'_1 = 2\pi/a(0, -1/\sqrt{3})$. One then expands the local density in mass density waves,

$$\langle \rho_j \rangle = \theta + \sum_{s=1}^{\ell} \rho(q_s) \exp(i q_s \cdot R_j), \tag{27}$$

R_j being the lattice vector of site j. If some g connects q_s and $-q_s$, the fourier component $\rho(q_s)$ is real and one has

$$\langle \rho_j \rangle = \theta + \sum_{s=1}^{\ell} \rho(q_s) \cos(q_s \cdot R_j), \tag{28}$$

while otherwise we split $\rho(q_s)$ in real and imaginary parts, $\rho(q_s) = \rho'(q_s) + i\rho''(q_s)$, where ρ', ρ'' are real, and

$$\langle \rho_j \rangle = \theta + 2 \sum_{s=1}^{\ell/2} \rho'(q_s) \cos(q_s \cdot R_j) + 2 \sum_{s=1}^{\ell/2} \rho''(q_s) \sin(q_s \cdot R_j) \qquad (29)$$

The first Landau rule states that for a second-order transition to be possible there should occur just a single star of q in the description of the ordered phase. Now the order parameter components ϕ_s can be identified as

$$\phi_s = \frac{1}{N} \sum_j \langle \rho_j \rangle \cos(q_s \cdot R_j), \qquad s = 1, \ldots, \ell, \qquad (30)$$

if eq. (28) holds, while in the case where eq. (29) holds, we have

$$\psi_s = \begin{cases} \dfrac{1}{N} \sum_j \langle \rho_j \rangle \cos(q_s \cdot R_j), & s = 1, \ldots, \dfrac{\ell}{2}, \\[2ex] \dfrac{1}{N} \sum_j \langle \rho_j \rangle \sin(q_s \cdot R_j), & s = 1, \ldots, \dfrac{\ell}{2}. \end{cases} \qquad (31)$$

As an illustration, we note for the square lattice that the lattice points $R_j = (m, n)a$ with m, n integers, and using $q_s = q_o = (\pi/a)(1, 1)$ for the c(2×2) structure in eq. (30), we recover the single order parameter component

$$\phi = \frac{1}{N} \sum_j \langle \rho_j \rangle (-1)^{m+n}. \qquad (32)$$

For the $(\sqrt{3} \times \sqrt{3})$R30° structure, on the other hand, we have $R_j = [(a/2)(m + n), (\sqrt{3}a/2)(m - n)]$ and we use $q_s = q_1 = (4\pi/3a)(1, 0)$ in eq. (31). This yields the X and Y-components of the order parameter

$$\phi_X = \frac{1}{N} \sum_j \langle \rho_j \rangle \cos \left[\frac{2\pi}{3} (m + n) \right],$$

$$\phi_Y = \frac{1}{N} \sum_j \langle \rho_j \rangle \sin \left[\frac{2\pi}{3} (m + n) \right]. \qquad (33)$$

The resulting free energy expansion is found to have the same symmetry (Alexander, 1975) as that of the 3-state Potts model, cf. eqs. (22)–(25). The general form of the "Ginzburg–Landau–Wilson"-Hamiltonian $F\{\phi(x)\}$ then is (see also Straley and Fisher, 1973; Stephanov and Tsypin, 1991)

$$
\frac{F\{\boldsymbol{\phi}(x)\}}{k_{\mathrm B}T} = \frac{F_o}{k_{\mathrm B}T} + \int d\mathbf{r}\Big\{ \tfrac{1}{2}r(\phi_X^2 + \phi_Y^2) + \tfrac{1}{3}w(\phi_X^3 - 3\phi_X\phi_Y^2)
$$

$$
+ \tfrac{1}{4}u(\phi_X^2 + \phi_Y^2)^2
$$

$$
+ \tfrac{1}{5}U_5(\phi_X^2 + \phi_Y^2)(\phi_X^3 - 3\phi_X\phi_Y^2)
$$

$$
+ \tfrac{1}{6}U_6(\phi_X^2 + \phi_Y^2)^3
$$

$$
+ \tfrac{1}{6}U_6'(\phi_X^6 - 15\phi_X^4\phi_Y^2 + 15\phi_X^2\phi_Y^4 - \phi_Y^6) + \cdots
$$

$$
+ \frac{R^2}{2d}\big[(\nabla\phi_X)^2 + (\nabla\phi_Y)^2\big]\Big\} \tag{34}
$$

The three ordered states of the Potts model correspond to a preferential occupation of one of the three sublattices a,b,c into which the triangular lattice is split in the $(\sqrt{3}\times\sqrt{3})R30^\circ$ structure. In the "order parameter" plane (ϕ_X, ϕ_Y), the minima of F occur at positions $(1, 0)M_s$, $(-1/2, \sqrt{3}/2)M_s$, $(-1/2, -\sqrt{3}/2)M_s$, where M_s is the absolute value of the order parameter, i.e. they are rotated by an angle of 120° with respect to each other. The phase transition of the three-state Potts model hence can be interpreted as spontaneous breaking of the (discrete) Z_3 symmetry. While Landau's theory implies [fig. 13 and eqs. (20), (21)] that this transition must be of first order due to the third-order invariant present in eq. (34), it actually is of second order in $d = 2$ dimensions (Baxter, 1982, 1973) in agreement with experimental observations on monolayer $(\sqrt{3}\times\sqrt{3})R30^\circ$ structures (Dash, 1978; Bretz, 1977). The reasons why Landau's theory fails in predicting the order of the transition and the critical behavior that results in this case will be discussed in the next section.

2.2. Critical and multicritical phenomena

In the previous section, we have seen that it cannot suffice to consider the order parameter alone. A crucial role is played by order parameter fluctuations that are intimately connected to the various singularities sketched in fig. 11. We first consider critical fluctuations in the framework of Landau's theory itself, and return to the simplest case of a scalar order parameter $\phi(x)$ with no third-order term, and $u > 0$ [eq. (14)], but add a weak wavevector dependent field $\delta H(x) = \delta H_q \exp(iq \cdot x)$ to the homogeneous field H. Then the problem of minimizing the free energy functional is equivalent to the task of solving the Ginzburg–Landau differential equation

$$
r\phi(x) + u\phi^3(x) - \left(\frac{R^2}{d}\right)\nabla^2\phi(x) = \frac{H + \delta H_q \exp(iq \cdot x)}{k_{\mathrm B}T}. \tag{35}
$$

We now treat the effect of $\delta H(x)$ in linear response, writing $\phi(x) = \phi_o + \delta\phi(x) = \phi_o + \delta\phi_q \exp(iq \cdot x)$, where ϕ_o is the solution of $r\phi_o + u\phi_o^3 = H/k_B T$ as previously (cf. eq. (16) for $H = 0$). Linearizing eq. (35) in $\delta\phi_q$ yields the wavevector-dependent order parameter response function $\chi(q)$

$$\chi(q) \equiv \frac{\delta\phi_q}{\delta H_q} = \left[k_B T \left(r + 3u\phi_o^2 + \frac{R^2 q^2}{d} \right) \right]^{-1}, \tag{36}$$

which can be rewritten in the well-known Ornstein–Zernike form

$$\chi(q) = \frac{\chi_T}{1 + q^2\xi^2}, \quad \chi_T = \left[k_B T (r + 3u\phi_o^2) \right]^{-1}, \quad \xi = R\sqrt{\frac{k_B T \chi_T}{d}} \tag{37}$$

Using now eqs. (15) and (16) yields eq. (6) with $\gamma = 1$, $\hat{C}^+ = T_c/r'$, $\hat{C}^- = \frac{1}{2}\hat{C}^+$ and an analogous law for the correlation length

$$\xi = \hat{\xi}^{\pm} \left| \frac{T}{T_c} - 1 \right|^{-\nu} \tag{38}$$

with $\nu = 1/2$, $\hat{\xi}^+ = R\sqrt{k_B/r'd}$, $\hat{\xi}^- = \hat{\xi}^+/\sqrt{2}$. The correlation function of fluctuations

$$G(x) = \langle [\phi(0) - \phi_o][\phi(x) - \phi_o] \rangle = \langle \phi(0)\phi(x) \rangle - \phi_o^2 \tag{39}$$

is related to $k_B T \chi(q)$ via the fluctuation relation by a Fourier transform,

$$k_B T \chi(q) = S(q) = \sum_x \exp(iq \cdot x) G(x). \tag{40}$$

Comparing eqs. (37) and (40) one easily shows by Taylor expansion in powers of q that

$$k_B T \chi_T = \sum_x G(x), \qquad \xi^2 = (2d)^{-1} \frac{\sum_x x^2 G(x)}{\sum_x G(x)}. \tag{41}$$

While for T different from T_c the asymptotic decay of the correlation function $G(x)$ corresponding to $\chi(q)$ written in eq. (37) is exponential,

$$G(x) \propto \left[\exp\left(-\frac{|x|}{\xi} \right) \right] |x|^{-(d-1)/2}, \qquad |x| \gg \xi, \tag{42}$$

right at T_c a power law decay is found. Defining an exponent η and an amplitude prefactor $\hat{G}$ via

$$G(x) = \hat{G}|x|^{-(d-2+\eta)}, \qquad T = T_c, \tag{43}$$

one finds from Landau's theory $\eta = 0$. This result is immediately found from Fourier transformation of eq. (36) for $T = T_c$ (i.e., $r = 0$), i.e. $\chi(q) = d/(k_B T_c R^2 q^2)$. If the exponent η defined in eq. (43) is non-zero, one also has a non-trivial power law for $\chi(q)$ as function of wavenumber q, namely

$$\chi(q) \propto q^{-(2-\eta)}, \qquad T = T_c. \tag{44}$$

Another critical exponent (δ) is defined considering the variation of the order parameter at T_c as a function of the conjugate field H, $\hat{D}$ being the associated critical amplitude,

$$\phi_o = \hat{D} H^{1/\delta}, \qquad T = T_c. \tag{45}$$

From $u\phi_o^3 = H/k_B T$ one concludes that in Landau's theory $\delta = 3$, $\hat{D} = (k_B T_c u)^{-1/3}$. Finally we return to the specific heat in zero field (which was already considered in eq. (7) for the general case) writing eq. (14) in the homogeneous case ($\triangledown \phi(x) \equiv 0$) as ($V$ is the volume of the system)

$$\frac{F - F_o}{k_B T V} = \phi_o^2 \left(\frac{r}{2} + \frac{u}{4} \phi_o^2 \right) = -\frac{r^2}{4u} = -\frac{r'^2}{(2k_B u)^2} (1 - T/T_c)^2, \tag{46}$$

where in the last step eq. (16) was used. This yields

$$C_{H=0} \equiv -T \left(\frac{\partial^2 F}{\partial T^2} \right)_{H=0} = \frac{r'^2 (T/T_c)}{2k_B u^2}, \qquad T < T_c, \tag{47}$$

while for $T > T_c$ the part of the specific heat associated with the ordering is identically zero, $C_{H=0} \equiv 0$. This jump singularity of $C_{H=0}$ at $T = T_c$ is formally compatible with the power law of eq. (7) if one puts $\alpha = \alpha' = 0$. Hence we can summarize the critical behavior of Landau's theory in terms of the following set of critical exponents

$$\alpha = \alpha' = 0, \quad \beta = \tfrac{1}{2}, \quad \gamma = \gamma' = 1, \quad \delta = 3, \quad \nu = \nu' = \tfrac{1}{2}, \quad \eta = 0. \tag{48}$$

As we shall see below, the Landau theory of critical phenomena fails badly for systems with short range interactions in the dimensionalities of physical interest ($d = 2$ and $d = 3$), and the critical exponents of physical systems differ considerably from the prediction eq. (48). As an example, we reproduce experimental results for the ordering of Oxygen on Ru(0001) at 1/4 monolayer coverage in fig. 15, taken from Piercy and Pfnür (1987).

In order to understand why Landau's theory is inaccurate, let us recall the justification of eq. (14) in terms of the coarse-graining eq. (13), where short wavelength fluctuations of a microscopic model [such as the Ising model, eq. (1)] are eliminated. In fact, if L in eq. (13) would be the lattice spacing a,

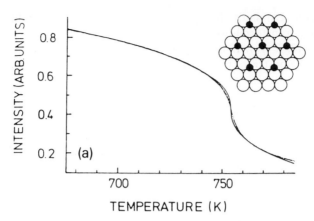

Fig. 15. (a) Integrated LEED intensity of an oxygen-induced second-order p(2×2) super-structure spot vs. temperature at an electron energy $E_{\text{prim}} = 65$ eV. Dots are data from a typical temperature sweep at 2 K/s; solid line is a fit by $\int d^3q\,S(q) \propto A - Ct \mp B_{\pm}|t|^{1-\alpha}$ with $T_{\text{c}} = 754$ K, $\alpha = 0.59$, $B_+ = B_-$, where A, C and $B_{\pm}$ are constants.

we would have $\phi(x) = \phi_i/a^d = \pm 1$ (measuring lengths in units of the lattice spacing), and for a disordered (or weakly ordered) Ising spin configuration $\phi(x)$ would be rapidly varying from one lattice site to the next; i.e. neither $|\phi(x)|$ nor $|\nabla \phi(x)|$ would be small. Obviously, the larger L the smaller the variations of $\phi(x)$ will be: but clearly L must be much smaller than the characteristic lengths which we want to study near T_{c}, such as the correlation length ξ. Consequently, we must have $a \ll L \ll \xi$ in order that eq. (14) makes sense: but even then one must consider that the coefficients r, u, R that result in eq. (14) from applying the coarse-graining to a microscopic Hamiltonian such as eq. (1) will depend somewhat on the size L of the coarse-graining cells. This fact is also demonstrated by explicit calculations (e.g. Kaski et al., 1984). On the other hand, critical amplitudes such as $\hat{C}^{\pm}$, $\hat{\xi}^{\pm}$, $\hat{B}$, $\hat{D}$ — which Landau's theory expresses in terms of these expansion coefficients r', u, R — cannot depend on the length L which to some extent is quite arbitrary.

The resolution of this puzzle is that $F[\phi(x)]$ in eq. (14) should not be confused with the actual Helmholtz free energy function of the system, but really plays the role of an effective Hamiltonian. The coarse graining leads from a microscopic Hamiltonian $\mathcal{H}\{\phi_i\}$ [such as eq. (1)] to $F\{\phi(x)\}$ by projecting out the short wavelength degrees of freedom and thus replace the Hamiltonian (defined on a discrete lattice) by a functional (defined in continuous space),

$$\exp\left[-\frac{1}{k_{\text{B}}T}\mathcal{F}_{(L)}\{\phi(x)\}\right] = \underset{\{\phi_i\}}{\text{Tr}}\; P_{(L)}(\{\phi(x), \{\phi_i\}\})\exp\left[-\frac{1}{k_{\text{B}}T}\mathcal{H}\{\phi_i\}\right].$$

$$(49)$$

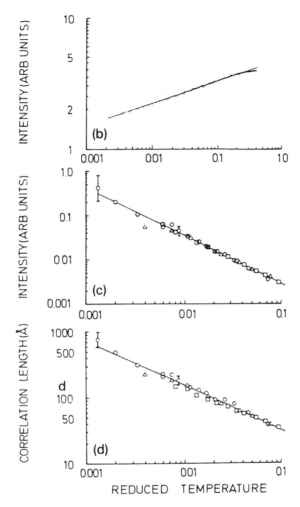

Fig. 15 (contd.) (h) Peak intensity of second-order spots at 65 eV as a log–log plot vs. reduced temperature $1 - T/T_c$ below T_c, after division by the Debye–Waller factor. Straight line corresponds to an exponent $\beta = 0.085$. (c) Log–log plot of peak intensity of fluctuations vs. reduced temperature $1 - T_c/T$ above T_c. Squares are first-order oxygen spots at 52 eV; circles and triangles are second-order spots at 36.5 and 65 eV, respectively. Straight line is a fit to eq. (6) with $\gamma = 1.08$. (d) Same as (c) but for correlation length ξ. Straight line is a fit to eq. (38) with $\nu = 0.68$. From Piercy and Pfnür (1987).

Here $P_{(L)}(\{\phi\}, \{\phi_i\})$ is a projection operator defined implicitly by eq. (13) or a similar procedure, and we emphasize in our notation that the resulting effective Boltzmann factor will depend on L. What needs to be done, is to obtain the free energy from the partition function Z which for the discrete lattice models means a trace over all Ising spins,

$$F = -k_B T \ln Z = -k_B T \ln \underset{\{\phi_i\}}{\mathrm{Tr}} \exp \left[-\frac{\mathcal{H}\{\phi_i\}}{k_B T} \right], \tag{50}$$

while for its continuum analog it means a functional integration

$$F = -k_B T \ln Z = -k_B T \ln \int D\{\phi(x)\} \exp \left[-\frac{\mathcal{F}_{(L)}\{\phi(x)\}}{k_B T} \right]. \tag{51}$$

Now the Landau theory [where simply $\mathcal{F}_{(L)}\{\phi(x)\}/k_B T$ is minimized, as discussed in eqs. (16) and (35)] results from eq. (51) only if one assumes that the path integral is dominated by this path $\phi(x)$ that minimizes the integrand $\exp[-\mathcal{F}_{(L)}\{\phi(x)\}/k_B T]$, and any fluctuations around this path yielding the largest contribution are neglected.

This neglect of fluctuations in general is not warranted. One can recognize this problem in the framework of Landau's theory itself. This criterion named after Ginzburg (1960) considers the mean square fluctuation of the order parameter in a coarse graining volume L^d and states that Landau's theory is selfconsistent if this fluctuation is much smaller than the square of the order parameter itself,

$$\langle [\phi(x) - \phi_o]^2 \rangle_{L=\xi} \ll \phi_o^2 \tag{52}$$

Here we use the maximum permissible choice for L, that is $L = \xi$. Using now the notation $N(t)$ for the number of microscopic degrees of freedom ϕ_i in a volume $V_\xi(x) \equiv [\xi(t)]^d$ centered at x where $-t \equiv 1 - T/T_c$, one can use eq. (13) in eq. (52) and make use of the translational invariance of the correlation function $\langle \phi_i \phi_j \rangle = \langle \phi_{i=0} \phi_{j-i} \rangle$ to find

$$\sum_{k \in V_\xi(x)} \left(\langle \phi_{i=0} \phi_k \rangle - \phi_o^2(t) \right) \ll N(t) \phi_o^2(t) \tag{53}$$

Comparing the left hand side of this inequality with eq. (41), $k_B T \chi_T(t) = \sum_k (\langle \phi_{i=0} \phi_k \rangle - \phi_o^2(t))$, we conclude that both expressions must be of the same order of magnitude, since the additional correlations over distances larger than ξ that contribute to $k_B T \chi_T(t)$ but not to eq. (53) are very small. Thus the inequality eq. (53) implies also

$$\chi_T(t) \ll N(t) \phi_o^2(t), \qquad \mathrm{const} \ll \left[\frac{\xi(t)}{a} \right]^d \phi_o^2(t) \chi_T^{-1}(t), \tag{54}$$

where in the last step we have used that the number $N(t)$ of degrees of freedom in a correlation volume $V_\xi(x)$ must be $[\xi(t)/a]^d$, since every elementary cell a^d contains one Ising spin. Using now the power laws eqs. (6), (8) and (38) we find (Als-Nielsen and Birgeneau, 1977)

$$\text{const} \ll |t|^{-\nu d + 2\beta + \gamma}. \tag{55}$$

Using the explicit findings of Landau's theory [eqs. (16), (37), ...], eq. (54) reads

$$\text{const} \ll \left(\frac{R}{a}\right)^d u^{-1} r'^{3-d} k_B^{3-d/2} T_c^{-2} |t|^{(4-d)/2} \propto \left(\frac{R}{a}\right)^d |t|^{(4-d)/2}, \tag{56}$$

where in the last steps constants of order unity have been suppressed. This condition for the Landau theory always breaks down for $d < 4$ as $t \to 0$. In fact, for $d < 4$ close enough to T_c a regime occurs where fluctuations dominate the functional integral, eq. (51). The "crossover" from the mean field regime, where the Landau description is essentially appropriate to the non-mean field regime occurs at a reduced distance $|t| = Gi$, the "Ginzburg number" (see e.g. Anisimov et al., 1992), which is described as (suppressing a numerical prefactor of order unity which is $(3/4\pi)^2$ in $d = 3$)

$$Gi = (\hat{C}_{MF}^+)^{2/(4-d)} (\hat{B}_{MF})^{-4/(4-d)} \left(\frac{a}{\hat{\xi}_+^{MF}}\right)^{2d/(4-d)} \tag{57}$$

where $\hat{C}_{MF}^+$, $\hat{B}_{MF}$ and $\hat{\xi}_+^{MF}$ are the critical amplitudes in the mean field critical regime. The condition $|t| = Gi$ amounts to treating eq. (56) as an equality. Thus mean-field like behavior occurs only as long as $Gi \ll |t| \ll 1$ which can occur only for systems with large interaction range R. It is also plausible from eq. (56) that deviations from mean field behavior set in earlier (and are stronger) in $d = 2$ than in $d = 3$ dimensions.

We now turn to the behavior in the regime $|t| \ll Gi$, where mean field theory has broken down and fluctuations dominate the critical behavior. In this regime of "non-classical" critical behavior the critical exponents have non-trivial values, and the development of methods for an accurate prediction of these exponents has found longstanding interest. While in $d = 3$ dimensions one has to rely mostly on renormalization group methods (Fisher, 1974; Wilson and Kogut, 1974; Ma, 1976; Domb and Green, 1976; Amit, 1984), exact series expansions (Domb and Green, 1974) and Monte Carlo methods (Binder, 1976, 1979b, 1984a, 1992a; Binder and Heermann, 1988), in $d = 2$ there are many exact solutions available (Baxter, 1982), and one can also predict critical exponents from conformal invariance (Cardy, 1987) and the finite size scaling analysis of transfer matrix calculations (Barber, 1983; Privman, 1990; Cardy, 1987). Even a very compact description of all these techniques would fill a whole book and thus must remain outside the scope of the present chapter: we only attempt now to summarize the main conclusions of all these studies.

It turns out that the critical behavior of physical systems is "universal" in the sense that for systems with finite range R of the interactions it depends only on the dimensionality of space (d), on the dimensionality of the order parameter (m), and on certain symmetry properties of the Hamiltonian. For example, a fully isotropic XY ferromagnet (where the order parameter M has full rotational invariance in the XY-plane) belongs to a different "universality class" than a "XY"-system with cubic anisotropy [eq. (26)] or with hexagonal (sixfold) anisotropy. But, for a given symmetry of the Hamiltonian and a given type of ordering, critical exponents normally (apart from so-called "marginal" cases, which will be discussed later) will not depend on other details of the model (e.g., the precise range and functional form of the interaction $J(x_i - x_j)$ does not affect the critical exponents, nor — in the case of XY- or Heisenberg ferro- or antiferromagnets — the spin quantum number, etc.). Such "details" of the system which do not show up in the critical exponents are called "irrelevant" (in the renormalization group sense: of course, these "details" do affect the critical temperature as well as critical amplitudes of the considered system). Only for the dimensionalities d exceeding the "upper critical dimensionality" d_u (where $d_u = 4$ for standard critical phenomena as occur for Ising ($m = 1$), XY ($m = 2$) or Heisenberg ($m = 3$) ferromagnets, for instance) does one recover the complete universality of Landau's theory, where critical exponents have the values listed in eq. (48) which for $d > d_u$ are independent of both d, m and any anisotropies of the model Hamiltonian (or free energy functional, respectively).

We have already mentioned that the effect of fluctuations is the stronger the lower the dimensionality d of the system. If d is low enough, fluctuations are so strong that the system no longer is able to maintain long range order in the system: there exists a "lower critical dimensionality" d_ℓ, such that for $d < d_\ell$ T_c is zero, long range order can exist in the ground state of the system only. As will be discussed below, $d_\ell = 1$ for Ising-models or Potts models (Baxter, 1982; Wu, 1982), while $d_\ell = 2$ for isotropic XY or Heisenberg models. Therefore surface layers due to their two-dimensional character are of particular interest, since they allow the experimental study of strongly fluctuating systems, which are at (or at least close to) their lower critical dimensionality. In addition, one must be aware that the above values for d_ℓ apply for "ideal" (pure) systems (with translationally invariant interactions) or systems where the frozen-in disorder (due to crystal defects of the substrate lattice, strongly chemisorbed impurities at the surface, etc.) is sufficiently *weak*. In this context, weak does not only mean that these defects are sufficiently dilute, but also that they produce only a perturbation in the local strength of pairwise interactions, which in the framework of the "Ginzburg–Landau–Wilson"-Hamiltonian (Wilson and Kogut, 1974; Ma, 1976) $F\{\phi(x)\}$, eq. (14), translates into a weak randomness of the parameter r. If the frozen-in defects lead to a random-field type term $H(x)\phi(x)$ in eq. (14), with

$$[H(x)]_{av} = \int dH\, P(H) H = 0,$$

$$[H(x)H(x')]_{av} = \int dH\, P(H) H(x) H(x') = \delta(x - x')h^2, \tag{58}$$

then an Ising-type order is also unstable and an arbitrarily weak random field amplitude h is sufficient to destroy long range order in the system in the sense that the system is split into an irregular configuration of (large) domains (fig. 16; Morgenstern et al., 1981). This means that the random field raises the lower critical dimensionality from $d_\ell = 1$ for the pure Ising model to $d_\ell = 2$ for the random field Ising model (RFIM) (Imry and Ma, 1975; Grinstein and Ma, 1982; Villain, 1982; Nattermann and Villain, 1988). In the following, we explicitly disregard all such effects of quenched random disorder, and assume the substrate is strictly ideal over large enough distances (e.g., of order 100 Å) to allow a meaningful study of critical phenomena in pure systems.

After these caveats, fig. 17 shows qualitatively the dimensionality dependence of the order parameter exponent β, the response function exponent γ, and correlation length exponent ν. Although only integer dimensionalities $d = 1, 2, 3$ are of physical interest (lattices with dimensionalities $d = 4, 5, 6$ etc. can be studied by computer simulation, see e.g. Binder, 1981a, 1985), in the renormalization group framework it has turned out useful to continue d from integer values to the real axis, in order to derive expansions for critical exponents in terms of variables $\epsilon = d_u - d$ or $\epsilon' = d - d_\ell$, respectively (Fisher, 1974; Domb and Green, 1976; Amit, 1984). As an example, we quote the results for η and ν (Wilson and Fisher, 1972)

$$\eta = \frac{m+2}{2(m+8)^2}\epsilon^2 + \ldots, \tag{59}$$

$$\nu = \frac{1}{2} + \frac{m+2}{4(m+8)}\epsilon + \left[\frac{(m+2)(m^2 + 23m + 60)}{8(m+8)^3}\right]\epsilon^2 + \ldots, \tag{60}$$

from which the expansions for all other exponents can be derived using the scaling laws which are discussed below. Figure 17 has been drawn qualitatively consistent with the results of these expansions. Also the limit $(m \to \infty)$ has been included which reduces (Stanley, 1968) to the exactly soluble spherical model (Berlin and Kac, 1952), which for $2 < d < 4$ has the exponents

$$\alpha = \frac{d-4}{d-2}, \quad \beta = \frac{1}{2}, \quad \gamma = \frac{2}{d-2}, \quad \eta = 0, \quad \nu = \frac{1}{d-2}. \tag{61}$$

It is seen that for $d \to d_\ell = 2$ both γ and ν diverge towards $+\infty$, while α diverges towards $-\infty$. This behavior is compatible with an exponential

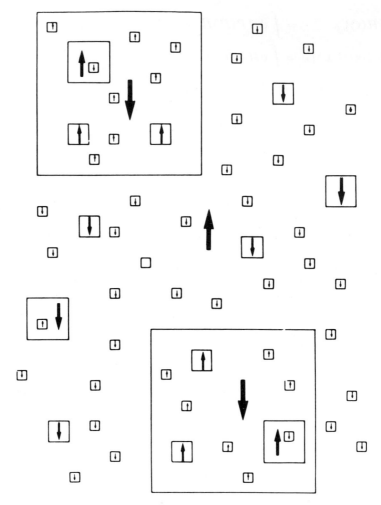

Fig. 16. Qualitative ground state domain pattern of the two-dimensional Ising lattice in a small random field. Note that this schematic picture neglects the random field induced roughness of the domain walls (Grinstein and Ma, 1982; Villain, 1982). Arrows indicate orientations of the domains, which are arranged such as to make optimum use of the local excess of one sign of the random field. From Morgenstern et al. (1981).

divergence of correlation length and response function at the lower critical dimension. This is most simply shown for the one-dimensional Ising model (for a pedagogic account, see e.g. Young, 1980a), where

$$\langle S_j S_{j+k} \rangle = \left[\tanh \left(\frac{J}{k_B T} \right) \right]^k = \exp \left(-\frac{k}{\xi} \right), \tag{62}$$

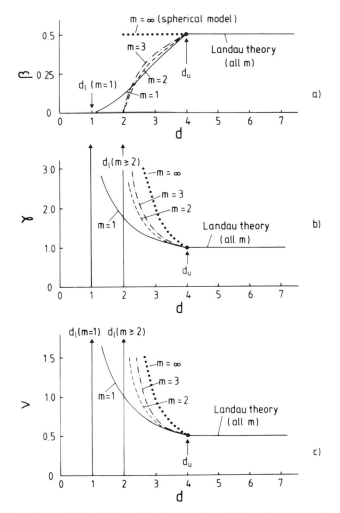

Fig. 17. Schematic variation of the critical exponents of the order parameter β (a), the order parameter response function γ (b), and the correlation length ν (c) with the spatial dimensionality, for the m-vector model. Upper (d_u) and lower (d_ℓ) critical dimensionalities are indicated. Here $m = 1$ corresponds to the Ising model, $m = 2$ to the XY model, $m = 3$ to the Heisenberg model of magnetism, while the limit of infinitely many order parameter components ($m \to \infty$) reduces to the exactly solved spherical model (Berlin and Kac; 1952, Stanley, 1968).

J being the exchange constant between nearest neighbors. Equation (62) implies

$$\xi = -\frac{1}{\ln\{\tanh(J/k_B T)\}} \approx \frac{1}{2} \exp\left(\frac{2J}{k_B T}\right), \qquad \text{for } T \to 0, \qquad (63)$$

and

$$k_B T \chi = \frac{1 + \tanh(J/k_B T)}{1 - \tanh(J/k_B T)} \approx \exp\left(\frac{2J}{k_B T}\right), \qquad \text{for } T \to 0, \qquad (64)$$

and a similar divergence can be proven for the m-vector model with $m \geq 3$ in $d = 2$. The XY-model in $d = 2$ exhibits a special behavior, Kosterlitz–Thouless (1973) finding a transition at a non-zero critical temperature T_c but with no long range order for $0 < T < T_c$. This will be briefly considered in sect. 2.4. On the other hand, the m-vector model (with a nearest neighbor exchange between "classical" spins) is soluble also in $d = 1$ (Baxter, 1982), and in this case one has $d < d_\ell$ and a zero-temperature phase transition with a simple power-law divergence of the correlation length occurs again,

$$\xi = \frac{2J}{(m-1)k_B T}, \qquad T \to 0, \ m \geq 2. \qquad (65)$$

It has already been mentioned in the introduction that due to fluctuations in $d = 2$ dimensions it is impossible to have at $T > 0$ long range order in strictly isotropic magnets (Mermin and Wagner, 1966) and in crystals (Mermin, 1968). Of course, crystalline order that is in registry with the substrate (fig. 2c) is not excluded, this order being stabilized by the corrugation potential maintained by the (three-dimensional) substrate. Long wavelength fluctuations do destabilize, however, the long range order of incommensurate two-dimensional solids (fig. 2c) at non-zero temperatures. Rather than discussing a general proof, we simply recall the well-known result of spin wave theory for the spontaneous magnetization of a ferromagnet with nearest neighbor exchange,

$$\phi_o(T) = g\mu_B \left\{ Ns - \sum_{k \in \text{1st Brillouin zone}} \left[\exp\left(\frac{\hbar\omega_k}{k_B T}\right) - 1 \right]^{-1} \right\}, \qquad (66)$$

where g is the gyromagnetic ratio, μ_B the Bohr magneton, s the spin quantum number, the wave vectors k of the spin waves with frequency ω_k being restricted to the first Brillouin zone. Using (δ is a vector to a nearest neighbor on the cubic lattice)

$$\hbar\omega_k = 2Jzs \left\{ 1 - z^{-1} \sum_\delta \exp(ik \cdot \delta) \right\} \approx 2Jsa^2 k^2, \qquad k \to 0, \qquad (67)$$

where z is the coordination number ($z = 4$ for the square lattice), one obtains, after transforming the sum in eq. (66) to an integral, in $d = 2$

$$\phi_o(T) \cong g\mu_B N \left\{ s - \frac{a^2}{2\pi} \int k \, dk \left[\exp\left(\frac{2Jsa^2 k^2}{k_B T}\right) - 1 \right]^{-1} \right\}. \qquad (68)$$

Clearly the integration $\int dk/k$ resulting from eq. (68) as $k \to 0$ would yield a logarithmic divergence, which already suggests that (without a stabilizing field or anisotropy that would induce a gap into the spin wave spectrum) there cannot be a spontaneous magnetization.

A similar argument applies to the crystalline order in the case of fig. 2d. Assuming that long range order exists, one would calculate the mean square displacement of an atom relative to its ideal position

$$\langle u^2 \rangle = \sum_{k \in 1.B.Z.} \langle u_k^2 \rangle = \left(\frac{L}{2\pi} \right)^2 \int dq \langle u_q^2 \rangle \tag{69}$$

Now the equipartion theorem implies that the elastic energy of long wavelength phonons is $(L^2/2)Bk^2 \langle u_k^2 \rangle = (1/2)k_B T$, where B is a constant, and hence eq. (69) becomes

$$\langle u^2 \rangle = \frac{k_B T}{2\pi} \int_{k_{min}}^{k_{max}} \frac{k \, dk}{Bk^2} = \frac{k_B T}{2\pi B} \ln \left(\frac{L}{a} \right), \tag{70}$$

where $k_{min} = 2\pi/L$ and $k_{max} = 2\pi/a$ was used. This logarithmic divergence of $\langle u^2 \rangle$ for $L \to \infty$ implies that the Debye–Waller factor vanishes; i.e. δ-function like Bragg peaks cannot exist. We calculate the structure factor (x_o, x_o' are positions of the perfect lattice)

$$S(k) = \frac{1}{N} \sum_{x_o, x_o'} \langle \exp[ik \cdot (u(x_o) - u(x_o'))] \rangle \tag{71}$$

For a harmonic crystal the Hamiltonian is (α, γ, ξ, η run here over the cartesian indices x, y)

$$\mathcal{H} = \mathcal{H}_o + \sum_q \sum_{\alpha\gamma\xi\eta} g_{\alpha\gamma}^{\xi\eta} q_\xi q_\eta u_q^\alpha u_{-q}^\gamma, \tag{72}$$

u_q being the Fourier transform of the displacement $u(x_o)$, and $g_{\alpha\gamma}^{\xi\eta}(q)$ is the "dynamical matrix". Now for harmonic phonons the variables $ik \cdot (u(x_o) - u(x_o'))$ are sums of independent gaussian variables and hence

$$S(k) = \frac{1}{N} \sum_{x_o, x_o'} \exp[ik \cdot (x_o - x_o')] \exp \left[-\frac{1}{2} \langle |k \cdot (u(x_o) - u(x_o'))|^2 \rangle \right]$$

$$= \sum_x \exp(ik \cdot x) \exp \left\{ -\frac{1}{N} \sum_q \langle |k \cdot u_q|^2 \rangle (1 - \exp(iq \cdot x)) \right\}, \tag{73}$$

where in the last step the translational invariance has been used, $\langle |k \cdot (u(x_o) - u(x_o'))|^2 \rangle = \langle |k \cdot (u(x) - u(0))|^2 \rangle$, $x = x_o - x_o'$, and $u(x) = \sum_q \exp(iq \cdot x)u_q$.

Concluding now from eq. (72) via the equipartition theorem

$$\left\langle u_q^\alpha u_{-q}^\gamma \right\rangle \approx \frac{k_B T}{g q^2},$$ (74)

where for simplicity the indices of $g_{\alpha\gamma}^{\xi\eta}$ in eq. (72) are suppressed, we obtain further (Jancovici, 1967)

$$
\begin{aligned}
S(k) &\approx \sum_x \exp(ik \cdot x) \exp\left\{-\frac{k_B T}{(2\pi)^2} \int d^2q \, \frac{k^2}{g q^2}[1 - \exp(iq \cdot x)]\right\} \\
&\approx \sum_x \exp(ik \cdot x) \exp\left\{-\frac{k_B T k^2}{(2\pi g)} \int_{q > x^{-1}} \frac{dq}{q}\right\} \\
&= \sum_x \exp(ik \cdot x) \exp\left\{-\frac{k_B T k^2}{(2\pi g)} \ln \frac{x}{a}\right\} \\
&= \sum_x \left(\frac{x}{a}\right)^{[k_B T k^2/2\pi g]} \exp(ik \cdot x) \propto |k - k_o|^{-(2 - k_B T k_o^2/2\pi g)}
\end{aligned}
$$ (75)

k_o being the position of the Bragg peak (at $T = 0$ we have $S(k) \propto \delta(k - k_o)$, of course).

The above calculation does not apply to any ordering that is commensurate with the periodic substrate potential $V(x, z)$ (fig. 1), since the periodic potential then removes the instability of the harmonic Hamiltonian, eq. (72). However, another instability does occur in adsorbed layers on stepped surfaces (fig. 3), if the layers on each terrace are independent from the neighboring terraces. Each terrace of width L can then be considered as a quasi-onedimensional infinite strip, which cannot maintain true long range order. Consider, for example, a c(2×2) ordering on the square lattice: there are two types of domain possible, depending which sublattice is preferentially occupied (see fig. 10, top part). Neither of these domains will be preferred by boundary effects at the steps, and hence the system does not develop infinite-range order at any non-zero temperature, but rather the strip is always spontaneously broken up in domains (fig. 18). The size of these domains in the direction parallel to the steps is very large, namely (Fisher, 1969)

$$\xi_D \propto \exp\left[\frac{2L\sigma_{\text{int}}}{k_B T}\right]$$ (76)

where σ_{int} is the interfacial free energy between the coexisting ordered phases.

After these remarks on situations where fluctuations are strong enough to destroy true long range order, we now assume that $d_\ell < d$ for the system of interest, and discuss the critical behavior. At T_c the power law decay of the correlation function, $G(x) = \hat{G}|x|^{-(d-2+\eta)}$ [eq. (43)] implies

Fig. 18. Snapshot pictures of a nearest-neighbor Ising ferromagnet on the square lattice with bulk field $H = 0$ and boundary fields $H_1 = H_L = 0$ (this model is isomorphic to the c(2×2) ordering at coverage $\theta = 0.5$) at the temperatures $T = 0.95\,T_c$ (a), $T = T_c$ (b) and $T = 1.05\,T_c$ (c), for a $L \times M$ system with $L = 24$, $M = 288$ and two free boundaries of length M, while periodic boundary conditions are used along the strip. Up spins (adatoms on sublattice 1) are shown in black, down spins (adatoms on sublattice 2) are shown in white. Domain formation at $T \lesssim T_c$ can be clearly recognized. From Albano et al. (1989b).

that the system is invariant against a transformation of the length scale: Choosing $x' = x/\Lambda$ changes only prefactors, but leaves the power law decay unchanged, $G(x') \propto |x'|^{-(d-2+\eta)}$. Thus critical fluctuations would look just the same when we study them with a light microscope, as they would look with an electron microscope! This "Gedankenexperiment" can be checked very nicely with computer experiments, of course (e.g. producing snapshot pictures of Ising model lattices at T_c such as in fig. 18 one can verify that a coarse-grained very large lattice looks just the same as a corresponding magnification of a smaller lattice).

This scale invariance which holds exactly at T_c on arbitrarily large scales (all scales must be much larger than the lattice spacing a, of course) is limited by the correlation length ξ at temperatures off T_c. But for distances $a \ll |x| \ll \xi$ the correlation function will still have precisely the same behavior as for $T = T_c$. Thus it is reasonable to assume that it is the ratio of the lengths $|x|$ and ξ that matters, i.e.

$$G(x, \xi) = \hat{G}|x|^{-(d-2+\eta)}\tilde{G}\left(\frac{|x|}{\xi}\right), \tag{77}$$

where the scaling function $\tilde{G}(z)$ behaves as $\tilde{G}(z \to 0) = 1$, $\tilde{G}(z \gg 1) \propto \exp(-z)$, disregarding preexponential power-law prefactors now.

From this "scaling hypothesis" eq. (77) one readily derives a relation between the exponents γ, η and ν. Using eq. (77) we find for the isothermal response function near T_c

$$k_B T \chi_T = \sum_x G(x, \xi) \approx \int dx\, G(x, \xi) = \hat{G} \int dx\, |x|^{-(d-2+\eta)} \tilde{G}\left(\frac{|x|}{\xi}\right)$$
$$= \hat{G} U_d \int_0^\infty dx\, x^{1-\eta} \tilde{G}\left(\frac{x}{\xi}\right) = \hat{G} U_d \xi^{2-\eta} \int_0^\infty dz\, z^{1-\eta} \tilde{G}(z), \quad (78)$$

U_d being the surface of a unit sphere in d dimensions. Equation (78) implies

$$\chi_T(t) \propto |t|^{-\gamma} \propto [\xi(t)]^{2-\eta} \propto |t|^{-\nu(2-\eta)}, \tag{79}$$

and hence we must have

$$\gamma = \nu(2 - \eta). \tag{80}$$

eq. (77) can also be interpreted as a "homogeneity postulate": we write $G(x, \xi)$ as a generalized homogeneous function, with an exponent κ

$$G(x, \xi) = \Lambda^\kappa G\left(\frac{x}{\Lambda}, \frac{\xi}{\Lambda}\right), \tag{81}$$

since for $\Lambda = \xi$

$$G(x, \xi) = \xi^\kappa G\left(\frac{x}{\xi}, 1\right) = \xi^\kappa \left(\frac{x}{\xi}\right)^\kappa G'\left(\frac{x}{\xi}, 1\right) \equiv x^\kappa \tilde{G}\left(\frac{x}{\xi}\right) \hat{G}, \tag{82}$$

which implies $\kappa = -(d - 2 + \eta)$.

A similar homogeneity postulate can be written for the singular part of the free energy,

$$F(T, H) = F_{reg}(T, H) + F_{sing}(T, H) = F_{reg}(T, H) + |t|^{2-\alpha} \tilde{F}(\tilde{H}), \tag{83}$$

where $\tilde{H}$ is the scaled ordering field,

$$\tilde{H} = \frac{H \hat{C} |t|^{-\Delta}}{\hat{B}}, \tag{84}$$

Δ being the "gap exponent" and $\tilde{F}(\tilde{H})$ the scaling function of the free energy. Note that the scaling power $|t|^{2-\alpha}$ was chosen for the sake of consistency with the assumed singular behavior of the specific heat, cf. eq. (7)

$$C = T\left(\frac{\partial^2 F}{\partial T^2}\right)_{H=0} \propto \frac{\partial^2 F}{\partial t^2} \propto |t|^{-\alpha}. \tag{85}$$

Equations (83) and (84) have been written such that the expected critical behavior of the order parameter follows,

$$\phi = -\left(\frac{\partial F}{\partial H}\right)_T = -\frac{\hat{C}|t|^{-\Delta}}{\hat{B}}|t|^{2-\alpha}\frac{\partial \tilde{F}}{\partial \tilde{H}} = -\frac{\hat{C}}{\hat{B}}|t|^{2-\alpha-\Delta}\tilde{F}'(\tilde{H}). \tag{86}$$

Defining [cf. eq. (8)]

$$\phi = \hat{B}|t|^{\beta}\tilde{M}(\tilde{H}) \tag{87}$$

with $\tilde{M}(0) = 1$ one finds $\hat{B} = -\hat{C}\tilde{F}'(0)/\hat{B}$, i.e. $\tilde{F}'(0) = -\hat{B}^2/\hat{C}$ and also

$$\beta = 2 - \alpha - \Delta. \tag{88}$$

Thus there exist relations between suitable ratios of critical amplitudes and derivatives of the scaling functions. We now consider the response function

$$\chi_T = \left(\frac{\partial \phi}{\partial H}\right)_T = \frac{\hat{C}|t|^{-\Delta}}{\hat{B}}(\hat{B}|t|^{\beta})\frac{d\tilde{M}}{d\tilde{H}} = \hat{C}|t|^{\beta-\Delta}\tilde{M}'(\tilde{H}) \tag{89}$$

The scale factor of the field relating $\tilde{H}$ to H in eq. (84) has been chosen such that $d\tilde{M}/d\tilde{H}|_{\tilde{H}=0} = \tilde{M}'(0) = 1$ and then eq. (89) yields eq. (6) if the scaling relations between critical exponents hold,

$$\beta - \Delta = -\gamma, \qquad \gamma + 2\beta = 2 - \alpha, \tag{90}$$

where in the second relation eq. (88) was used.

We now consider also the critical isotherm, $\phi = \hat{D}H^{1/\delta}$, $t = 0$ [eq. (45)]. In order to obtain this relation from eq. (87), the scaling function $\tilde{M}(\tilde{H} \gg 1)$ must behave as a power law $\tilde{M} \propto \tilde{H}^x$, with $|t|^{\beta}|t|^{-\Delta x} = |t|^0$, since then the powers of t cancel each other for large $\tilde{H}$ and the limit $t \to 0$ can be taken in a meaningful way. This implies, using also eq. (90), $x = \beta/\Delta = \beta/(\gamma+\beta)$ and hence

$$\phi_{T=T_c} \propto H^x = H^{\beta/(\gamma+\beta)} = H^{1/\delta}, \qquad \delta = 1 + \frac{\gamma}{\beta}. \tag{91}$$

One easily verifies that all these scaling relations hold for the critical exponents of Landau's theory, eq. (48), as well as for the spherical model [eq. (61)] and Ising and Potts models (see sect. 2.3 below).

Another consequence that follows from the homogeneity postulate for the free energy is the fact that exponents of corresponding quantities above and below T_c are identical, $\alpha = \alpha'$, $\gamma = \gamma'$, $\nu = \nu'$.

A very interesting scaling law which does not hold in Landau's theory for $d > d_u$ but which does hold for $d \leq d_u$ is the so-called hyperscaling law, which relates the critical exponent ν of the correlation length ξ to the exponent $2 - \alpha$ of the singular part of the free energy. To motivate it, we present a plausibility argument: consider a non-interacting Ising spin system at $H = 0$. Its total free energy would simply be $F^{\text{tot}} = NF = k_B T \ln 2$;

next we argue that near T_c we can divide the system into blocks of volume ξ^d, and that within a block the spins are very strongly correlated with each other, while different blocks can be treated as uncorrelated, as far as the free energy is concerned. In analogy with the non-interacting system of single spins we conclude

$$F_{\text{sing}}^{\text{tot}} \propto k_B T_c N_{\text{blocks}} \ln 2, \qquad F_{\text{sing}} \propto \frac{k_B T_c N_{\text{blocks}}}{N} \ln 2, \tag{92}$$

apart from factors of order unity. Since each block contains $[\xi(t)/a]^d$ spins, we have $N = N_{\text{blocks}}[\xi(t)/a]^d$, and thus

$$F_{\text{sing}} \propto \left[\frac{\xi(t)}{a}\right]^{-d} \propto |t|^{\nu d} \implies \nu d = 2 - \alpha, \tag{93}$$

comparing to eq. (83) for $\tilde{H} = 0$. Obviously, for Landau exponents [eq. (48)] this holds at the upper critical dimensionality d_u only.

At this point, we mention a further consequence of the "universality principle" alluded to above. For each "universality class" (such as that of the Ising model or that of the XY model, etc.) not just the critical exponents are universal, but also the scaling function $\tilde{F}(\tilde{H})$, apart from non-universal scale factors for the occurring variables (a factor for H we have expressed via the ratio $\hat{C}/\hat{B}$ in eq. (84), for instance). A necessary implication then is the universality of certain critical amplitude ratios, where all scale factors for the variables of interest cancel out. In particular, ratios of critical amplitudes of corresponding quantities above and below T_c, $\hat{A}^+/\hat{A}^-$ [eq. (7)], $\hat{C}^+/\hat{C}^-$ [eq. (6)] and $\hat{\xi}^+/\hat{\xi}^-$ [eq. (38)] are universal (Privman et al., 1991). A further relation exists between the amplitude $\hat{D}$ and $\hat{B}$ and $\hat{C}^-$. Writing $\tilde{M}(\tilde{H} \to \infty) = X\tilde{H}^{1/\delta}$, cf. eqs. (87) and (91), the universality of $\tilde{M}(\tilde{H})$ states that X is universal. But since $\phi = \hat{B}|t|^\beta \tilde{M}(\tilde{H}) = \hat{B}|t|^\beta X \tilde{H}^{1/\delta} = \hat{B}^{1-1/\delta}\hat{C}^{1/\delta}H^{1/\delta}X$, a comparison with eq. (45) yields

$$X = \hat{D}\hat{B}^{1/\delta}\hat{C}^{-1/\delta} = \text{universal}. \tag{94}$$

While there is reasonable experimental evidence for the universality of scaling functions, the experimental evidence for the universality of amplitude relations such as eq. (94) is not very convincing. One reason for this problem is that the true critical behavior can be observed only asymptotically close to T_c, and if experiments are carried out not close enough to T_c the results for both critical amplitudes and critical exponents are affected by systematic errors due to corrections to scaling. For example, eq. (6) must be written more generally as

$$\chi = \hat{C}^\pm |t|^{-\gamma}\left\{1 + \hat{C}_1^\pm|t|^{x_1} + \hat{C}_2^\pm|t|^{x_2} + \ldots\right\}, \qquad \text{with} \quad 0 < x_1 < x_2 \ldots \tag{95}$$

Very close to T_c these corrections $\hat{C}_1^{\pm}|t|^{x_1} \ll 1$, and higher order terms are also negligible, but how close one has to get to T_c to see unambiguously the leading behavior depends both on the (universal) correction-to-scaling exponent x_1 and the associated (non-universal) correction-to-scaling amplitude $\hat{C}_1^{\pm}$.

It is also interesting to consider the critical behavior of non-ordering fields near critical points. Such phenomena occur since there is a coupling of the order parameter field $\phi(x)$ to other variables. Consider for instance the transition of a monolayer held at fixed ambient gas pressure (or chemical potential, respectively) from the disordered state to the c(2×2) structure. In an Ising spin representation, the c(2×2) order parameter corresponds to a staggered magnetization, while the coverage translates into the uniform magnetization $M(x)$ of these pseudo-spins. In the Landau free energy, we expect a lowest-order coupling term of the form

$$\Delta\mathcal{F} = \int d^d x \frac{c}{2} \phi^2(x) M^2(x), \qquad c = \text{const}; \tag{96}$$

there cannot be a term linear in $\phi(x)$ (in the absence of a "staggered field" coupling to the order parameter) due to the invariance of the Hamiltonian against an interchange of the two sublattices. Similarly, in the absence of a field h conjugate to M there is a symmetry against an interchange of all the spins (the state $h = 0$, $M = 0$ corresponds to the coverage $\theta = \frac{1}{2}$, around this coverage the lattice gas has particle–hole symmetry if there are pairwise additive interactions only.) In the homogeneous case, the relevant part of the free energy then becomes, taking only the terms containing M into account

$$\Delta F = V \left(\frac{c}{2} \phi^2 M^2 + \frac{1}{2} \chi_o^{-1} M^2 - Mh \right), \tag{97}$$

where χ_o^{-1} is the expansion coefficient of M^2. From $\partial(\Delta F)/\partial M = 0$ we find that in thermal equilibrium we have $h = \chi_o^{-1} M + c\phi^2 M$, i.e.

$$M = \frac{\chi_o h}{1 + c\chi_o\phi^2} \approx \chi_o h - c\chi_o^2 h\phi^2 = M_{\phi=0} + \Delta M, \qquad \phi \to 0. \tag{98}$$

Obviously, for $T > T_c$ where $\phi \equiv 0$ we have $\Delta M \equiv 0$, while for $T < T_c$

$$\Delta M \propto \langle\phi\rangle^2 \propto (-t)^{2\beta} = (-t)^1. \tag{99}$$

Of course, eqs. (97)–(99) are just mean-field results. A more accurate theory yields

$$\Delta M \propto \langle\phi^2\rangle_L^{\text{sing}} = \frac{1}{L^{2d}} \sum_{i,j \in L^d} \langle\phi_i\phi_j\rangle_T^{\text{sing}} \propto |t|^{1-\alpha}. \tag{100}$$

We now assume $\alpha > 0$ so the temperature dependence of eq. (100) is more important than the regular term ($\propto t$) that is also present. Here we used the fact that the singular part of short range correlations has the same singularity as the internal energy $U_{\text{sing}} = \langle \mathcal{H} \rangle_{\text{sing}} = -\sum_{i \neq j} J_{ij} \langle \phi_i \phi_j \rangle_{\text{sing}} \propto |t|^{1-\alpha}$. The same correlation function singularity $\langle S_i S_j \rangle_{\text{sing}} \propto |t|^{1-\alpha}$ is picked up by the electrical resistivity at phase transitions in conducting materials, by the refractive index etc. This energy singularity is also seen if one studies the intensity $I\,(k_I, t)$ of scattering carried out at finite wavevector resolution characterized by a wavenumber k_I of the instrument (Bartelt et al., 1985a–c). For the wavevector $q = q_{\text{B}}$ where the superlattice Bragg spots appear in the ordered phase, we may write a scaling form, with $\tilde{I}_{\pm}(z)$ a scaling function,

$$I\,(q_{\text{B}}, k_I, t) = |t|^{2\beta}\,\tilde{I}_{\pm}(k_I \xi) \tag{101}$$

One can motivate eq. (101) as follows: In the limit of $k_I \to 0$ (perfect resolution) one would simply see the Bragg delta function peak superimposed by diffuse scattering, cf. eq. (37), $k \equiv q - q_{\text{B}}$,

$$S_T\,(q) = k_{\text{B}} T \chi\,(q) \propto \left\{ \phi^2(t) \delta(k) + \frac{k_{\text{B}} T \chi_T}{1 + k^2 \xi^2} \right\}, \qquad k\xi \ll 1; \tag{102}$$

thus from a scattering experiment with k_I sufficiently small both exponents β (of the order parameter $\phi(t)$), γ (of the ordering susceptibility χ_T), and ν (of the correlation length ξ) can be extracted (see fig. 15 for a practical example). Note that $\tilde{I}_{-}(0) = \hat{B}^2$ while $\tilde{I}_{+}(0) \equiv 0$, of course, since (for simplicity, we consider a step function-like resolution function)

$$I\,(q_{\text{B}}, k_I, t) = \int_{k < k_I} d^d k\, S_T\,(q) \propto k_I^d \chi_T \propto k_I^d |t|^{-\gamma}, \quad \text{for } k_I \xi \ll 1, \tag{103}$$

and writing $\tilde{I}_{+}(z) \propto z^d$ we see that eq. (101) is indeed compatible with eq. (103),

$$I\,(q_{\text{B}}, k_I, t) \propto |t|^{2\beta} k_I^d \xi^d \propto k_I^d |t|^{2\beta} |t|^{-d\nu} = k_I^d |t|^{-\gamma}, \tag{104}$$

if one invokes the hyperscaling relation, eq. (93), $d\nu = 2 - \alpha = \gamma + 2\beta$. Now for $k\xi \gg 1$ $\chi(q)$ no longer has the simple Ornstein–Zernike form, but rather an expansion compatible with eq. (44) (note that $q_{\text{B}} \equiv 0$ was assumed there and thus q and k need not be distinguished then)

$$\chi\,(q) = k^{-(2-\eta)} \tilde{\chi}(k\xi) \propto k^{-(2-\eta)} + \hat{\chi}_{\infty}(k\xi)^{-(1-\alpha)/\nu} + \cdots, \tag{105}$$

so that upon Fourier transformation of eq. (105) indeed a $|t|^{1-\alpha}$ behavior of the correlation $\langle \phi_i \phi_j \rangle_{\text{sing}}$ results. Using eq. (105) in eq. (103), we expect a behavior (adding now also regular terms in t)

$$I\,(q_{\text{B}}, k_I, t) \propto 1 + I_1 t \pm I_2 |t|^{1-\alpha} + \cdots, \tag{106}$$

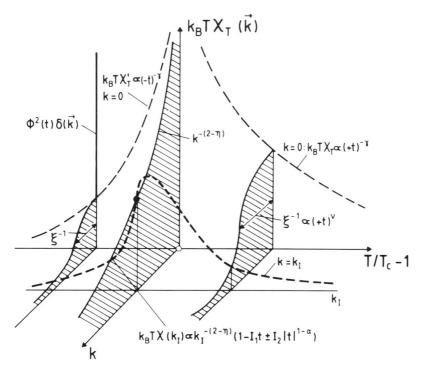

Fig. 19. Schematic plot of the scattering intensity as a function of temperature distance from a critical point and as a function of the distance k from a superstructure Bragg spot in reciprocal space. For $k = 0$ one has a critical divergence according to the ordering susceptibility $k_B T \chi_T \propto (-t)^{-\gamma}$ and an analogous divergence below T_c, superimposed by a delta-function whose weight is given by the order parameter. For a non-zero wavenumber k_I the scattering intensity has a maximum slightly above T_c, reflecting the smooth crossover from Ornstein–Zernike behavior for $k\xi \leq 1$ to the critical decay $k^{-(2-\eta)}$ at T_c (Fisher and Burford, 1967). At T_c for $k \neq 0$, the intensity exhibits a singular temperature derivative due to the term $\pm|t|^{1-\alpha}$, as indicated in the figure.

which was also used in fig. 15 to estimate α from scattering data. For clarity, fig. 19 summarizes schematically this behavior of the scattering function $k_B T \chi(\mathbf{q})$.

A further property related to the energy singularity induced in non-ordering fields is the so-called "Fisher renormalization" (Fisher, 1968) of critical exponents. Consider again the order–disorder transition of the $c(2\times2)$ structure (fig. 10) as a function of chemical potential μ controlling the coverage θ. Since $T_c = T_c(\mu)$ depends on μ, we can also probe the transition varying μ at constant T across $\mu_c(T)$, the inverse function of $T_c(\mu)$. Thus $\phi \propto [\mu_c(T) - \mu]^\beta$, and invoking that the coverage change $\Delta\theta$ is proportional to $\Delta M \equiv M - M_c(T) \propto [\mu_c(T) - \mu]^{1-\alpha}$ via the Ising magnet–lattice translation, we also find

$$\phi \propto [\theta - \theta_c(T)]^{\beta/(1-\alpha)} . \tag{107}$$

Thus if one studies the phase transitions as a function of coverage, one finds (remembering that $\alpha > 0$) an enhanced value of the critical exponent, $\beta/(1 - \alpha)$ instead of β.

As a final topic of this section, we return to the simple Landau theory, eq. (14), and consider the special case that by the variation of a non-ordering field h one can reach a special point h_t, T_t where the coefficient $u(h_t) = 0$: while for $h < h_t$ one has $u(h) > 0$ and thus a standard second-order transition occurs, for $h > h_t$ one has $u(h) < 0$ and thus the Landau theory implies a first-order transition. This behavior actually occurs for Ising antiferromagnets in a field or the related order–disorder transitions of the corresponding lattice gas models of adsorbed monolayers. Figure 20 shows a phase diagram and the corresponding order parameter behavior of a related model as obtained by Monte Carlo simulation (Binder and Landau, 1981).

We thus write $u(h) = u'(h_t - h) + \ldots$ and obtain from eq. (17)

$$\frac{1}{k_B T V} \left(\frac{\partial F}{\partial \phi_o} \right)_T \bigg|_{H=0} = \phi_o(r + v\phi_o^4) = 0,$$

$$\phi_o = (-r/v)^{1/4} = (r'/k_B v)^{1/4}(-t)^{1/4}. \tag{108}$$

Thus we conclude that the order parameter exponent at the point h_t, T_t, which is called a tricritical point (Sarbach and Lawrie, 1984), has a different

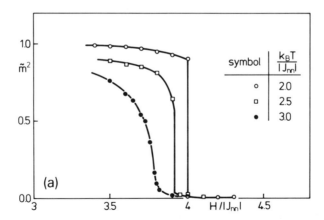

Fig. 20. (a) Square of the order parameter ψ (staggered magnetization) of the square lattice gas model (or corresponding Ising antiferromagnet, respectively) with a ratio $R = J_{nnn}/J_{nn} = -1$ of the interaction energies between next nearest and nearest neighbors plotted versus the non-ordering field $H/|J_{nn}|$ at three temperatures. Highest temperature corresponds to a second-order transition while for the two lower temperatures the transition is first order.

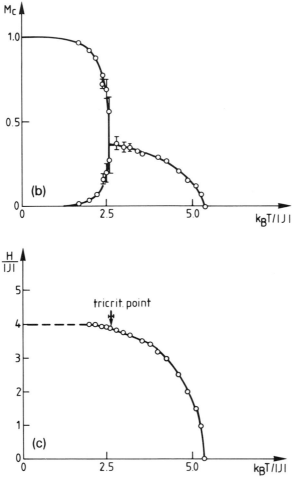

Fig. 20 (contd.). (b) Critical magnetization m_c of the Ising antiferromagnet plotted vs. temperature. Two values m_c^-, m_c^+ for m_c indicate the magnetization jump at the transition, which translates into the two phase coexistence regions for the coverage ($\theta_{coex}^{(1)} = (1/2)(1 - m_c^+)$), $\theta_{coex}^{(2)} = (1/2)(1 - m_c^-)$, see sect. 2.3). (c) Critical (non-ordering) magnetic field (for $R = -1$) plotted versus temperature. The transition is second order for temperatures higher than the tricritical temperature T_t while for $T < T_t$ it is of first order. From Binder and Landau (1981).

value, $\beta_t = 1/4$, instead of $\beta = 1/2$. Similarly, the equation for the critical isotherm differs from normal critical points, where $\delta = 3$,

$$v\phi_o^5 = \frac{H}{k_B T_c}, \qquad \phi_o = (k_B T_c v)^{-1/5} H^{1/5}, \qquad \delta_t = 5. \qquad (109)$$

It is easy to see that the above treatment of the critical scattering and correlations in terms of the wavevector dependent susceptibility goes through
as previously, i.e. we still have $\gamma_t = 1$, $\nu_t = 1/2$, $\eta_t = 0$ as in the standard
Landau theory. But the behavior of the specific heat changes, since [cf.
eq. (46)]

$$\frac{F(\phi) - F(0)}{V k_B T} = \phi_o^2 \left(\frac{r}{2} + \frac{v}{6} \phi_o^4 \right) = \frac{r}{3} \sqrt{\frac{-r}{v}} \propto (-t)^{3/2},$$
(110)

$$C_{H=0,h_t} \propto (-t)^{-1/2}, \qquad T < T_t.$$

Thus Landau's theory predicts the following set of tricritical exponents
(Griffiths, 1970; Sarbach and Lawrie, 1984)

$$\alpha_t = \tfrac{1}{2}, \quad \beta_t = \tfrac{1}{4}, \quad \gamma_t = 1, \quad \delta_t = 5, \quad \nu_t = \tfrac{1}{2}, \quad \eta_t = 0$$
(111)

One easily verifies that these exponents also satisfy the scaling relations
[eqs. (90), (91)] $2 - \alpha_t = \beta_t(\delta_t + 1) = \gamma_t + 2\beta_t$ and that the hyperscaling
relation [eq. (93)] $\nu_t d = 2 - \alpha_t$ is satisfied for $d_t = 3$. Returning to the
Ginzburg criterion, eqs. (52)–(55), we conclude that the Landau theory of
tricritical phenomena is selfconsistent if

$$\text{const} \ll |t|^{-\nu_t d + 2\beta_t + \gamma_t} = |t|^{(3-d)/2},$$
(112)

i.e. for $d > 3$, $d_t^* = 3$ being the upper critical dimensionality for tricritical
points. Renormalization group theory (Ma, 1976; Domb and Green, 1976;
Amit, 1984) predicts logarithmic correction factors to the Landau-type
power laws in $d = 3$ for tricritical points, just as it does at $d = d_u = 4$ for
ordinary critical points.

An additional exponent β_2 describing the shape of the phase diagram in
fig. 20b is defined as $M_c^\pm - M_{ct} \propto (T_t - T)^{\beta_2}$, where M_{ct} is the value of
$M_c(T)$ at the tricritical point, and $M_c^\pm$ denote the two magnetizations of
the coexisting phases in the first order region. One can show that Landau's
theory yields $\beta_2 = 1$ (Sarbach and Lawrie, 1984), while in $d = 2$ one has
$\beta_2 = 1/4$.

Not every case where a critical line in a phase diagram turns into a
first-order transition implies the occurrence of a tricritical point: other possible phase diagram scenarios involve critical end points or bicritical points
(Fisher and Nelson, 1974). Instructive examples for phase diagrams involving
such special points can be found in antiferromagnets with uniaxial anisotropy
of weak or intermediate strength (fig. 21). A critical end point (CEP) occurs
if a line of critical points terminates at a first-order line that describes a
transition involving other degrees of freedom than those involved in the
ordering at the critical line. For example, in fig. 21b at the critical line

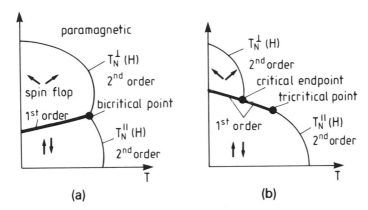

Fig. 21. Schematic phase diagrams of antiferromagnets with uniaxial anisotropy in an applied uniform magnetic field $H_\parallel$ in the direction parallel to the easy axis: case (a) shows the case of weak anisotropy, case (b) shows the case of intermediate anisotropy. In addition to the antiferromagnetic ordering of the spin components in the direction of the easy axis, a spin–flop ordering of the transverse components also occurs. In case (a) both transitions $T_\parallel(H_\parallel)$ and $T_\perp(H_\parallel)$ are of second order and meet in a bicritical point. For intermediate strength of the anisotropy the line $T_\perp(H_\parallel)$ does not end at a bicritical point, but rather at a critical end point (CEP) at the first-order transition line. Then a tricritical point also appears where the antiferromagnetic transition $T_\parallel(H_\parallel)$ becomes first order. For very strong anisotropy, the spin–flop phase disappears altogether, and a phase diagram as shown in fig. 20c results.

$T_N^\perp(H_\parallel)$ the transverse (xy) components of the spins order antiferromagnetically, while the z-components in that phase are disordered (and thus have a uniform magnetization, induced by the field $H_\parallel$: the resulting spin arrangement then is called the "spin flop" (SF) phase). The line $T_N^\parallel(H_\parallel)$, on the other hand, describes the transition into the antiferromagnetically (AF) ordered phase of the z-component of the spins. When the line $T_N^\perp(H_\parallel)$ hits the first-order line, this means that in configuration space a completely different minimum of the free energy hypersurface (describing the AF state) takes over. At the point (CEP) where the disordered and SF phases coexist with the AF phase, no special critical behavior is expected, however, the critical behavior at the CEP being still the same as along the whole line $T_N^\perp(H_\parallel)$. A different situation, however, occurs when the anisotropy is weak enough such that the SF phase exists for high enough temperatures, and the line $T_N^\perp(H_\parallel)$ meets the AF ordering in a part of the phase diagram where the line $T_N^\parallel(H_\parallel)$ is still describing a second-order transition: then the phase transition topology changes (fig. 21a) and one encounters a bicritical point (Fisher and Nelson, 1974). We can describe this situation in the framework of Landau's theory by writing down the Landau free energy functional for an anisotropic m-vector model, i.e., instead of eq. (14)

we write

$$\frac{1}{k_B T} \mathcal{F}[\boldsymbol{\phi}(x)] = \frac{F_o}{k_B T} + \int dx \left\{ \frac{1}{2} r_1(p)\phi_1^2(x) + \ldots \frac{1}{2} r_m(p)\phi_m^2(x) \right.$$
$$\left. \frac{u}{4}[\boldsymbol{\phi}^2(x)]^2 + \frac{R^2}{2d}[(\nabla\phi_1)^2 + \ldots (\nabla\phi_m)^2] \right\},$$

$$(113)$$

where for simplicity both the fourth order term and the gradient terms have been taken fully isotropic. If we have $r_1(p) = r_2(p) = \ldots = r_m(p)$, we would have the fully isotropic m-vector model whose exponents were mentioned in eqs. (59) and (60). We now consider the case where some of these coefficients may differ from each other, but depend on a parameter p which has the character of a non-ordering field (such as a longitudinal uniform field $H_\parallel$ has for antiferromagnets, fig. 21). The nature of the ordering will be determined by the term $r_i\phi_i^2$ for which the coefficient r_i changes sign at the highest temperature. Suppose this is the case for $i = 1$ for $p < p_b$; we then have a one-component ordering which sets in at $T_{c1}(p)$ given by $r_1(p, T) = 0$ [this corresponds to the line $T_N^\parallel(H_\parallel)$!]. The other components ϕ_i for $i > 1$ are then "secondary order parameters" just as the uniform magnetization would be or other non-ordering fields, cf. eqs. (96)–(100). If, however, for $p > p_b$ the coefficients $r_2(p, T) = r_3(p, T)$ vanish at a higher critical temperature $T_{c2}(p)$, the components ϕ_2 and ϕ_3 drive the transition as primary order parameters (here we have anticipated that the SF-ordering at $T_N^\perp(H_\parallel)$ in fig. 21 is a two-component ordering.) The point $p = p_b$, $T_{c1}(p) = T_{c2}(p) = T_b$ then is called a bicritical point.

While in Landau's theory the exponents do not depend on the number of components m of the order parameter, and hence are the same along both lines $T_{c1}(p)$ and $T_{c2}(p)$ and at T_b, this is no longer true if one considers fluctuations. Renormalization group theory (Fisher and Nelson, 1974; Fisher, 1975; Mukamel et al., 1976) shows that apart from this change of critical exponents there is also one additional exponent, the "crossover exponent" φ, describing the singular approach of phase transition lines towards the multicritical point, as well as the change of critical behavior from one universality class to the other. It is advisable to define scaling axes (t, g) which are perpendicular (t) or parallel (g) to the critical line $T_c(p)$ at the multicritical point (p_m, T_m), see fig. 22. (One calls all such special points like bicritical, tricritical etc. "multicritical".) For all $p < p_m$, the same type of critical behavior occurs (as it should do according to the universality principle!) but the region in the $T-p$-plane where it actually can be observed shrinks to zero smoothly as $p \to p_m$. Both the critical line and the center of the crossover region can be expressed in terms of the crossover exponent φ (Riedel and Wegner, 1969),

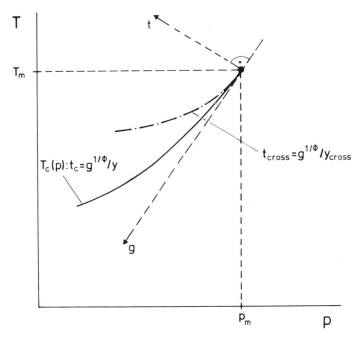

Fig. 22. Schematic phase diagram of a system exhibiting crossover between "ordinary" critical phenomena along the line $T_c(p < p_m)$ and the multicritical point $p = p_m$, $T_m = T_c(p_m)$. Considering the approach to the critical line along an axis parallel to the T-axis one will observe multicritical behavior as long as one stays above the dash-dotted curve $t_{cross} = g^{1/\varphi}/y_{cross}$ describing the center of the crossover region. Only in between the dash-dotted curve and the critical line $T_c(p)$ (full curve) the correct asymptotic behavior for $p < p_m$ can be seen.

$$t_c = \frac{g^{1/\varphi}}{y_c}, \qquad t_{cross} = \frac{g^{1/\varphi}}{y_{cross}}, \tag{114}$$

y_c, y_{cross} being constants. The singular part of the free energy then becomes near the multicritical point

$$F^{sing}(T, H, p) = t^{2-\alpha_m} \tilde{F}\left(Ht^{-(\beta_m+\gamma_m)}, g^{-1/\varphi}t\right), \tag{115}$$

where α_m, β_m, γ_m, ... are the exponents at the multicritical point (T_m, p_m) and $\tilde{F}(x, y)$ is a scaling function. This function has at $y = y_c$ a singularity described by critical exponents α, β, γ, ... characteristic of the universality class at the critical line, e.g. $\tilde{F}(0, y) \propto (y - y_c)^{2-\alpha}$, while for $y \ll y_c$ the y-dependence of $\tilde{F}(x, y)$ can be neglected, $F_{sing} \propto t^{2-\alpha_m}$ for $t \gg t_{cross}$. For a tricritical point in Landau's theory, $\varphi = 1/2$.

At this point, we note that while the phase diagram of fig. 21(a) with a bicritical point where lines of $m = 2$-component ordering and $m = 1$-

component ordering meet is perfectly meaningful at $d = 3$ (the bicritical point then has Heisenberg character, $m = 3$), it is unclear whether this phase diagram with $T_b > 0$ exists in two dimensions (Landau and Binder, 1981). Indeed, one believes that for $m \geq 3$ $T_c \equiv 0$ in $d = 2$ dimensions, the system being at its lower critical dimensionality. On the other hand, bicritical phenomena where two critical lines $T_{c1}(p)$, $T_{c2}(p)$ both in the Ising universality class ($m = 1$) meet, clearly should be possible, as well as bicritical points where lines $T_c(p)$ with $m = 2$ and suitable higher-order anisotropies (e.g. cubic anisotropy) meet with a line with $m = 1$.

A completely different type of multicritical point arises in Landau's theory when the coefficient of the gradient term vanishes in eq. (14). Just as in the case where u changes sign we need a higher order term $\frac{1}{6}v\phi^6$ to stabilize the free energy, we now need a higher order gradient term $\frac{1}{4}K_2(\nabla^2\phi)^2$ when the coefficient K_1 of the term $\frac{1}{2}K_1[\nabla\phi]^2$ is allowed to become negative,

$$\frac{1}{k_B T}\mathcal{F}\{\phi(x)\} = \frac{F_o}{k_B T} + \int dx \left\{\frac{1}{2}r\phi^2(x) + \frac{1}{4}u\phi^4(x) + \frac{1}{2}K_1[\nabla\phi(x)]^2 \right.$$
$$\left. + \frac{1}{4}K_2[\nabla^2\phi(x)]^2\right\}. \tag{116}$$

As in eqs. (35)–(44), we obtain from eq. (116) the wavevector-dependent susceptibility which now be comes [$r = r't = r'(T/T_c - 1)$]

$$k_B T \chi(q) = \left(r + K_1 q^2 + K_2 q^4\right)^{-1} \tag{117}$$

If $K_1 < 0$ the first divergence of $\chi(q)$ no longer occurs for $T = T_c(t = 0)$ and $q = 0$, but rather a divergence occurs at a higher temperature $t_c = K_1^2/4K_2 r'$ at $q^* = \sqrt{-K_1/2K_2}$, where $\chi(q)$ is maximal. Writing $t' = t - t_c$, eq. (117) can be rewritten as

$$k_B T \chi(q) = \left[r't' + K_2(q^2 - q^{*2})^2\right]^{-1} \approx (r't')^{-1}\left[1 + (q - q^*)^2\xi^2\right]^{-1},$$
$$q \to q^*, \tag{118}$$

where $\xi^2 = 4q^{*2}K_2/r't'$. Thus one finds that $k_B T\chi(q^*)$ has again a Curie–Weiss-like divergence, $k_B T\chi(q^*) \sim t'^{-1}$, and long range order characterized by a wavevector q^* develops, i.e. a structure modulated with a wavelength $\lambda^* = 2\pi/q^*$. The correlation function corresponding to eq. (118) is, apart from power law prefactors,

$$\langle\phi(0)\phi(x)\rangle \propto \exp\left(-\frac{x}{\xi}\right)\cos(q^* \cdot x). \tag{119}$$

Now another multicritical point arises for the special case where $K_1 = 0$ (cf. fig. 23), and then eq. (117) yields a Lifshitz point (Hornreich et al., 1975)

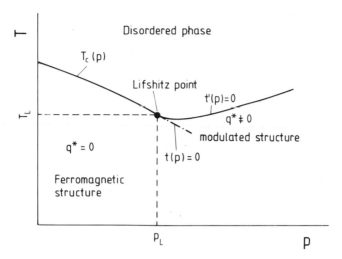

Fig. 23. Schematic phase diagram of a system where by a variation of a parameter p the coefficient $K_1(p)$ of the gradient energy $(1/2)K_1(p)(\nabla\phi)^2$ vanishes at a Lifshitz point $K_1(p_L) = 0$, $T_c(p_L) = T_L$. For $p < p_L$ one has a ferromagnetic structure, while for $p > p_L$ where $K_1(p) < 0$ one has a modulated structure, with a characteristic wavenumber $q^\star$ describing the modulation. For $p \to p_L$ from above one has $q^\star \to 0$ along the critical line $t'(p) = 0$.

$$k_B T \chi(\boldsymbol{q}) = (r + K_2 q^4)^{-1} = r^{-1}(1 + \xi^4 q^4)^{-1}, \quad \xi = \left(\frac{K_2}{r}\right)^{1/4} \propto t^{-1/4},$$

$$(120)$$

while at $T = T_L$ we have $r = r't = 0$ and hence $k_B T \chi_L(\boldsymbol{q}) = K_2^{-1} q^{-4}$. Thus the Landau theory predictions for the critical exponents of an (isotropic) Lifshitz point are (thermal properties remain identical to the normal critical behavior)

$$\alpha_L = 0, \quad \beta_L = \tfrac{1}{2}, \quad , \gamma_L = 1, \quad \delta_L = 3, \quad \nu_L = \tfrac{1}{4}, \quad \eta_L = -2. \quad (121)$$

Again one concludes that the scaling relations eqs. (80), (90) and (91) are satisfied, while the hyperscaling relation [eq. (93)] would only be satisfied at $d_o = 8$. Indeed, using the Lifshitz exponents in the Ginzburg criterion [eqs. (52)–(55)] one does find that the Landau description of Lifshitz points becomes self-consistent only for $d > 8$. Thus it is no surprise that the behavior at physical dimensionalities ($d = 2, 3$) is very different from the above predictions. In fact, in $d = 2$ one does not have Lifshitz points at non-zero temperature (Selke, 1992).

    Now a further complication that often arises in solids is that there exists an uniaxial anisotropy, and then one does not have an isotropic gradient energy term $\frac{1}{2} K_1(p)[\nabla \phi(x)]^2$ but rather one has

$$\text{gradient energy} = \frac{1}{2} K_{1\parallel}(p) \left[\frac{\partial \phi(x)}{\partial x_1} \right]^2$$
$$+ \frac{1}{2} K_{1\perp}(p) \left\{ \left[\frac{\partial \phi(x)}{\partial x_2} \right]^2 + \cdots \left[\frac{\partial \phi(x)}{\partial x_d} \right]^2 \right\} \quad (122)$$

For ordinary critical phenomena, such a spatial anisotropy is not very important — it gives rise to an anisotropy of the critical amplitude $\hat{\xi}_\parallel$, $\hat{\xi}_\perp$ of the correlation length in different lattice directions ($\xi_\parallel = \hat{\xi}_\parallel |t|^{-\nu}$, $\xi_\perp = \hat{\xi}_\perp |t|^{-\nu}$), while the critical exponent clearly is the same for all spatial directions. Of course, this is no longer necessarily true at Lifshitz points: There is no reason to assume that both functions $K_{1\parallel}(p)$, $K_{1\perp}(p)$ in eq. (122) vanish for $p = p_L$. Let us rather assume that only $K_{1\parallel}(p_L) = 0$ while $K_\perp(p_L) > 0$: this yields the uniaxial Lifshitz point (Hornreich et al., 1975). We then have to add a term $\frac{1}{4} K_{2\parallel}(p)[\partial^2 \phi(x)/\partial x_1^2]^2$ to eq. (122) to find

$$k_B T \chi(q) = [r + K_\perp q_\perp^2 + K_{2\parallel} q_\parallel^4]^{-1} = r^{-1}[1 + \xi_\perp^2 q_\perp^2 + \xi_\parallel^4 q_\parallel^4]^{-1} \quad (123)$$

with

$$\xi_\parallel = \left(\frac{K_{2\parallel}}{t} \right)^{1/4}, \qquad \xi_\perp = \left(\frac{K_{1\perp}}{t} \right)^{1/2}. \quad (124)$$

In this case the correlation lengths in parallel and perpendicular directions diverge with different exponents, $\xi_\parallel \propto t^{-\nu_\parallel}$, $\xi_\perp \propto t^{-\nu_\perp}$, with $\nu_\parallel = 1/4$, $\nu_\perp = 1/2$ in Landau's theory. One can generalize this assuming that the gradient energy coefficients $K_{1i}(p)$, $i = 1, \ldots, d$ vanish at $p = p_L$ in $k \leq d$ directions [$k = d$ is the isotropic Lifshitz point considered in eqs. (120) and (121), $k = 1$ the uniaxial Lifshitz point of eqs. (123) and (124)]. One can show with renormalization group methods that the lower critical dimensionality is $d_\ell = 2 + k/2$ for order parameter dimensionality $m \geq 3$ (Grest and Sak, 1978): thus for $d = 2$ and $m \geq 3$ one not only has $T_L = 0$ but at the same time the system is always below its lower critical dimensionality, i.e. one expects a power-law growth of correlations as $T \to 0$ similar to the case of one-dimensional isotropic spin models [eq. (65)].

We conclude this section by remarking that an additional critical exponent of interest here characterizes the vanishing of $q^\star$ as $p \to p_L$, i.e.

$$q^\star \propto \left(\frac{p}{p_L} - 1 \right)^{\beta^\star}. \quad (125)$$

Assuming $K_1(p) \propto p_L - p$ near $p = p_L$ we conclude from our above result $q^\star = \sqrt{-K_1/K_2}$ that $\beta^\star = 1/2$ in Landau's theory.

At $T < T_L$, one can also consider the transition from the ferromagneti-cally ordered structure to the structure with modulated order, or, more gen-erally speaking, consider commensurate–incommensurate (CI) transitions: rather than considering a modulation around the center of the Brillouin zone ($q = 0$), one can consider now more general orderings characterized by a superstructure Bragg spot at q_B in reciprocal space, assuming that q_B is commensurate with the substrate lattice (fig. 2c). A modulation of this com-mensurate structure now is described by two order parameter components ϕ_1, ϕ_2 in terms of an amplitude A and phase ρ, $\phi_1 = Ae^{i\rho}$, $\phi_2 = Ae^{-i\rho}$. Assuming that the amplitude A is constant while $\rho(x)$ may vary in x-direction in the considered uniaxial system, the free energy contribution is (Dzyaloshinskii, 1964; de Gennes, 1968; Bak and Emery, 1976; Selke, 1992)

$$\frac{\Delta\mathcal{F}}{k_BT} = \int dx \left\{ \gamma A^2 \left(\frac{d\rho}{dx}\right)^2 + 2\sigma A^2 \frac{d\rho}{dx} + 2\omega A^n \cos[n\rho(x)] \right\}, \quad (126)$$

where γ, σ and ω are phenomenological coefficients, and different cases $n = 1, 2, \ldots$ can be distinguished. Minimizing $\Delta\mathcal{F}\{\rho(x)\}$ leads to the Euler–Lagrange equation

$$\frac{d^2(n\rho)}{dx^2} + v \sin[n\rho(x)] = 0, \qquad v = \frac{n^2\omega A^{n-2}}{\gamma}. \quad (127)$$

One can show that the CI transition occurs at $v_c = n^2\pi^2\sigma^2/16\gamma^2$. The incommensurate phase (for $v < v_c$) consists of a periodic arrangement of regions where the phase is nearly constant, separated by "walls" where $\rho(x)$ increases by $2\pi/n$ (see fig. 24 for $n = 1$). One describes this structure as a domain wall lattice or "soliton lattice", whose lattice constant ℓ_d

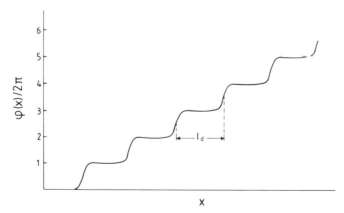

Fig. 24. Variation of the phase $\varphi(x)$ of an incommensurately modulated structure, character-ized by a lattice of domain walls periodically spaced at a distance ℓ_d.

diverges logarithmically on approaching the IC transition, $\ell_d \propto |\ln(v_c - v)|$. However, a consideration of fluctuations (domain wall meandering!) rather implies $\ell_d \propto |q^* - q_B|^{-1} \propto (v_c - v)^{-1/2}$ (Pokrovskii and Talapov, 1978).

2.3. Basic models: Ising model, Potts model, clock model, ANNNI model, etc.

While in the previous section we have emphasized the general phenomenological descriptions of phase transitions to work out concepts that hold for wide classes of systems, we are here concerned with the more specific modelling of adsorbate monolayers at crystal surfaces, having in mind the situation sketched in figs. 1 and 2. The simplest situation arises if the minima of the corrugation potential in fig. 1 are rather deep and the barriers in between them steeply rising. The locations of the minima of the corrugation potential thus form a well-defined lattice, at which the occupation probability density of the adatoms is sharply peaked. Then we may neglect deviations of the adatom positions from the sites of this "preferred lattice" altogether, introducing the *lattice gas model* which has as single degree of freedom, an occupation variable c_i with $c_i = 1$ if at site i there is an adatom while $c_i = 0$ if site i is empty. Multiple occupancy of the lattice sites is forbidden. The coverage θ of the monolayer then is given by a thermal average $\langle \ldots \rangle_T$ summed over all N lattice sites,

$$\theta = \frac{1}{N} \sum_{i=1}^{N} \langle c_i \rangle_T. \tag{128}$$

In addition to the binding energy ϵ (fig. 1) which we assume to be independent of temperature and coverage, there will be lateral interactions between adatoms (fig. 25). Although a pairwise interaction between adatoms at short distances is the simplest description of the energetics, the need for non-pairwise interactions may also arise. For example, in fig. 25 it is assumed that the energy of the occupied triangle shown there is $-2\rho_{nn} - 2\rho_{nnn} - \rho_t$, ρ_t being a three-body interaction term. Again these interaction parameters are assumed to be independent of temperature and coverage. The total configurational energy of the system then is

$$\mathcal{H} = -\epsilon \sum_{i=1}^{N} c_i - \sum_{i \neq j} \rho_{ij} c_i c_j - \sum_{i \neq j \neq k} \rho_t c_i c_j c_k \tag{129}$$

Here $\rho_{ij} = \rho_{nn}$, ρ_{nnn}, ρ_3, etc., when i, j are nearest, next nearest, third nearest neighbors, etc. The second sum on the right hand side of eq. (129) runs over all these pairs once, while the third sum runs over all appropriate triangles once. Of course, one could consider four-body interactions along

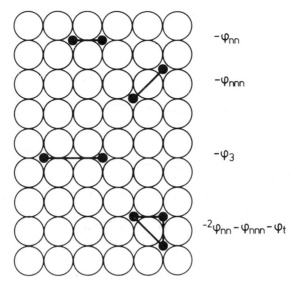

Fig. 25. Interaction energies on the square lattice, as were used for modelling of H on Pd(100). Pairwise interactions are considered between nearest (φ_{nn}), next nearest (φ_{nnn}) and third nearest neighbors (φ_3). In addition, three-body interactions (φ_t) around a nearest-neighbor triangle also are considered. From Binder and Landau (1981).

an elementary plaquette of the square lattice as well, and one may find other choices of interaction parameters relevant, or other lattice structures (see e.g. Roelofs, 1982 or Binder and Landau, 1989). We treat eq. (129) and fig. 25 merely as an illustrative example which shows how one proceeds in the general case. The average adsorption energy per lattice site U_{ads} for eq. (129) then is

$$U_{ads} = \frac{\langle \mathcal{H} \rangle_T}{N} = -\epsilon\theta - \frac{1}{N}\sum_{i\neq j}\rho_{ij}\langle c_i c_j\rangle_T - \frac{1}{N}\sum_{i\neq j\neq k}\rho_t\langle c_i c_j c_k\rangle_T \quad (130)$$

It is often convenient to transform from the canonical ensemble (T,θ being the fixed independent variables) to the grand-canonical ensemble, where T, μ are the independent variables, μ being the chemical potential of the adsorbed layer. In a physisorption experiment at high enough temperatures, the adsorbed layer is in thermal equilibrium with surrounding gas, and hence $\mu = \mu_{gas}$. The chemical potential of the layer can then be controlled by choosing the appropriate gas pressure (see fig. 4c). We transform to the grand-canonical ensemble by subtracting a term $\mu \sum_i c_i$, i.e.

$$\mathcal{H}' = \mathcal{H} - \mu\sum_i c_i = -(\epsilon + \mu)\sum_i c_i - \sum_{i\neq j}\rho_{ij}c_i c_j - \sum_{i\neq j\neq k}\rho_t c_i c_j c_k \quad (131)$$

Coverage and chemical potential are related via the adsorption isotherm,

$$\frac{\mu}{k_B T} = \frac{1}{(N k_B T)} \left(\frac{\partial F_{tot}}{\partial \theta} \right)_T, \tag{132}$$

where F_{tot} is the total free energy of the system which is given here as

$$F = -k_B T \ln \mathop{\mathrm{Tr}}_{c_i=0,1} \exp\left[-\frac{\mathcal{H}\{c_i\}}{k_B T} \right], \tag{133}$$

the trace (Tr) being taken over all configurations of the occupation variables $\{c_i\}$.

As is well known, the lattice gas model can be rewritten in terms of an equivalent Ising Hamiltonian $\mathcal{H}_{Ising}$ by the transformation $c_i = (1 - S_i)/2$, which maps the two choices $c_i = 0, 1$ to Ising spin orientations $S_i = \pm 1$. In our example this yields (Binder and Landau, 1981)

$$\mathcal{H}' = -\frac{1}{2} N(\mu + \epsilon) - \frac{1}{4} \sum_{i \neq j} \rho_{ij} - \frac{1}{8} \sum_{i \neq j \neq k} \rho_t + \mathcal{H}_{Ising}, \tag{134}$$

with

$$\mathcal{H}_{Ising} = -H \sum_{i=1}^{N} S_i - \sum_{i \neq j} J_{ij} S_i S_j - \sum_{i \neq j \neq k} J_t S_i S_j S_k. \tag{135}$$

The "magnetic field" H is related to the chemical potential μ as

$$H = -\left[\frac{\epsilon + \mu}{2} + \frac{\sum\limits_{j(\neq i)} \rho_{ij}}{4} + \sum_{j \neq k(\neq i)} \rho_t \right], \tag{136}$$

and the effective two- and three-spin exchange constants J_{ij}, J_t are given by

$$J_{ij} = \frac{\rho_{ij}}{4} + \sum_{k(\neq i, j)} \rho_t, \qquad J_t = -\frac{\rho_t}{8}. \tag{137}$$

The coverage θ then is simply related to the magnetization m of the Ising magnet,

$$\theta = \frac{1 - \langle m \rangle_T}{2}, \qquad \langle m \rangle_T = \frac{1}{N} \sum_{i=1}^{N} \langle S_i \rangle_T. \tag{138}$$

The transformation to the (generalized) Ising model is useful since it clearly brings out the symmetries of the problem: eq. (135) is invariant under the transformation

$$H, \quad J_t, \quad \{S_i\} \rightarrow -H, \quad -J_t, \quad \{-S_i\} \tag{139}$$

which transforms θ into $1 - \theta$ [via eq. (138)]. Thus the phase diagrams for positive and negative values of J_t are related: the phase diagram for $-J_t$ is obtained from the phase diagram for $+J_t$ by taking its mirror image around the axis $\theta = 1/2$ in the (T, θ) plane. If $J_t = 0$, i.e. for a model with only pairwise interactions, the phase diagram must possess therefore perfect mirror symmetry around the line $\theta = 1/2$. The adsorption isotherm, eq. (132) then is antisymmetric around the point $\theta = 1/2$, $\mu = \mu_c$, where μ_c is the chemical potential corresponding to $H = 0$ [cf. eq. (136)].

In the non-interacting case (or for "infinite temperature") the model is analyzed very simply: $\langle c_i c_j \rangle_{i \neq j} = \theta^2$, $\langle c_i c_j c_k \rangle = \theta^3$, and thus a simple polynomial results for U_{ads},

$$U_{ads}(T \to \infty) = -\epsilon\theta - \frac{1}{2}\theta^2 \sum_{i(\neq j)} \rho_{ij} - \frac{1}{3}\theta^3 \sum_{i \neq j (\neq k)} \rho_t. \tag{140}$$

Since in this limit the "magnetization process" of the Ising model is just given by the Brillouin function ($h \equiv H/k_B T$ remains non-zero)

$$m = \tanh h, \qquad h = \frac{1}{2}\ln\left[\frac{1+m}{1-m}\right] = \frac{1}{2}\ln\left[\frac{1-\theta}{\theta}\right] \tag{141}$$

one obtains the well-known Langmuir isotherm (Zangwill, 1988)

$$\frac{\mu + \epsilon}{k_B T} = \ln\left[\frac{\theta}{1-\theta}\right]. \tag{142}$$

In the presence of the lateral interactions, both adsorption isotherms and adsorption energies can be calculated conveniently and accurately from Monte Carlo simulations (Binder and Landau, 1981, 1989; Binder 1976, 1979a, b, 1984a, 1992a; Binder and Heermann, 1988). Figures 26 and 27 give some examples, and fig. 28 shows some examples of phase diagrams computed for such models.

We now discuss the critical behavior of the various transitions that one encounters in the lattice gas model. Only the unmixing critical point that arises for purely attractive interactions and the order–disorder transitions to the c(2×2) structure on the square lattice (fig. 28a, b) have a one-component order parameter and thus belong to the same universality class as the nearest-neighbor Ising model solved by Onsager (1944). In fact, once the consideration of the ground states of a lattice gas model with a specific set of interactions has yielded insight into the ordered structures that need to be considered (e.g. for the lattice gas model on the square lattice fig. 29 presents such ground state phase diagrams), one can then use the mass density wave method described in eqs. (27)–(31) to construct the corresponding structure that the "Ginzburg–Landau–Wilson"-Hamiltonian for this model must have, and thus arrive at the corresponding assignment

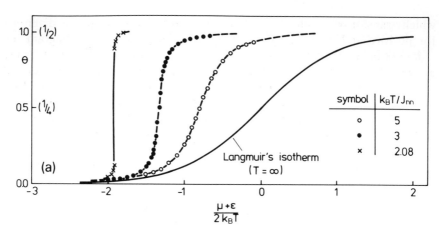

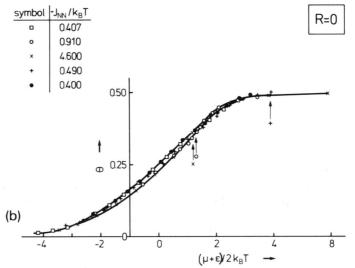

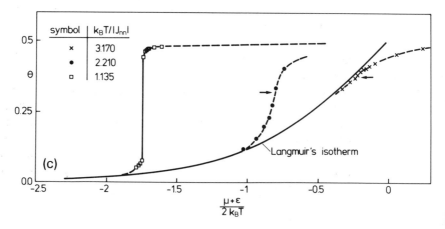

Table 1

Classification of continuous order–disorder transitions of commensurate superstructures in adsorbed monolayers at surfaces (from Schick, 1981).

Universality class and critical exponents	Ising	XY with cubic anisotropy	3-state Potts	4-state Potts
α	0 (log)		1/3	2/3
β	1/8	non-universal	13/9	7/6
γ	7/4	non-universal	13/9	7/6
ν	1		5/6	2/3
δ	15	15	14	15
η	1/4	1/4	4/15	1/4
Substrate symmetry				
Skew (p 1 mm) or rectangular (p 2 mm)	(2×1) (1×2) $c(2\times2)$			
Centered, rectangular (c 2 mm) or square (p 4 mm)	$c(2\times2)$	(2×2) (1×2) (2×1)		
Triangular (p 6 mm)			$(\sqrt{3}\times\sqrt{3})$	(2×2)
Honeycomb (p 6 mm)	(1×1)			(2×2)
Honeycomb in a crystal field (p 3 ml)			$(\sqrt{3}\times\sqrt{3})$	(2×2)

whether a second-order transition is possible and to which universality class it belongs (Schick, 1981). Table 1 presents a catalogue of transitions, which are of second order, and also lists the corresponding predictions for the values of the critical exponents which one believes are known exactly (Baxter, 1982; den Nijs, 1979; Cardy, 1987; Nienhuis, 1987). It is seen that only a relatively small number of structures qualify as candidates for second-order disorder transitions. Other structures with larger unit cells (fig. 29) readily can be obtained from lattice gas models, but one expects that either they show direct first-order transitions into the disordered phase, or they

Fig. 26. Adsorption isotherms of the lattice gas model on the square lattice with (a) only nearest neighbor attractive interaction, and (b) only nearest neighbor repulsive interaction, and (c) nearest neighbor repulsion and next-nearest neighbor attraction of the same strength. Temperature is always measured in units of the (absolute value) of the nearest neighbor exchange energy of the corresponding Ising Hamiltonian [eqs. (135), (137)]. The Langmuir isotherm [eq. (142)] is included for comparison. Second-order phase transitions from the disordered phase to the $c(2\times2)$ structure are indicated by arrows. Two-phase coexistence regions between island of the $c(2\times2)$ structure and the lattice gas show up as vertical positions of the adsorption isotherms in (a) and (c), respectively. From Binder and Landau (1980, 1981).

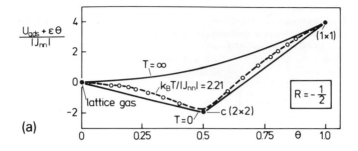

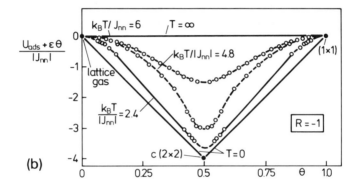

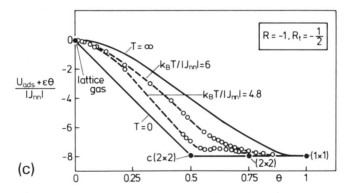

Fig. 27. Adsorption energy plotted versus coverage for $R \equiv J_{nnn}/J_{nn} = -1/2$, $R_t \equiv J_t/J_{nn} = 0$ (a), $R = -1$, $R_t = 0$ (b) and $R = -1$, $R_t = -1/2$ (c). From Binder and Landau (1981)

Fig. 28. Phase diagram for the lattice gas on the square lattice with (a) only nearest neighbor attractive interaction and (b) only nearest neighbor repulsive interaction and (c) with $R = -1$, $R_t = -1/2$ corresponding to case (c) of fig. 27. From Binder and Landau (1980, 1981).

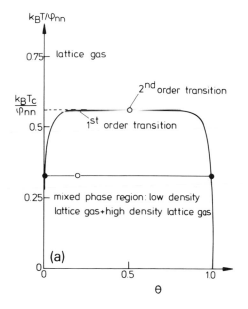

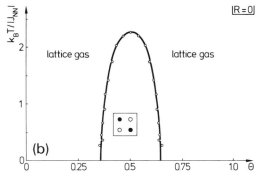

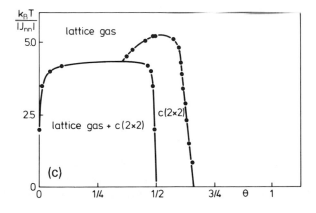

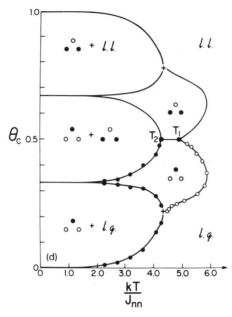

Fig. 28 (contd.). Case (d) shows the phase diagram for the lattice gas on the triangular lattice with nearest neighbor repulsion and next-nearest neighbor attraction, for $J_{nnn}/J_{nn} = -1$, in the coverage-temperature plane. For $\theta = 0.5$ a Kosterlitz–Thouless transition occurs at T_1 and a commensurate–incommensurate transition at T_2. Two commensurate $\sqrt{3} \times \sqrt{3}$ phases, with ideal coverages of 1/3 and 2/3, respectively, occur whose order–disorder transition belongs to the class of the three-states Potts model. The crosses denote Potts tricritical points, where two-phase regions between these commensurate phases and lattice gas ($l.g.$) or lattice liquid ($l.l.$) form. From Landau (1983).

may also in certain circumstances exhibit commensurate–incommensurate transitions.

The latter situation in fact is predicted for the so-called ANNNI (axial next-nearest neighbor Ising) model (Selke, 1988, 1992). This Ising model has a competing interaction $J_2 < 0$ in one lattice direction only, and thus the Hamiltonian is

$$\mathcal{H}_{ANNNI} = -J_o \sum_{i_x, i_y} S(i_x, i_y) S(i_x + 1, i_y) - J_1 \sum_{i_x, i_y} S(i_x, i_y) S(i_x, i_y + 1)$$
$$- J_2 \sum_{i_x, i_y} S(i_x, i_y) S(i_x, i_y + 2) \tag{143}$$

We assume here both $J_o > 0$ and $J_1 > 0$. Then the model has a ferromagnetic ground state for $\kappa \equiv -J_2/J_1 < 1/2$, while for $\kappa > 1/2$ the ground state is a structure where two rows of up-spins (the rows are oriented in x-direction) alternate with two rows of down spins. Hence along the y-axis

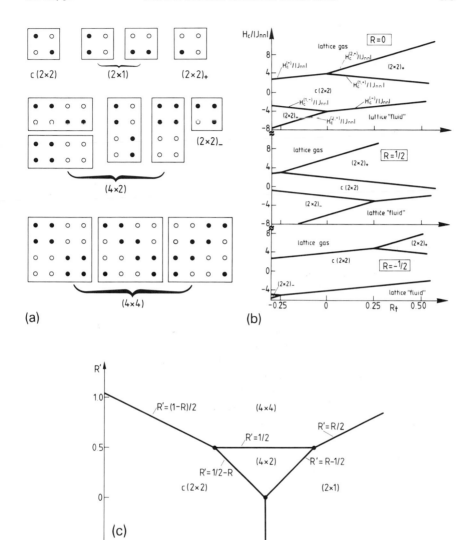

Fig. 29. (a) Unit cells of various overlayer structures on the square lattice with coverage $\theta = 1/2$ [c(2×2), (2×1), (4×2) and (4×4)], $\theta = 1/4$ [(2×2)$_+$] and $\theta = 3/4$ [(2×2)$_-$]. (b) Ground state phase diagram of the square lattice gas for $J_3 = 0$ and three choices of $R = J_{nnn}/J_{nn}$. Here $R_t \equiv J_t/J_{nn}$. (c) Ground state phase diagram for $J_t = 0$, $H = 0$ (i.e., for $\theta = 1/2$) in the plane of variables $R' = J_3/J_{nn}$ and $R = J_{nnn}/J_{nn}$. From Binder and Landau (1981).

one has a spin sequence $\ldots + + - - + + - - \ldots$ This structure is abbreviated as $\langle 2 \rangle$ in the literature. Figure 30 depicts the phase diagram as it was obtained from transfer matrix calculations (Beale et al., 1985). The

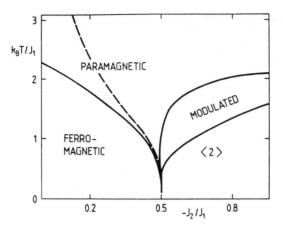

Fig. 30. Phase diagram of the two-dimensional ANNNI model. The broken curve in the paramagnetic phase is the "disorder line": below this line, the correlation function has a simple ferromagnetic exponential decay (uniform in sign), while above this line an oscillatory decay of the type of eq. (119) is found. From Beale et al. (1985).

transition from the paramagnetic to the modulated phase is believed to be of Kosterlitz–Thouless (1973) character, see sect. 2.4, while the transition from $\langle 2 \rangle$ to the modulated phase is a commensurate–incommensurate transition of the Pokrovskii–Talapov (1978) type.

While the standard lattice gas and Ising models (including the ANNNI model) start out from a two-state description of each lattice site ($S_i = \pm 1$), we have already mentioned the Potts model (Potts, 1952; Wu, 1982) where each site may be in one of q discrete states, with $q = 3, 4, \ldots$, see eq. (22). For a nearest-neighbor ferromagnetic interaction, the critical temperature is known exactly as (Wu, 1982; Kihara et al., 1954)

$$\frac{J}{k_B T_c} = \ln(1 + \sqrt{q}), \qquad q = 2, 3, 4, \ldots \tag{144}$$

For $q > 4$, however, the transition is known to be of first order and one can obtain exactly the latent heat at the transition (Baxter, 1973)

$$\frac{U_c^+ - U_c^-}{J} = 2(1 + \sqrt{q}^{-1}) \tanh \frac{\theta}{2} \prod_{n=1}^{\infty} (\tanh n\theta)^2, \qquad \theta = \text{arccosh} \frac{\sqrt{\theta}}{2}, \tag{145}$$

while (Kihara et al., 1954)

$$\frac{U_c^- + U_c^+}{2J} = -1 - \frac{1}{\sqrt{q}}. \tag{146}$$

A variant of the Potts model [eq. (22)] is the "vector Potts model" or "clock model" (Wu, 1982; José et al., 1977)

$$\mathcal{H}_{\text{clock}} = -J \sum_{\langle i,j \rangle} \cos\left[\frac{2\pi}{q}(S_i - S_j)\right], \qquad S_i = 1, 2, \ldots, q \qquad (147)$$

While for $q = 2$ this is still identical to the Ising model and for $q = 3$ identical to the standard Potts model, a different behavior results for $q \geq 4$. In particular, the exponents for $q = 4$ can be non-universal for variants of this model (Knops, 1980). For $q > 4$ there occurs a Kosterlitz–Thouless phase transition to a floating phase at higher temperatures, where the correlation function decays algebraically, and a IC-transition to a commensurate phase with q-fold degenerate ordered structure at lower temperatures (Elitzur et al., 1979; José et al., 1977).

The clock model with $q = 4$ is also called the Z(4) model. It can be represented in terms of *two* Ising spins s_i, τ_i associated with each lattice site. In this form it is known as the Ashkin–Teller model (1943)

$$\mathcal{H} = -J \sum_{\langle i,j \rangle} (s_i s_j + \tau_i \tau_j) - \Lambda \sum_{\langle i,j \rangle} s_i \tau_i s_j \tau_j \qquad (148)$$

It is believed to have non-universal critical exponents, depending on the ratio Λ/J. A related exactly solvable model, the 8-vertex model (Baxter, 1971, 1972) can be written rather similarly in spin representation as (Kadanoff and Wegner, 1971)

$$\mathcal{H} = -J \sum_{\langle s_i, s_j \rangle_{\text{nnn}}} s_i s_j - \Lambda \sum_{\langle i,j,k,l \rangle_{\text{plaquettes}}} s_i s_j s_k s_l \qquad (149)$$

eq. (149) leads to a singular free energy (Baxter, 1971)

$$\frac{F_{\text{sing}}}{k_{\text{B}} T} \propto \begin{cases} \cot\dfrac{\pi^2}{2\mu} |T - T_{\text{c}}|^{\pi/\mu}, & \dfrac{\pi}{\mu} = \text{non-integer} \\[3mm] \dfrac{1}{2\pi} |T - T_{\text{c}}|^{2m} \ln |T - T_{\text{c}}|, & \dfrac{\pi}{\mu} = m = \text{integer}, \end{cases} \qquad (150)$$

where $\cos\mu = \{[1 - \exp(-4\Lambda)/k_{\text{B}}T)]/[1 + \exp(-4\Lambda/k_{\text{B}}T)]\}$.

This is a celebrated example of non-universal critical behavior, since the specific heat exponent depends via μ on the coupling constant Λ. In the framework of the renormalization group theory of critical phenomena, one can understand this case as follows: if $\Lambda = 0$, eq. (149) splits into two uncoupled Ising models with two sublattices A,B with lattice spacing $\sqrt{2}a$ (the next-nearest neighbor distance). Denoting by ϵ^A the energy density of sublattice A and by ϵ^B the energy density of sublattice B, the four spin interaction term in eq. (149) is simply written as an energy–energy coupling, $\Lambda \sum_r \epsilon_r^A \epsilon_r^B$. One can show that such a perturbation is a "marginal" operator in a renormalization group sense [unlike "irrelevant" operators which simply would yield a correction to scaling, cf. eq. (95)]. Marginal operators lead to non-universal critical behavior (Domb and Green, 1976).

While it is not clear whether the Ashkin–Teller-model [eq. (148)] or the 8-vertex model [eq. (149)] have an experimental realization in adsorbed layers, the lattice gas model with repulsive interactions between nearest and next-nearest neighbors for $R = J_{nnn}/J_{nn} > 1/2$ also has non-universal behavior (Krinsky and Mukamel, 1977; Domany et al., 1978). This model has an ordered structure of (2×1) type (fig. 19, fig. 29a) and belongs to the class of the XY model with cubic anisotropy, which also acts as a "marginal operator". As pointed out in table 1, this structure is expected to be realized in many cases. Consequently, prediction of the R-dependence of the exponents of this model (which cannot be solved exactly) has become a challenge for real space renormalization group methods (Nauenberg and Nienhuis, 1974), high temperature series extrapolation (Oitmaa, 1981), transfer matrix techniques (Nightingale, 1977), Monte Carlo renormalization group (MCRG) techniques (Swendsen and Krinsky, 1979) and finite size scaling analyses of Monte Carlo data (Binder and Landau, 1980; Landau and Binder, 1985). Figure 31 reproduces an example taken from Landau and Binder (1985). We do not go into the details of these computational methods here, as they have been reviewed in detail elsewhere (Binder and Heermann, 1988; Binder and Landau, 1989; Privman, 1990).

We now return to the clock model [eq. (147)] and mention a variant called the "chiral clock model". Particularly the 3-state chiral clock (CC_3) model has been studied in detail (Ostlund, 1981; Huse, 1981; Schulz, 1983; Haldane et al., 1983). Its Hamiltonian is

$$\mathcal{H} = -J_o \sum_{\substack{\text{intralayer}}} \cos\left[\frac{2\pi}{3}(s_i - s_j)\right] - J_1 \sum_{\substack{\text{interlayer}}} \cos\left[\frac{2\pi}{3}(s_i - s_j + \Delta)\right],$$

$$s_i = 0, 1, 2, \qquad (151)$$

The phase diagram of this model has some similarity with the ANNNI model: varying the chirality parameter Δ in the range $0 \leq \Delta \leq 1/2$ the ground state is ferromagnetic, while for $1/2 < \Delta \leq 1$ the chiral ordering (in interlayer direction) $\ldots 012012012\ldots$ is stable. At $\Delta = 1/2$ the ground state is highly degenerate. For $0 < \Delta < 1/2$ the ferromagnetic phase melts into an incommensurate floating phase before a Kosterlitz–Thouless (1973) transition to the disordered phase occurs. For a more detailed review of this model and related models we refer to Selke (1992). But we wish to draw attention to a different concept for describing incommensurate phases, where one does not invoke a lattice description in terms of Ising or Potts spins as in the ANNNI model or CC_3 model but takes the description of fig. 2c and d more literally and allows for displacements r_i of the ith particle

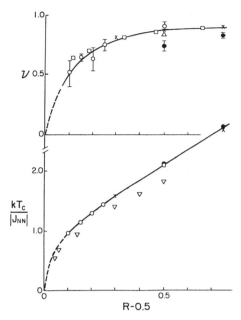

Fig. 31. Variation of the correlation length exponent ν (upper part) and the critical temperature T_c (lower part) of the lattice gas model with $\theta = 0.5$ and interactions between nearest and next-nearest neighbors with $R - 0.5$ (note that for $R \equiv J_{nnn}/J_{nn} < 0.5$ the structure of the model is the c(2×2) structure, while for $R > 0.5$ it is the (2×1) structure). Results of phenomenological finite size scaling renormalization group (Binder, 1981a) are shown by open circles (Landau and Binder, 1985), and the "data collapsing" finite size scaling method by an open triangle in the upper part of the figure (Binder and Landau, 1980). Crosses denote MCRG (Swendsen and Krinsky, 1979), open squares transfer matrix renormalization (Nightingale, 1977), solid circles series extrapolations (Oitmaa, 1981). Open circles in the lower part are due to real space renormalization (Nauenberg and Nienhuis, 1974). From Landau and Binder (1985).

away from the ith lattice site. The simplest model of this type is the Frenkel–Kontorova (1938) model ("FK model"). We discuss here its one-dimensional version only: a harmonic chain of particles in an external sinusoidal potential (which may represent the corrugation potential due to the substrate acting on the adatoms). Thus the potential energy U is (A,k are constants)

$$U = \frac{1}{2}k \sum_i (r_{i+1} - r_i - b)^2 + \frac{1}{2}A \sum_i \left(1 - \cos \frac{2\pi r_i}{a}\right) \qquad (152)$$

The harmonic potential (described by the spring constant k) favors an interparticle spacing b while the sinusoidal potential favors an interparticle spacing a: to balance these competing interactions, the particles may choose non-trivial positions already in the ground state. Requiring that the force

$\partial U / \partial r_i = 0$ for each particle yields

$$u_{i+1} - 2u_i + u_{i-1} = \frac{\pi}{2\ell_o^2} \sin 2\pi u_i, \quad u_i = \frac{r_i}{a}, \quad \ell_o = \left(\frac{ka^2}{2A} \right)^{1/2}. \quad (153)$$

Replacing differences by differentials eq. (153) is reduced to the sine-Gordon-equation (Frank and Van der Merwe, 1949a, b),

$$\frac{d^2 u}{dn^2} = \frac{\pi}{2\ell_o^2} \sin(2\pi u), \quad (154)$$

which is solved in terms of elliptic integrals. One finds that the commensurate phase is the ground state for small enough misfit, $\delta \equiv |b - a|/a \leq \delta_c = 2/(\ell_o \pi)$; otherwise the ground state of eq. (153) is described by a solution for $u(n)$ which closely resembles the picture drawn in fig. 24 for $\rho(x)/2\pi$, as expected, since eq. (154) is identical with eq. (127). For $\delta = \delta_c$, the solution of eq. (154) reduces to the well-known "domain-wall" or "kink" or "soliton" solution

$$u(n) = \frac{2}{\pi} \arctan \left[\exp \frac{\pi n}{\ell_o} \right], \quad (155)$$

to create such a wall, one has to imagine that one particle has to be deleted (if b is less than a) or added (if b is larger than a). The thickness of the wall is ℓ_o. For larger misfits, $\delta > \delta_c$, the ground state consists of a lattice ("soliton lattice") of regularly spaced domain walls of thickness $h\ell_o$, where h is given by $\ell_o \delta = 2E(h)/(\pi h)$, $E(h)$ being the complete elliptic integral of the second kind. The spacing ℓ_d of domain walls is given by $\ell_d = 2\ell_o h K(h)/\pi$, with $K(h)$ being the complete elliptic integral of the first kind, fig. 24. The separation between particles is then on average $a\ell_d/(\ell_d - 1)$. One can consider the quantity $\ell_d/(\ell_d - 1)$ as a "winding number". Since ℓ_d changes continuously with δ, the ground states in general are incommensurate with respect to the sinusoidal potential. At $\delta = \delta_c$ a continuous commensurate to incommensurate (CI) transition takes place, with $\ell_d \propto |ln(\delta - \delta_c)|$.

At non-zero temperatures, of course there is no sharp phase transition in an one-dimensional model with short-range interactions. One finds that already in the commensurate region ($\delta < \delta_c$) kink and antikink excitations appear via spontaneous thermal fluctuations (Brazovskii et al., 1977; Burkov and Talapov, 1980).

A generalization of the Frank and Van der Merwe model to two dimensions has been given by Pokrovskii and Talapov (1979). The incommensurate phase is described by an array of parallel domain walls running along a given axis (say, the y-axis) and crossing the whole sample from top to bottom (i.e., wall crossing is forbidden). At $T = 0$, the ground state is identical to the one-dimensional case as described above. At finite T, it behaves very differently because the interaction between walls decays exponentially with distance ℓ_d between walls. While this exponential tail governs the critical behavior at $T = 0$, it can be neglected for $T > 0$, replacing the interaction by a hard core repulsion. The problem then depends on the chemical potential μ (controlling coverage of the adsorbed layer) and the line tension $\gamma^\star$ of the walls; each wall is described by a harmonic Hamiltonian

$$\mathcal{H} = \gamma^\star \sum_{y=1}^{N_y} [X(y+1) - X(y)]^2 \tag{156}$$

where we have assumed that there are N_y particles in y-direction at positions $y = 1, 2, \ldots, N_y$ and $X(y)$ is the abscissa of the wall at ordinate y. One has to add to eq. (156) a term $\mu^\star N_w$, N_w being the number of walls, and $\mu^\star$ being the chemical potential of the walls, which is $\mu^\star = w_1 \pm ka(a-b) + \mu$, where w_1 is the excitation energy of a wall ($w_1 \approx \frac{4}{\pi} a \sqrt{kA}$), μ the chemical potential per adatom, and the two signs refer to light ($+$) or heavy ($-$) walls, depending whether the mass excess of the incommensurate structure relative to the commensurate one is positive ("heavy wall") or negative ("light wall"). As an example, we mention the CI-transition of Kr on graphite (Chinn and Fain, 1977; Larher, 1978) created by increasing the gas pressure where heavy walls are introduced into the commensurate $\sqrt{3} \times \sqrt{3}$R30° structure. The critical behavior at constant temperatures as a function of $\mu^\star$ is then given by (Pokrovskii and Talapov, 1979)

$$\frac{\pi}{\ell_d} \approx \sqrt{\frac{\mu_c^\star(T)}{k_B T} - \frac{\mu^\star}{k_B T}} \exp\left[\frac{\gamma^\star}{2k_B T}\right], \quad \frac{\mu_c^\star(T)}{k_B T} = 2\exp\left(-\frac{\gamma^\star}{k_B T}\right). \tag{157}$$

Equation (157) shows a square root divergence of the distance between walls at the CI-transition. For more details (including a discussion of the Novaco–McTague (1977) orientational instability on hexagonal substrates) we refer the reader to Villain's (1980) beautiful review.

As a final point of this section, we return to tricritical phenomena in $d = 2$ dimensions. The tricritical exponents are known exactly from conformal invariance (Cardy, 1987). For the Ising case, the results are (Pearson, 1980; Nienhuis, 1982)

$$\alpha_t = \frac{8}{9}, \quad \beta_t = \frac{1}{24}, \quad \gamma_t = \frac{37}{36}, \quad \delta_t = \frac{77}{3}, \quad \nu_t = \frac{5}{9}, \quad \eta_t = \frac{3}{20},$$

$$\varphi = \frac{4}{9}, \quad \beta_2 = \frac{1-\alpha_t}{\varphi} = \frac{1}{4}. \tag{158}$$

This set of exponents agrees with those of the hard square model (Huse, 1982). This model is defined as follows: consider placing hard squares of linear dimension $\sqrt{2}a$ on a square lattice of lattice spacing a, such that squares are allowed to touch but not overlap. At a critical coverage $\theta^\star \approx 0.37$ a second-order phase transition occurs from a disordered lattice gas of these hard squares to a structure with long range order of c(2×2) type (figs. 10, 29a). This phase transition can be considered as the $T \to 0$ limit of the lattice gas problem considered in eqs. (128)–(139) where one chooses a nearest-neighbor repulsive interaction $\rho_{nn} > 0$ only: in the limit

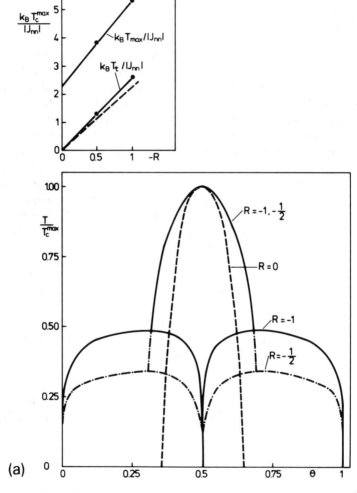

(a)

Fig. 32. (a) Phase diagram of the square lattice gas with nearest neighbor repulsion $J_{nn} > 0$ and next-nearest neighbor attraction $J_{nnn} < 0$, in the plane of variables temperature and coverage, for three choices of $R = J_{nnn}/J_{nn}$. Insert shows the variation of the maximum transition temperature (at $\theta = 1/2$) and of the tricritical temperature T_t with R. From Binder and Landau (1981).

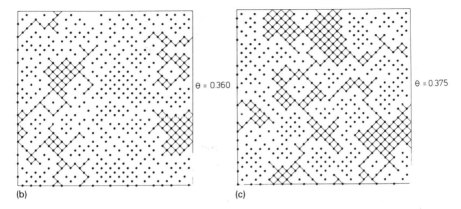

Fig. 32 (contd.). (b) Typical configurations of the hard square model at two values of θ, $\theta = 0.36$ (b) and $\theta = 0.375$ (c), for a lattice of linear dimension $L = 40$ and periodic boundary conditions. Points show the centers of the hard squares. The largest cluster of the c(2×2) structure is indicated by connecting the points. From Binder and Landau (1980).

$\rho_{nn}/T \to \infty$, μ/T finite an occupation of nearest neighbor sites becomes strictly forbidden, and a hard-square exclusion results. Thus this transition is the end-point of the phase diagram shown in fig. 28a. But at the same time, it is the end-point of a line of tricritical transitions obtained in the lattice gas model when one adds an attractive next-nearest neighbor interaction ρ_{nnn} and considers the limit $R \equiv \rho_{nnn}/\rho_{nn} \to 0$ (Binder and Landau, 1980, 1981; fig. 32).

Also the tricritical 3-state Potts exponents (for a phase diagram, see fig. 28c) can be obtained from conformal invariance (Cardy, 1987). But in this case the standard Potts critical exponents are related to an exactly solved hard core model, namely the "hard hexagon model" (Baxter, 1980), and not the tricritical ones. The latter have the values $\alpha_t = 5/6$, $\beta_t = 1/18$, $\gamma_t = 19/18$, $\delta_t = 20$, $\nu_t = 7/12$, $\eta_t = 4/21$, $\varphi = 1/3$, and $\beta_2 = 1/2$. Note that for $q = q_c = 4$ critical and tricritical exponents coincide (den Nijs, 1979).

2.4. Kosterlitz–Thouless transitions

In this section, we follow Young (1980b) and first focus on a variant of the two-dimensional XY model, namely the plane rotator model where each lattice site i carries an unit vector $(\cos\theta_i, \sin\theta_i)$ and the Hamiltonian depends only on the relative orientations of these vectors,

$$\mathcal{H} = -J \sum_{\langle i,j \rangle} \cos(\theta_i - \theta_j). \tag{159}$$

At very low temperatures spins at neighboring sites are strongly correlated, and hence one may expand the cosine keeping only the quadratic term. For long wavelength fluctuations one may also make a continuum approximation, replacing the θ_i by $\theta(x)$ and hence

$$\mathcal{H} = \frac{J}{2} \int dx \, [\nabla \theta(x)]^2 \tag{160}$$

If, finally, one neglects the fact that $\theta + 2\pi n$ is equivalent in eq. (159) to θ for n integer, one finds extending the range of integration over θ from $-\infty$ to $+\infty$ that the partition function can be written as functional integration involving gaussian integrals,

$$Z = \int_{-\infty}^{+\infty} \mathcal{D}\theta(x) \exp\left\{ -\frac{J}{2k_B T} \int dx \, [\nabla \theta(x)]^2 \right\}. \tag{161}$$

From this spin wave approximation (Wegner, 1967) one can also obtain the correlation function

$$G(x) = \langle \exp[i(\theta(x) - \theta(0))] \rangle = \exp\left\{ -\tfrac{1}{2} \langle [\theta(x) - \theta(0)]^2 \rangle \right\}, \tag{162}$$

where the harmonic character of the Hamiltonian eq. (160) was used, cf. eqs. (72), (73). Using equipartition as in eqs. (69)–(75), one concludes that for large x

$$\langle [\theta(x) - \theta(0)]^2 \rangle = \frac{2k_B T}{J} \int \frac{dk}{(2\pi)^2} \frac{1 - \exp(ik \cdot x)}{k^2} \approx \frac{k_B T}{\pi J} \ln \frac{x}{a}, \tag{163}$$

where the wavevector integral was cut off at $k \approx 1/a$, a being the lattice spacing. Equation (163) shows that there is no long range order (Mermin, 1968). Equations (162) and (163) imply a power law decay of the spin correlation function,

$$G(x) \approx x^{-\eta}, \quad x \to \infty, \qquad \eta = \frac{k_B T}{2\pi J}. \tag{164}$$

Thus the spin wave approximation predicts a line of critical points at all $T > 0$, each temperature being characterized by its own (non-universal) critical exponent η.

Of course, the approximations made by the spin wave theory are reasonable at very low temperatures only, and thus it is plausible that this line of critical temperatures terminates at a transition point T_{KT}, the Kosterlitz–Thouless (1973) transition, while for $T > T_{KT}$ one has a correlation function that decays exponentially at large distances. This behavior is recognized when singular spin configurations called vortices (fig. 33; Kawabata and Binder, 1977) are included in the treatment (Berezinskii, 1971, 1972). Because $\theta(x)$ is a multivalued function it is possible that a line integral such

t=1000

Fig. 33. Spin configuration of a 30×30 XY model at $k_B T / J = 0.01$ exhibiting various frozen-in vortices. This (non-equilibrium) configuration was prepared by choosing an initial state where all spins $S_i = (0, 0, 1)$ and then quenching the system to the considered temperature and following the time evolution for 1000 MCS per spin. From Kawabata and Binder (1977).

as $\oint \nabla \theta \cdot d\ell$ around a closed contour is non-zero and equal to $2\pi n$, where the (integer) n is called the winding number. Such a contour with $n \neq 0$ encloses at least one vortex. For such configurations, $\nabla \theta = 1/\rho$ where ρ is the distance from the core of the vortex. The energy of a single isolated vortex would be

$$E_{1 \text{ vortex}} = \frac{1}{2} J \int d\mathbf{x} \left[\nabla \theta(\mathbf{x}) \right]^2 = \pi J \int_a^L \rho \, d\rho \left(\frac{1}{\rho} \right)^2 = \pi J \ln \left(\frac{L}{a} \right), \quad (165)$$

where L is the linear dimension of the system. A single vortex therefore would cost an infinite amount of energy in the thermodynamic limit. However, the energy of a pair of opposite vortices (i.e., one with $n = +1$, one with $n = -1$) is finite, and is given by

$$E_{\text{vortex pair}} = 2\pi J \ln \left(\frac{\rho}{a} \right), \quad (166)$$

ρ being the separation of the pair. Thus one expects at low temperatures a small but non-zero density of tightly bound vortex–antivortex pairs. Kosterlitz and Thouless (1973) argue that at T_c these vortex pairs can unbind due to the gain in entropy. The entropy of a single isolated vortex in a $L \times L$ lattice is $2k_B \ln(L/a)$ and thus the free energy of an isolated vortex would be

$$F_{1 \text{ vortex}} = (\pi J - 2k_B T) \ln \left(\frac{L}{a} \right), \tag{167}$$

which is negative for $k_B T > \pi J/2$. Thus one estimates the critical temperature as $k_B T_{KT} = \pi J/2$.

Another quantity of interest is the stiffness S characterizing the free energy increase against a twist of the angle $\theta(x)$,

$$\Delta F = \frac{1}{2} S \int [\nabla \theta(x)]^2 \, dx. \tag{168}$$

For an XY-model at dimensionalities $d > d_\ell$ this coefficient vanishes at T_c with a power law $S \propto |t|^{2\beta - \eta\nu}$. Although in $d = 2$ the magnetization $\langle \cos \theta(x) \rangle \equiv 0$, the stiffness S is non-zero in the spin wave regime: in fact, comparison of eqs. (160) and (168) suggests $S = J$, independent of T. The Kosterlitz–Thouless (1973) theory implies that S is reduced from J at non-zero temperatures due to vortex–antivortex pairs, and that the equation that yields the transition temperature rather is

$$k_B T_{KT} = \frac{\pi S(T_{KT}^-)}{2}. \tag{169}$$

Therefore the ratio of the stiffness as $T \to T_{KT}^-$ to the transition temperature T_{KT} has the universal value $\pi/2$ (Nelson and Kosterlitz, 1977) and the exponent η as $T \to T_{KT}^-$ is also universal (Kosterlitz, 1974), $\eta(T_{KT}^-) = 1/4$.

We now recall that the classical planar rotator model may be used as a model of superfluid He^4, θ being the phase of the condensate wave function, S being related to the superfluid density ρ_s as $S = \rho_s(\hbar/m)^2$, m being the mass of a He^4 atom. Thus one can have superfluid–normal fluid transition in $d = 2$ dimensions, despite the lack of conventional long range order! This conclusion seems to be corroborated by experiments on He^4 films (Bishop and Reppy, 1978).

At this point we also note the relation to the Halperin–Nelson (1978)– Young (1978) theory of continuous melting in two dimensions via an unbinding of dislocation pairs. Writing the Hamiltonian in terms of the strain tensor $\epsilon_{\alpha\beta} = \frac{1}{2}[\partial u_\beta/\partial x_\alpha + \partial u_\alpha/\partial x_\beta]$,

$$H = \int dx \left[\frac{\lambda}{2}(\epsilon_{\alpha\alpha})^2 + \mu \epsilon_{\alpha\beta} \epsilon_{\alpha\beta} \right], \tag{170}$$

where λ is a Lamé coefficient, μ the shear elastic constant, and it is understood that indices occurring twice are summed over. Now the structure factor $S(k)$ is related to the displacement $u(x)$ by introducing Fourier components ρ_k of the density, $\rho_k = \sum_j \exp(ik \cdot x_j)$, $x_j = R_j + u_j$, R_j being the position of the jth lattice site at zero temperature,

$$S(k) = \frac{1}{N} \langle \rho_k \rho_{-k} \rangle = \sum_x \exp(ik \cdot x) C_k(x),$$

$$C_k(x) = \langle \exp\{ik \cdot [u(x) - u(0)]\} \rangle \tag{171}$$

The correlation $C_k(x)$ is the analogue of the correlation $G(x)$ in eq. (162). Within continuum elasticity theory $C_k(x)$ can be evaluated as treated already in eqs. (69)–(75). Thus $\langle [u(x) - u(o)]^2 \rangle$ diverges logarithmically with x, as eq. (163), and there is no true (positional) long range order, $\langle \rho_G \rangle = 0$ for all reciprocal lattice vectors G except $G = 0$ (Mermin, 1967). One can show (Halperin and Nelson, 1978; Nelson and Halperin, 1979) that $C_G(x) \propto r^{-\eta_G}$ with $\eta_G = [k_B T |G|^2/(4\pi)](3\mu + \lambda)/[\mu(2\mu + \lambda)]$.

Just as continuum elasticity theory is the analogue of the continuum version of the spin wave approximation, so dislocations are the analogue of vortices. The multivaluedness of $\theta(x)$ corresponds to replacing displacements $u(x)$ by $u(x)+na$, where a is a lattice vector. Consequently, the integral $\oint (\partial/\partial x_i) u \, d\ell_i$ around a closed loop can equal a lattice vector, so that it is not necessarily zero. One calls the resulting vector the "Burger's vector" of the dislocation. The energy of an isolated dislocation in the lattice would again be described by eq. (165) if one replaces J by $\mu(\mu + \lambda)a^2/[(2\pi^2)(2\mu + \lambda)]$, and at large distances the energy of a pair of dislocations with opposite Burger's vectors is given by eq. (166). The analogue of the stiffness S, which has a universal value as $T \rightarrow T_{KT}^-$ and is identically zero for $T > T_{KT}^-$, is the shear modulus here, which controls transverse fluctuations of the strain. Thus the low temperature phase although it lacks positional long range order is "solid" since the shear modulus is finite. The high temperature phase at $T > T_{KT}$ has a vanishing shear modulus but is no true liquid yet since it displays a power law decay of bond orientational correlation functions (Halperin and Nelson, 1978). This "hexatic phase" (as it is called for solids ordering at $T = 0$ in a triangular lattice structure) melts at a higher temperature (via disclination-pair unbinding) by a second Kosterlitz–Thouless transition into a true liquid where both positional and orientational correlations decay with finite correlation lengths. This Halperin–Nelson (1978) scenario of two-dimensional melting as a sequence of two (continuous) Kosterlitz–Thouless transitions, with a hexatic phase in between, is still debated since in most circumstances one finds instead a single first-order transition from solid (without positional long range order) to liquid.

We now return to the critical behavior of the Kosterlitz–Thouless (1973) transition in the XY model. One assumes that the small oscillations (spin waves) superimposed on top of any vortex configuration have the same energy as when there are no vortices. In this approximation the Hamiltonian splits in two independent parts, a spin wave part and a vortex part. This decoupling is rigorous for a variant of the XY model, the Villain (1975) Hamiltonian $\mathcal{H}_V$, which has the statistical weight

$$\exp\left[-\frac{\mathcal{H}_V}{k_BT}\right] \equiv \prod_{\langle ij\rangle} \sum_{m_{ij}=-\infty}^{+\infty} \exp\left\{-\frac{J}{2k_BT}(\theta_i - \theta_j - 2\pi m_{ij})^2\right\}. \qquad (172)$$

Equation (172) has the same periodicity as eq. (159) does, $\mathcal{H}\{\theta + 2\pi n\} = \mathcal{H}\{\theta\}$, and one can justify the replacement of eq. (159) by eq. (172) with renormalization group arguments (José et al., 1977). The vortex Hamiltonian is [cf. eq. (166)]

$$\mathcal{H}_{\text{vortex}} = 2\pi J \sum_{\langle i,j\rangle} \ln\left(\frac{|\boldsymbol{r}_i - \boldsymbol{r}_j|}{a}\right) n_i n_j + \ln y \sum_i n_i^2, \qquad (173)$$

where $n_i = \pm 1$ is the winding number of the ith vortex and $y = \exp(-E_c/k_BT)$ involves the "core energy" of a vortex. To avoid divergencies the condition $\sum_i n_i = 0$ is imposed. Since eq. (173) can be reinterpreted as the Hamiltonian of a two-dimensional classical Coulomb gas (Poisson's equation would yield a logarithmic interaction in $d = 2!$), this condition can be interpreted as charge neutrality. The number of vortices (or charges, respectively) is not fixed, so y is equivalent to the fugacity in a grand-canonical ensemble.

It turns out that concepts of dielectric media are helpful to describe such a system: so the effective interaction between a pair of opposite charges at distance r is not the bare interaction $2\pi J \ln(r/a)$, but rather it is screened by a distance-dependent "dielectric constant" $\epsilon(r)$ which must be calculated self-consistently. A renormalization group treatment (Kosterlitz, 1974; Young, 1978) shows that the quantity $K\{\ln(r/a)\} \equiv J/(k_BT\epsilon(r))$ vanishes above T_{KT} for $r \to \infty$ but behaves as

$$K(\infty) = \frac{2}{\pi} + c(-t)^{\bar{\nu}}, \qquad \bar{\nu} = \frac{1}{2}, \qquad (174)$$

for $T \lesssim T_{KT}(t \equiv T_{KT}/T - 1)$, where c is a non-universal quantity. Thus $k_BTK(\infty)$ is the stiffness S mentioned above. One also finds that the specific heat C and correlation length ξ have essential singularities,

$$C \propto \exp\left[-\frac{A_\pm}{|t|^{\bar{\nu}}}\right], \qquad \xi(t > 0) \propto \exp\left[\frac{B}{t^{\bar{\nu}}}\right], \qquad (175)$$

where $A_\pm$, B are other (non-universal) constants.

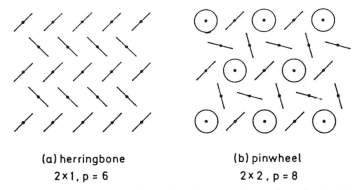

(a) herringbone
2×1, p = 6

(b) pinwheel
2×2, p = 8

Fig. 34. Herringbone (a) and pinwheel (b) orientational ordering of uniaxial diatomic molecules on a triangular lattice. The heavy bars represent planar rotators and the circles denote vacancies. p is the degeneracy of the ordering. From Mouritsen (1985).

There exists many generalizations and variants of the isotropic planar rotator model, eq. (159). Here we only mention the anisotropic planar rotor model (Mouritsen and Berlinsky, 1982; Harris et al., 1984)

$$\mathcal{H}_{APR} = -J \sum_{\langle i,j \rangle} \cos(2\theta_i + 2\theta_j - 4\phi_{ij}), \tag{176}$$

where $0 \leq \theta_i < \pi$ describe the rotor orientations, and ϕ_{ij} describes the angle of the vector connecting lattice points i, j. This model exhibits long range orientational order (fig. 34) and can be used to model N_2 molecules physisorbed in a commensurate $(\sqrt{3} \times \sqrt{3})$ overlayer on graphite.

2.5. Interfacial phenomena

We return here to the simple mean field description of second-order phase transitions in terms of Landau's theory, assuming a scalar order parameter $\phi(x)$ and consider the situation $T < T_c$ for $H = 0$. Then domains with $\phi_o = +\sqrt{-r/u}$ can coexist in thermal equilibrium with domains with $-\phi_o$ [eq. (16)]. We wish to consider the case where a domain with $\phi(x) = -\phi_o$ exists in the halfspace with $z < 0$ and a domain with $\phi(x) = +\phi_o$ in the other halfspace with $z > 0$ (fig. 35a), the plane $z = 0$ hence being the interface between the coexisting phases. While this interface is sharp on an atomic scale at $T = 0$ for an Ising model, with $\phi_i = -1$ for sites with $z < 0$, $\phi_i = +1$ for sites with $z > 0$ (assuming the plane $z = 0$ in between two lattice planes), we expect near T_c a smooth variation of the (coarse-grained) order parameter field $\phi(z)$, as sketched in fig. 35a. Within Landau's theory (remember $|\phi(x)| \ll 1$, $|\nabla \phi(x)| \ll 1$) the interfacial profile is described by

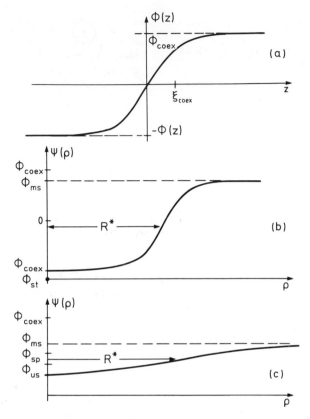

Fig. 35. (a) Order parameter profile $\phi(z)$ across an interface between two coexisting phases $\pm\phi_{coex}$, the interface being oriented perpendicular to the z-direction. (b) The radial order parameter profile for a marginally stable "critical droplet" in a metastable state which is close to the coexistence curve. (c) Same as (b) but for a state close to the spinodal curve, ϕ_{sp}. In (a) and (b) the intrinsic "thickness" of the interface is of the order of the correlation length ξ_{coex} whereas in (c) it is of the order of the critical droplet radius R^*. From Binder (1984b).

the equation [cf. eq. (35)]

$$r\phi(z) + u\phi^3(z) - \frac{R^2}{d}\frac{\mathrm{d}^2\phi}{\mathrm{d}z^2} = 0, \tag{177}$$

with boundary conditions

$$\phi(z \to \pm\infty) \to \pm\phi_o, \qquad \lim_{z \to \pm\infty}\left(\frac{\partial\phi}{\partial z}\right) = 0. \tag{178}$$

This problem is formally analogous to a problem in classical mechanics, namely the motion of a point particle in a potential $U(x) = -(rx^2/2 + ux^4/4)$, $m\ddot{x} = -\mathrm{d}U/\mathrm{d}x$, if we identify x with ϕ, z with t, and m with

R^2/d. The energy E of this problem is then chosen as $E = r^2/(4u)$ so the particle starts at $t = -\infty$ at the left potential well with $\dot{x} = 0$ and comes again to rest for $t \to +\infty$ at the right potential well. The velocity dx/dt (corresponding to the slope $d\phi/dz$ of the interfacial profile) is maximal for $x = 0$ ($\phi = 0$, respectively). Since the conservation of energy implies $E = U + m\dot{x}^2/2 = const$, multiplication of Newton's law by $\dot{x}$ and integration over time from $t = -\infty$ to t yields $m\dot{x}^2/2 = r(x^2 - x_o^2)/2 + u(x^4 - x_o^4)/4$. Analogously, we find the rescaled order parameter profile $\psi(Z) = \phi(z)/\phi_o$, $Z = z/\xi$, $\xi = R/\sqrt{-2rd}$ [eq. (37)] being here the correlation length at phase coexistence for $T < T_c$ $\{\phi_o = \phi_{coex} = \sqrt{-r/u}, k_B T \chi_T = (-2r)^{-1}\}$:

$$[1 - \psi^2(Z)]^2 = 4\left(\frac{d\psi}{dZ}\right)^2, \qquad \phi(z) = \phi_o \tanh\left(\frac{z}{2\xi}\right). \qquad (179)$$

Thus the thickness of the interfacial profile diverges in the same way as the correlation length does, when T approaches the critical temperature T_c.

The interfacial free energy is then defined as the excess contribution of a system such as that considered in fig. 35a, containing one interface, and a homogeneous system where $\phi(z) = \phi_o$ everywhere. Denoting the surface area of the interface by A, we thus obtain the interfacial tension f_{int} [eq. (4)] as

$$f_{int} = \frac{F_{int}}{k_B T A} = \lim_{L \to \infty} \int_{-L/2}^{+L/2} dz \left\{ \frac{1}{2} r \left[\phi^2(z) - \phi_o^2\right] + \frac{1}{4} u \left[\phi^4(z) - \phi_o^4\right] \right.$$
$$\left. + \frac{R^2}{2d}\left(\frac{d\phi}{dz}\right)^2 \right\} =$$
$$= 2u\phi_o^4 \xi \int_{-\infty}^{+\infty} dZ \left(\frac{d\psi}{dZ}\right)^2 = \frac{4}{3} u\phi_o^4 \xi \qquad (180)$$

Using the results for the critical behavior [eqs. (16), (38)] we now find $f_{int} = const\, R(1 - T/T_c)^{3/2}$. Thus the interfacial tension vanishes at T_c. We define an associated critical amplitude $\hat{f}_{int}$ and exponent μ as

$$f_{int} = \hat{f}_{int}(-t)^\mu, \qquad (181)$$

with $\mu = 3/2$ in Landau's theory. Using a generalized Landau theory one can show (Fisk and Widom, 1969) that $f_{int} \propto f_{sing}(\phi_o)\xi$, and since the singular part of the free energy scales as $f_{sing}(\phi_o) \propto (-t)^{2-\alpha}$, one obtains Widom's (1972) scaling law, $\mu = 2 - \alpha - \nu = (d - 1)\nu$. One can understand this relation by a similar plausibility argument as was used for justifying the hyperscaling relation, eq. (93): again we divide our systems in cells of size ξ^d, to obtain quasi-independent degrees of freedom describing the "cell-spin" orientations. While the total free energy was $F_{bulk}^{sing} \propto L^d/\xi^d$, an excess free energy due to an interface is expected only in a layer of thickness ξ, which contains the interface and contributes $(L/\xi)^{(d-1)}$ cells. Since in this L^d geometry $A = L^{d-1}$, we have $F_{int}/k_B T \propto L^{d-1}/\xi^{d-1}$, $f_{int} = F_{int}/(k_B T A) \propto \xi^{-(d-1)} \propto (-t)^{(d-1)\nu}$.

This coarse-graining can also be used to justify the "drumhead model" of an interface, cf. fig. 6, where on a more macroscopic scale the internal structure of the interface is disregarded, and one is more interested in large scale fluctuations of the local position $z = h(x, y)$ of this interface. In this "sharp kink" approximation the interface is described similarly to an elastically deformable membrane.

A basic concept is then the "interfacial stiffness" and the description in terms of the "capillary wave Hamiltonian" (Privman, 1992). To introduce these terms, we consider the one-dimensional interface $z = h(x)$ of a two-dimensional system for simplicity. Noting that in lattice systems the interfacial energy E_{int} will depend on the angle θ between the tangent to the interface and the x-axis, we write $[\theta = \arctan(dh/dx)]$

$$\frac{E_{int}}{k_B T} = \int d\ell\, f_{int}(\theta) = \int dx\, f_{int}(\theta)\sqrt{1 + \left(\frac{dh}{dx}\right)^2} \tag{182}$$

using the fact that the line element $d\ell$ along the interface satisfies $d\ell^2 = (dh)^2 + (dx)^2$. Of course, in this coarse grained description of the interface both overhangs and bubbles are deliberately ignored, cf. fig. 6, and we even assume that this coarse grained interface is rather flat, such that $(dh/dx) \ll 1$ and we can expand $\sqrt{1 + (dh/dx)^2} \approx 1 + \frac{1}{2}(dh/dx)^2$, $f_{int}(\theta) \approx f_{int}(0) + f'_{int}(0)(dh/dx) + \frac{1}{2}f''_{int}(0)(dh/dx)^2 + \ldots$. The linear term in dh/dx yields only boundary terms to the integral eq. (182), and can thus be omitted. Thus one obtains

$$\frac{E_{int}}{k_B T} = f_{int}(0) \int dx + \frac{\kappa}{2} \int dx \left(\frac{dh}{dx}\right)^2, \tag{183}$$

where the interfacial stiffness κ is defined as

$$\kappa = f_{int}(0) + f''_{int}(0) \tag{184}$$

While $k_B T f_{int}(0)$ tends to a finite constant for $T \to 0$ (in the nearest-neighbor Ising model $f_{int}(0) = 2J/k_B T - \ln\{[1 + \exp(-2J/k_B T)]/[1 - \exp(-2J/k_B T)]\}$, cf. Onsager (1944), and hence $k_B T f_{int} \to 2J$ as $T \to 0$), $k_B T \kappa \to \infty$ as $T \to 0$, reflecting considerable rigidity of the interface at low temperatures. Figure 36 summarizes the situation qualitatively in both $d = 2$ and $d = 3$ dimensions.

In $d = 3$ dimensions (fig. 6) an analogous treatment yields

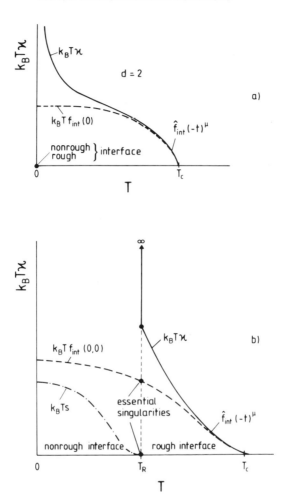

Fig. 36. Schematic temperature variation of interfacial stiffness $k_B T \kappa$ and interfacial free energy, for an interface oriented perpendicularly to a lattice direction of a square (a) or simple cubic (b) lattice, respectively. While for $d = 2$ the interface is rough for all non-zero temperatures, in $d = 3$ it is rough only for temperatures T exceeding the roughening transition temperature T_R (see sect. 3.3). For $T < T_R$ there exists a non-zero free energy $k_B T s$ of surface steps, which vanishes at $T = T_R$ with an essential singularity. While κ is infinite throughout the non-rough phase, $k_B T \kappa$ reaches a universal value as $T \to T_R^+$. Note that κ and f_{int} to leading order in their critical behavior become identical as $T \to T_c^-$.

$$
\begin{aligned}
\frac{E_{int}}{k_B T} &= f_{int}(0, 0) \int dx \, dy + \frac{\mathcal{H}_{cw}}{k_B T} \\
&= f_{int}(0, 0) \int dx \, dy + \frac{\kappa}{2} \int dx \, dy \left[\left(\frac{dh}{dx} \right)^2 + \left(\frac{dh}{dy} \right)^2 \right],
\end{aligned} \tag{185}
$$

where $\mathcal{H}_{cw}$ stands for the "capillary wave Hamiltonian". As will be discussed in sect. 3.3, in lattice systems eq. (185) applies only for temperatures T exceeding the roughening transition temperature T_R, while $\kappa = \infty$ for $T < T_R$. Here we only discuss properties of the rough phase and note by Fourier transformation in $d - 1$ dimensions

$$\frac{\mathcal{H}_{cw}}{k_B T} = \frac{\kappa}{2} \frac{1}{(2\pi)^{d-1}} \int d^{d-2}q \, q^2 |h_q|^2, \tag{186}$$

where h_q is the Fourier component of the height variable $h(x, y)$. From equipartition we conclude

$$\frac{k_B T \kappa}{2} \frac{1}{(2\pi)^{d-1}} q^2 \langle |h_q|^2 \rangle = \frac{1}{2} k_B T \tag{187}$$

and hence

$$\langle h^2(x) \rangle - \langle h(x) \rangle^2 = \frac{1}{(2\pi)^{d-1}} \int d^{d-2}q \, \langle |h_q|^2 \rangle \propto \kappa^{-1} \int q^{d-2} \, dq \, q^{-2}, \tag{188}$$

which yields in $d = 3$ for the interfacial width $W(L)$ due to capillary wave fluctuations

$$W^2(L) \equiv \langle h^2(x) \rangle - \langle h(x) \rangle^2 \propto \kappa^{-1} \int_{2\pi/L}^{2\pi/\xi} \frac{dq}{q} = \kappa^{-1} \ln\left(\frac{L}{\xi}\right), \tag{189}$$

while in $d = 2$ a power law divergence of $W^2(L)$ with the linear dimension L along the interface results,

$$W^2(L) \propto \kappa^{-1} \int \frac{dq}{q^2} \propto \kappa^{-1} L. \tag{190}$$

The latter result can be simply interpreted by the "random walk" picture of an one-dimensional fluctuating interface (Fisher, 1984).

As a final topic of this section, we consider the free energy of droplets in metastable phases. Metastable phases are very common in nature, and also readily predicted by approximate theories such as the Landau theory. Consider e.g. the transition of an Ising model at $T < T_c$ as a function of magnetic field H (fig. 37). From eq. (14) one obtains

$$[k_B T V]^{-1} \left(\frac{\partial F}{\partial \phi}\right)_T = r u + u\phi^3 - \frac{H}{k_B T} = 0, \tag{191}$$

and hence one finds the stability limit where $\chi_T = (\partial\phi/\partial H)_T$ diverges as follows

$$r + 3u\phi_{sp}^2 = (k_B T \chi_T)^{-1} = 0,$$

$$\phi_{sp} = \sqrt{-\frac{r}{3u}} = \frac{\phi_o}{\sqrt{3}}, \qquad H_c = -\frac{2r}{3}\sqrt{-\frac{r}{3u}} \tag{192}$$

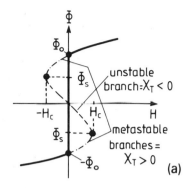

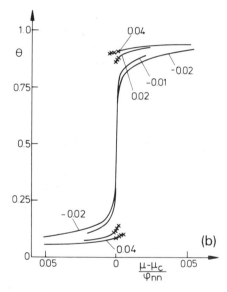

Fig. 37. (a) Order parameter ϕ vs. conjugate field H according to the phenomenological Landau theory for a system at a temperature T less than the critical temperature T_c of a second-order phase transition (schematic). At $H = 0$, a first order transition from ϕ_o to $-\phi_o$ occurs (thick straight line). The metastable branches (dash-dotted) end at the "limit of metastability" or "spinodal point" (ϕ_s, H_s), respectively, and are characterized by a positive order parameter susceptibility $\chi_T > 0$, whereas for the unstable branch (broken curve) $\chi_T < 0$. (b) Order parameter (coverage) vs. conjugate field (chemical potential difference) for the nearest neighbor lattice gas model in $d = 2$. Crosses denote metastable states. Parameter of the curves is $1 - T/T_c$. From Binder and Müller-Krumbhaar (1974).

Since in the metastable states the susceptibility can be written as

$$\chi_T = (3k_B T u)^{-1}(\phi^2 - \phi_{sp}^2)^{-1} = (3k_B T u)^{-1}(\phi - \phi_{sp})^{-1}(\phi + \phi_{sp})^{-1}$$

$$(193)$$

one sees that $\chi_T \to \infty$ as $\phi \to \pm\phi_{sp}$. The "spinodal curve" $\phi = \phi_{sp}(T)$ in the (ϕ, T) plane thus plays in mean field theory the role of a line of critical points (see fig. 39).

Similar behavior occurs in many other theories: e.g. the van der Waals equation of state describing gas–liquid condensation exhibits an analogous loop of one-phase states in the two-phase coexistence region. However, it must be emphasized that metastable states are in general not well-defined in statistical mechanics for systems with short range forces (Binder, 1984b, 1987; Penrose and Lebowitz, 1971). A simple argument to see this concerns the decay of metastable states via nucleation and growth, which is a problem of great practical interest (Zettlemoyer, 1969). For detailed expositions of nucleation theory we refer to various reviews (Binder and Stauffer, 1976a; Gunton et al., 1983). Here we summarize a few key points only. Let us compare in fig. 37 the thermodynamic potential of a stable state ($\phi_{stable} = \phi_o + \chi H$) and of a metastable state ($\phi_{ms} = -\phi_o + \chi H$), noting $F(T, H) = G(T, \phi) - H\phi$, and since at phase coexistence we have (cf. fig. 13a) $G(T, \phi_o) = G(T, -\phi_o)$, we conclude for H small

$$F(T, H)_{stable} = G(T, \phi_o) - H\phi_o, \qquad F(T, H)_{ms} = G(T, \phi_o) + H\phi_o, \tag{194}$$

thus a spherical droplet of "volume" (in d dimensions) $V = V_d \rho^d$ (V_d is the volume of a d-dimensional unit sphere, ρ is the droplet radius) involves a volume energy of order $\Delta F = F_{stable} - F_{ms} = -2HV\phi_o$. On the other hand, there is also an interfacial free energy contribution associated with the surface area of this droplet, $F_{int} = S_d \rho^{d-1} f_{int}$, S_d being the surface area of a d-dimensional unit sphere, and hence the "formation free energy" of a spherical droplet of radius ρ is

$$\Delta F(\rho) = S_d \rho^{d-1} f_{int} - 2H V_d \rho^d \phi_o \tag{195}$$

Obviously, $\Delta F(\rho)$ increases for small ρ (where the "surface term" $S_d \rho^{d-1} f_{int}$ dominates), reaches a maximum $\Delta F^\star$ at a critical droplet radius $R^\star$, and then decreases again due to the negative volume term. In this "classical" nucleation theory, it is straightforward to obtain the critical droplet radius $R^\star$ from

$$0 = \frac{\partial(\Delta F(\rho))}{\partial \rho}\bigg|_{\rho=R^\star} = (d-1)S_d \rho^{d-2} f_{int} - 2d H V_d \rho^{d-1} \phi_o, \tag{196}$$

$$R^\star = \frac{(d-1)S_d f_{int}}{2d V_d H \phi_o}, \qquad \Delta F^\star = \frac{(S_d f_{int}/d)^d}{[2V_d H \phi_o/(d-1)]^{d-1}}. \tag{197}$$

Thus for $H \to 0$ the free energy barrier $\Delta F^\star$ diverges as $H^{-(d-1)}$, which means that the lifetime of metastable states can get very large. Since for a large droplet the radial order parameter profile across a droplet is similar to that of a flat interface (fig. 35a,b), the interfacial tension f_{int} that enters in eqs. (195)–(197) is taken to be that of a flat interface ("capillarity approximation"; Zettlemoyer, 1969).

The classical nucleation theory can be used only when the droplet radius ρ^* is much larger than the interfacial width (which is of the same order as the correlation length $\xi(\phi)$). Since $\xi(\phi) \rightarrow \infty$ as one approaches the limit of metastability ϕ_{sp} [eq. (192)],

$$\xi = \frac{R}{[6du\phi_s(T)]^{1/2}}[\phi - \phi_{sp}(T)]^{-1/2}, \tag{198}$$

the classical theory cannot be used when ϕ is close to this spinodal ϕ_{sp}. There ρ^* becomes comparable to ξ (fig. 35c), as an extension of Landau's theory to this problem due to Cahn and Hilliard (1959) shows. One now solves the Ginzburg–Landau equation [cf. eqs. (35), (177)] but instead of an one-dimensional geometry one chooses a spherical geometry where only a radial variation of $\phi(\rho)$ with radius ρ is permitted, and a boundary condition $\phi(\rho \rightarrow \infty) = \phi_{ms}$ is imposed (fig. 35b, c). Whereas for ϕ_{ms} near $\phi_{coex} = \phi_o$ this treatment agrees with the classical theory, eqs. (195)–(197), it differs significantly from it for ϕ near $\phi_{sp}(T)$: then the critical droplet radius R^* is of the same order as the (nearly divergent!) correlation length ξ, eq. (198), and the profile is extremely flat, reaching in the droplet center only a value slightly below ϕ_{sp} rather than the other branch of the coexistence curve. One obtains for ΔF^* and temperatures T near T_c (Klein and Unger, 1983; Binder, 1984b)

$$\frac{\Delta F^\star}{k_B T_c} \propto R^d \left(1 - \frac{T}{T_c}\right)^{(4-d)/2} \left(\frac{\phi - \phi_{sp}}{\phi_{coex}}\right)^{(6-d)/2}, \tag{199}$$

whereas near the coexistence curve the result is (using eq. (197), $\phi_{coex} - \phi \approx \chi_{coex} H$, and the mean field critical behavior of $\phi_{coex} = \phi_o$ [eq. (16)], χ_{coex} [eq. (37)] and f_{int} [eq. (181)])

$$\frac{\Delta F^\star}{k_B T_c} \propto R^d \left(1 - \frac{T}{T_c}\right)^{(4-d)/2} \left(\frac{\phi_{coex} - \phi}{\phi_{coex}}\right)^{-(d-1)}. \tag{200}$$

In both eqs. (199) and (200) all prefactors of order unity are omitted. In a system with a large but finite range R of interaction, the nucleation barrier is very high in the mean-field critical region, in which $R^d(1 - T/T_c)^{(4-d)/2} \gg 1$ [cf. eq. (56)]. This factor, which controls the Ginzburg criterion, also controls the scale of the nucleation barrier as a prefactor (see fig. 38). In this region, the condition for the actual breakdown of the metastable state due to fast formation of many droplets (in gas to liquid nucleation this is called

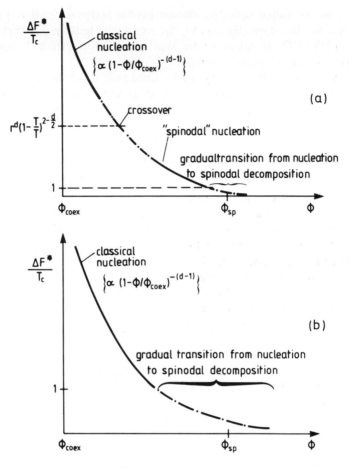

Fig. 38. Schematic plots of the free energy barrier for (a) the mean field critical region, i.e. $R^d (1 - T/T_c)^{(4-d)/2} \gg 1$, and (b) the non-mean field critical region, i.e. $R^d (1 - T/T_c)^{(4-d)/2} \ll 1$. When $\Delta F^*/T_c$ is of order unity, a gradual transition from nucleation to "spinodal decomposition" (in a phase-separating mixture) or "spinodal ordering" (in a system undergoing an order-disorder transition with non-conserved order parameter distinct from ϕ) occurs. From Binder (1984b).

the "cloud point"), $\Delta F^*/k_B T \approx 1$, is located very close to the mean-field spinodal. Then the description of nucleation phenomena close to the spinodal in terms of the diffuse droplets described by fig. 35c is meaningful ("spinodal nucleation"). On the other hand, for a system with short range interactions where R [measured in units of the lattice spacing in eqs. (199) and (200)] is unity, the free energy barrier becomes of order unity long before the spinodal curve is reached. The singularity at the spinodal then completely lacks any physical significance, as the metastable state decays to the stable phase long before the spinodal is reached. This is the situation usually encountered for phase transitions in two-dimensional systems.

2.6. Kinetics of fluctuations and domain growth

Second-order phase transitions also show up via the "critical slowing down" of the critical fluctuations (Hohenberg and Halperin, 1977). In structural phase transitions, one speaks about "soft phonon modes" (Blinc and Zeks, 1974; Bruce and Cowley, 1981); in isotropic magnets, magnon modes soften as T approaches T_c from below; near the critical point of mixtures the interdiffusion is slowed down; etc. This critical behavior of the dynamics of fluctuations is characterized by a dynamic critical exponent z: one expects that some characteristic time τ exists which diverges as $T \to T_c$,

$$\tau \propto \xi^z \propto \left| 1 - \frac{T}{T_c} \right|^{-\nu z}. \tag{201}$$

Many concepts developed for static critical phenomena (scaling laws, universality, etc.) can be carried over to dynamic critical phenomena. Hohenberg and Halperin (1977) discuss the various "dynamic universality classes": each static universality class is split into several dynamic classes, depending on which conservation laws apply, and whether mode coupling terms occur in the basic dynamic equations, etc. For example, anisotropic magnets such as $RbMnF_3$, ordering alloys such as β-brass (CuZn), unmixing solid mixtures such as ZnAl-alloys, unmixing fluid mixtures such as lutidine–water, and the gas–fluid critical point all belong to the same static universality class as the $d = 3$ Ising model, but each of these systems belongs to a different dynamic universality class! Thus, in the anisotropic antiferromagnet, no conservation law needs to be considered, whereas the conservation of concentration matters for all mixtures (where it means that the order parameter is conserved) and for ordering alloys (where the order parameter is not conserved but coupled to the conserved concentration, a "non-ordering density"). Whereas in solid mixtures the local concentration relaxes simply by diffusion, in fluid mixtures hydrodynamic flow effects matter and also play a role at the liquid–gas critical point. For the latter case, energy conservation also needs to be considered — but it does not play a role, of course, for phase transitions in solid mixtures where the phonons act as a "heat bath" to the considered configurational degrees of freedom.

These considerations apply analogously to adsorbed monolayers at surfaces — if one considers equilibrium with a surrounding (three-dimensional) gas, the coverage of the monolayer is not conserved, while for an adsorbed layer at very low temperatures, or for a chemisorbed layer in ultrahigh vacuum, no evaporation/condensation processes of adatoms occur, while configurations may still relax via surface diffusion. In fact, the interplay between ordering of layers and surface diffusion is quite subtle even in grossly simplified lattice gas models (Sadiq and Binder, 1983).

We shall not discuss critical dynamics in any depth here, but outline only the simple van Hove (1954) phenomenological approach, the so-called "conventional theory" of critical slowing down. First we consider a scalar, non-conserved order parameter, and ask how a deviation $\Delta\phi(x, t)$ from equilibrium occurring at a space point x and time t relaxes. According to Landau's theory, we have in equilibrium $\delta\mathcal{F}\{\phi(x)\}/\delta\phi(x) = 0$ for $\phi(x) = \phi_o$; the standard assumption of irreversible thermodynamics is now a generalized friction ansatz — the generalized velocity is proportional to the generalized force. Thus for this model ("model A" in the Hohenberg–Halperin (1977)-classification)

$$\frac{\partial}{\partial t}\Delta\phi(x, t) = -\Gamma_o \frac{\partial\mathcal{F}\{\phi(x, t)\}}{\partial(\Delta\phi(x, t))}. \tag{202}$$

Using eq. (14) and expanding $\phi(x, t) = \phi_o + \Delta\phi(x, t)$, $\phi^3(x, t) \approx \phi_o^3 + 3\phi_o^2\Delta\phi(x, t)$ we obtain for $H = 0$

$$\Gamma_0^{-1}\frac{\partial}{\partial t}\Delta\phi(x, t) = -\{r + 3u\phi_o^2\}\Delta\phi(x, t) + \frac{R^2}{d}\nabla^2[\Delta\phi(x, t)]. \tag{203}$$

Let us assume [as in eqs. (35)–(40)] that the deviation from equilibrium $\Delta\phi(x, t)$ has been produced by a field $\delta H(x) = \delta H_q \exp(iq \cdot x)$, which had been switched on at $t \to -\infty$ but was switched off suddenly at $t = 0$. This treatment is hence the dynamic counterpart of the linear response theory presented in sect. 2.2. Writing $\Delta\phi(x, t) = \Delta\phi_q(t)\exp(iq \cdot x)$, eq. (203) is solved by

$$\Gamma_0^{-1}\frac{d}{dt}\Delta\phi_q(t) = -\left\{r + 3u\phi_o^2 + \frac{R^2}{d}q^2\right\}\Delta\phi_q(t)$$
$$= -[k_B T\chi(q)]^{-1}\Delta\phi_q(t), \tag{204}$$

where eq. (35) was used. Since for $t \leq 0$ we have equilibrium with $\Delta\phi(x, t) = \Delta\phi_q \exp(iq \cdot x)$ with $\Delta\phi_q = \chi(q)H_q$, eq. (204) amounts to an initial value problem which is solved by

$$\frac{\Delta\phi_q(t)}{\Delta\phi_q(0)} = \frac{\Delta\phi_q(t)}{\chi(q)H_q} = \exp[-\omega(q)t], \tag{205}$$

the characteristic frequency $\omega(q)$ being

$$\omega(q) = \frac{\Gamma_o}{k_B T\chi(q)} = \frac{\Gamma_o(1 + q^2\xi^2)}{k_B T\chi_T}. \tag{206}$$

Thus $\omega(q = 0)$ vanishes as $\omega(q = 0) \propto \chi_T^{-1} \propto \xi^{-\gamma/\nu} = \xi^{-(2-\eta)}$, and eq. (201) hence implies the classical value $z_{\mathrm{cl}} = 2 - \eta$. Although eq. (206) thus suggests a relationship between the dynamic exponent and static ones, this is not true if effects due to non-mean-field critical fluctuations are taken into account. In fact, for the kinetic Ising model (Kawasaki, 1972) extensive numerical calculations imply that $z \approx 2.18$ in $d = 2$ dimensions (Dammann and Reger, 1993; Stauffer, 1992; Landau et al., 1988) rather than $z_{\mathrm{cl}} = 2 - \eta = 1.75$. Note also [this is already evident from eq. (206)] that not all fluctuations slow down as T_{c} is approached but only those associated with long wavelength order parameter variations. One can express this fact in terms of a dynamic scaling principle

$$\omega(q) = q^z \tilde{\omega}(q\xi), \qquad \tilde{\omega}(\mathcal{Z} \to 0) \to \mathcal{Z}^{-z}, \qquad \tilde{\omega}(\mathcal{Z} \gg 1) \to \text{const} \quad (207)$$

As a second system, we consider "model B" which has a conserved order parameter. E.g., we may consider an adsorbed monolayer at constant coverage θ, assuming that below some critical temperature T_{c} there occurs a phase separation, in a phase of low coverage $\theta_{\mathrm{coex}}^{(1)}$ and a phase of high coverage $\theta_{\mathrm{coex}}^{(2)}$ (e.g., as occurs in the lattice gas model if one assumes attractive interactions between nearest neighbors only, cf. figs. 25 and 26a). The order parameter is now the deviation from the critical coverage θ_{crit}, and since $\theta = \theta_{\mathrm{crit}} + V^{-1} \int \mathrm{d}x \phi(x, t) = \text{const}$ (V is the volume or, in $d = 2$, the area available for adsorption, respectively), the conservation of the order parameter is expressed by a continuity equation

$$\frac{\partial \phi(x, t)}{\partial t} + \nabla \cdot j(x, t) = 0, \tag{208}$$

where $j(x, t)$ is the current density. Irreversible thermodynamics (De Groot and Mazur, 1962) relates this current density to a gradient of the local chemical potential $\mu(x, t)$,

$$j = -M \nabla \mu(x, t), \tag{209}$$

M being a mobility. Just as in thermal equilibrium $\mu = (\partial F/\partial \theta)_T$, we have a functional derivative in the inhomogeneous case,

$$\mu(x, t) \equiv \frac{\delta(\mathcal{F}\{\phi(x, t)\})}{\delta \phi(x, t)}, \tag{210}$$

and using again the Landau expansion eq. (14) we conclude, since H equals μ here,

$$\mu(x, t) = r\phi(x, t) + u\phi^3(x, t) - \frac{R^2}{d} \nabla^2 \phi(x, t) \tag{211}$$

eqs. (208)–(210) yield, again substituting $\phi(x, t) = \phi_o + \Delta\phi(x, t)$ and linearizing in $\Delta\phi(x, t)$, an equation proposed by Cahn and Hilliard (1958) to describe the phase separation of binary mixtures (where $\phi(x, t)$ represents a concentration difference and μ a chemical potential difference between the two species A, B forming the mixture),

$$\frac{\partial \Delta\phi(x, t)}{\partial t} = M \nabla^2 \left\{ [r + 3u\phi_o^2] \Delta\phi(x, t) - \frac{R^2}{d} \nabla^2 [\Delta\phi(x, t)] \right\}. \quad (212)$$

This is analogous to eq. (203) but the rate factor Γ_o is now replaced by the operator $-M\nabla^2$. Using again $\Delta\phi(x, t) = \Delta\phi_q(t) \exp(iq \cdot x)$ yields

$$\frac{d}{dt} \Delta\phi_q(t) = -Mq^2 \left(r + 3u\phi_o^2 + \frac{R^2 q^2}{d} \right) \Delta\phi_q(t)$$

$$= -Mq^2 [k_B T \chi(q)]^{-1} \Delta\phi_q(t), \quad (213)$$

and hence

$$\frac{\Delta\phi_q(t)}{\Delta\phi_q(0)} = \exp[-\omega(q)t], \qquad \omega(q) = \frac{Mq^2}{k_B T \chi(q)} = \frac{Mq^{4-\eta}}{k_B T \tilde{\chi}(q\xi)} \quad (214)$$

where in the last step we used the scaling relation, eq. (105), for the static scattering function. Equation (214) thus implies for the case of conserved order parameters a stronger slowing down,

$$z = 4 - \eta \quad \text{(model B)}. \quad (215)$$

This result holds also beyond Landau's theory, as a renormalization group treatment shows (Halperin et al., 1974).

So far, we have considered the dynamics of fluctuations (with small amplitudes!) in equilibrium states. But it also is of great interest to study dynamic processes far from equilibrium, as occur in the context of phase transitions when we treat (Binder, 1981b) the kinetics of ordering or the kinetics of phase separation (fig. 39). Suppose we bring the system at a time $t = 0$ suddenly from a state in the disordered region above T_c by rapid cooling into the region below T_c. This disordered state now is unstable, and we expect to see ordered domains grow out of the initially disordered configuration. The growth of the size of these domains, and the magnitude of the scattering function describing this ordering, is a problem of great interest (Gunton et al., 1983; Binder, 1991). Such a process can be observed e. g. for the c(2×2) structure when the ordering occurs via a second-order transition (see e.g. fig. 28b).

A second problem that we consider is the kinetics of unmixing of a binary system AB (fig. 39). Quenching the system at time $t = 0$ from an equilibrium

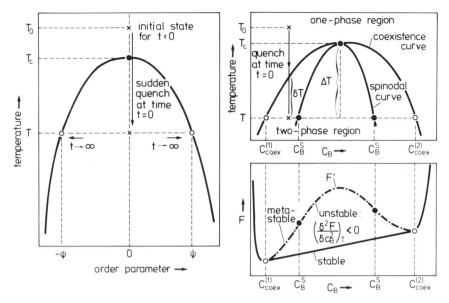

Fig. 39. Order parameter ψ of a second-order transition plotted vs. temperature (left part), assuming a two-fold degeneracy of the ordered state (described by the plus and minus sign of the order parameter; a physical realization in an adsorbed monolayer would be the c(2×2) structure). The quenching experiment is indicated. The right part shows the phase diagram of a binary mixture with a miscibility gap ending in a critical point (T_c, c_B^{crit}) of unmixing, in the temperature–concentration plane. (Alternatively, one can interpret c_B as the coverage of a monolayer that undergoes phase separation in a phase of low density $c_{coex}^{(1)}$ and a second phase at high density $c_{coex}^{(2)}$, respectively.) Again the quenching experiment is indicated, and the quenching distances from the coexistence curve (δT) and from the critical point (ΔT) are indicated. Lower part shows a schematic free energy curve plotted vs. c_B at constant T. The dash-dotted part represents metastable and unstable homogeneous one-phase states in the two-phase region. From Binder (1981b).

state in the one-phase region to a state underneath the coexistence curve leads to phase separation; in thermal equilibrium, macroscopic regions of both phases with concentrations $c_{coex}^{(1)}$ and $c_{coex}^{(2)}$ coexist.

Of course, the two processes considered schematically in fig. 39 are only the basic "building blocks" of much more complex processes that are expected to occur in real systems. Consider e.g. the phase diagram of fig. 28c: If we quench the system from the disordered phase to low temperatures for $\theta < 1/2$, the system separates into a disordered low-density lattice gas and ordered islands exhibiting the c(2×2) structure and $\theta \approx 1/2$; similarly, for the system of fig. 28d quenching experiments would produce simultaneously phase separation and ordering in ($\sqrt{3} \times \sqrt{3}$) structures. Here we do not aim at a detailed description of such processes, but rather sketch the main ideas only. Zinke-Allmang et al. (1992) provide a much more thorough review on the kinetics of clustering at surfaces and related processes.

As discussed already in the previous section, one basic concept is the idea of distinguishing between metastable and unstable homogeneous one-phase states in the two-phase region, described by the dash-dotted double-well free energy F' in fig. 39: one assumes that immediately after the quench some sort of local equilibrium in a homogeneous state is established, which is described by F'. Of course, this idea is rather questionable, because the system in its unstable part is predicted to decay immediately after the quench. The decay process is believed to be qualitatively different from nucleation, as considered in the previous section, since arbitrarily weak long-wavelength fluctuations grow spontaneously as the time after the quench elapses (fig. 40). This is easily recognized from eqs. (212)–(214), since the time constant $\omega(q)$ is negative for states inside the two branches of the spinodal curve and small enough q, since $r + 3u\phi_o^2 < 0$

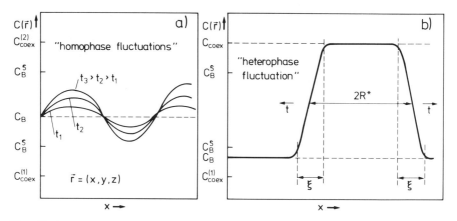

Fig. 40. Schematic description of unstable thermodynamic fluctuations in the two-phase regime of a binary mixture AB at a concentration c_B (a) in the unstable regime inside the two branches c_B^s of the spinodal curve and (b) in the metastable regime between the spinodal curve c_B^s and the coexistence curve $c_{coex}^{(1)}$. The local concentration $c(r)$ at a point $r = (x, y, z)$ in space is schematically plotted against the spatial coordinate x at some time after the quench. In case (a), the concentration variation at three distinct times t_1, t_2, t_3 is indicated. In case (b) a critical droplet is indicated, of diameter $2R^*$, the width of the interfacial regions being the correlation length ξ. Note that the concentration profile of the droplet reaches the other branch $c_{coex}^{(2)}$ of the coexistence curve in the droplet center only for weak "supersaturations" of the mixture, where $c_B - c_{coex}^{(1)} \ll c_B^s - c_B$ and $R^* \gg \xi$; for the sake of clarity, the figure therefore is not drawn to scale. Note that the same description also holds for homophase systems, e.g. lattice gas models where A corresponds to the state with $\theta = 0$, and B to a phase with non-zero coverage (e.g., $\theta = 1$ in system with attractive interactions only, or $\theta = 1/2$ in the case of systems like shown in fig. 28c, or $\theta = 1/3$ for fig. 28d). From Binder (1981b).

for $-\phi_{sp} < \phi < \phi_{sp}$ [for the simple Landau ϕ^4-model $\phi_{sp} = \sqrt{-r/3u}$, cf. eq. (192)]. The condition $\omega(q_c) = 0$ defines a critical wavenumber,

$$q_c^2 = \frac{d(-r - 3u\phi_o^2)}{R^2}, \quad \lambda_c = \frac{2\pi}{q_c} = \frac{2\pi R}{\sqrt{3du(\phi_{sp} - \phi_o)(\phi_{sp} + \phi_o)}}. \quad (216)$$

All fluctuations with wavelengths $\lambda > \lambda_c$ thus get spontaneously amplified, since for them $\omega(q)$ is negative. The divergence of λ_c as $\phi_o \to \pm\phi_{sp}$ again expresses critical slowing down, in mean field theory the spinodal curve is a line of critical points. The maximum growth rate of these unstable fluctuations occurs for $\lambda_m = \sqrt{2}\lambda_c$. It must be emphasized, however, that for systems of physical interest the transition between nucleation and spinodal decomposition is gradual and not sharp (Binder, 1981b, 1984b, 1991), and also the growing unstable waves in the region in between the spinodal (fig. 40a) do not show exponential growth, since fluctuations and non-linear effects need to be taken into account immediately after the quench (Gunton, 1983; Binder, 1991). The details of this behavior shall not be discussed here; rather we draw attention to the behavior of late stages after the quench. Then the typical linear dimension $\ell(t)$ of the ordered domains that are formed after the quench (or of the islands of high the density [or concentration] phase in the case of phase separation) grows with a power law of time (Lifshitz, 1962; Allen and Cahn, 1979; Lifshitz and Slyozov, 1961; Binder and Stauffer, 1974; Binder, 1977; Ohta et al., 1982; Furukawa, 1985; Binder and Heermann, 1985; Komura and Furukawa, 1988; Mouritsen, 1990)

$$\ell(t) \propto t^{1/2} \text{ (non-conserved order parameter)},$$
$$\ell(t) \propto t^{1/3} \text{ (conserved order parameter)} \quad (217)$$

Figures 41–43 give some examples from a model calculation (Sadiq and Binder, 1984) for the ordering process of the (2×1) structure of a monolayer at coverage $\theta = 1/2$ on the square lattice (fig. 10). One can recognize the steady growth of the four kinds of domains, and at the same time the excess energy $\Delta E(t)$ due to the domain walls decreases, and a diffuse peak grows at the Bragg spots (e.g. $q = \pi(1, 0)$, cf. fig. 43a). Binder and Stauffer (1974, 1976b) have extended the dynamic scaling principle to such phenomena far from equilibrium, by postulating that in the late stages where $\ell(t)$ is much larger than the lattice spacing the equal-time structure factor $S(q, t)$ describing the scattering from the growing domains can be scaled with $\ell(t)$ as

$$S(q, t) = [\ell(t)]^d \tilde{S}\{(q - q_B)\ell(t)\}, \quad (218)$$

q_B being the position in reciprocal space where long range order leads to a Bragg peak, and $\tilde{S}$ is a scaling function. This idea, which has found great

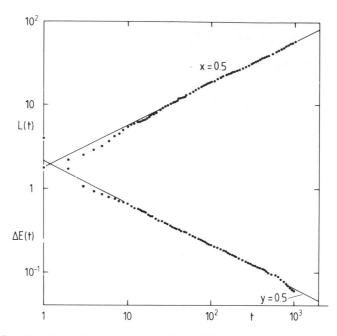

Fig. 41. Log–log plot of characteristic domain size $\ell(t)$ (upper part) and excess energy $\Delta E(t)$ (lower part) versus time t after the quench from a random initial configuration at $\theta = 1/2$ on the square lattice to a temperature $k_B T/|J_{nn}| = 1.33$, for the lattice gas model with repulsive interactions between both nearest and next nearest neighbors ($J_{nnn} = J_{nn} < 0$, ordering temperature then is at $k_B T_c/|J_{nn}| \approx 2.07$). The system evolves according to the Glauber (1963) kinetic Ising model, simulating random condensation–evaporation events of adatoms at the chemical potential corresponding to $\langle\theta\rangle = 1/2$. Time is measured in units of Monte Carlo steps (MCS) per lattice site. Straight lines indicate exponents $\ell(t) \propto t^x$ with $x = 1/2$ and $\Delta E(t) \propto t^{-y}$ with $y = 1/2$. These data were obtained from averages over 45 runs of 80×80 lattices. From Sadiq and Binder (1984).

theoretical interest (e.g. Komura and Furukawa, 1988), has been verified by simulations (e.g. fig. 43b) and in favorable cases has also been established experimentally: fig. 44 gives an example for oxygen in the p(2×1) structure on W(110) surfaces (Wu et al., 1989), and a similar behavior was also shown for the ($\sqrt{3}\times\sqrt{3}$) structure of Ag adsorbed on Ge(111) surfaces (Henzler

Fig. 42. A series of snapshot pictures of a time evolution of the model system described in fig. 41, for a 120×120 lattice with periodic boundary conditions. Occupied sites are denoted by a circle, if they belong to a domain of type 1, by a triangle if they belong to a domain of type 2, by a cross ($\times$) if they belong to a domain of type 3, and by a standing cross ($+$) if they belong to a domain of type 4 (cf. fig. 10). Atoms belonging to walls are not shown. Times shown are $t = 20$ (a), 60 (b) and 100 (c). From Sadiq and Binder (1984).

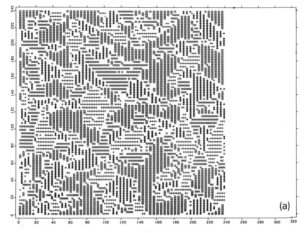

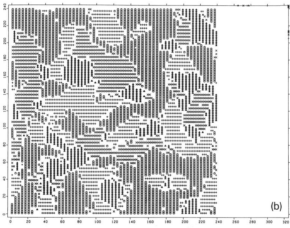

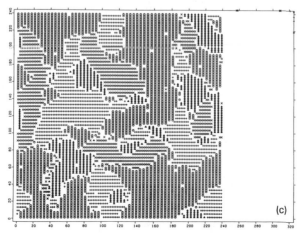

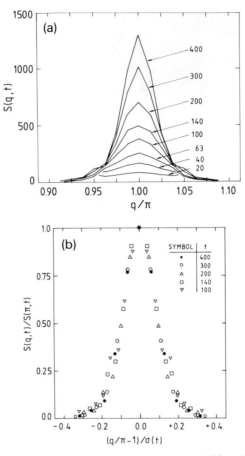

Fig. 43. (a) Structure factor $S(q, t)$ plotted vs. q for the model of figs. 41 and 42, using a lattice size of 160×160 and averaging over 250 runs. Parameter of the curves is the time t (measured in units of Monte Carlo steps per site), and lattice spacing is chosen as unit of length. Note that q is oriented in x-direction and q is then only defined for integer multiples of $2\pi/L = \pi/80$. Thus these discrete values of $S(q, t)$ were connected by straight lines in between. (b) Structure factor of part (a) replotted in scaled form, normalizing $S(q, t)$ by its peak value $S(\pi, t)$ and normalizing $q/\pi - 1$ by the halfwidth $\sigma(t)$. From Sadiq and Binder (1984).

and Busch, 1990). In reality, it is difficult to establish this behavior of eqs. (217) and (218) because the growth is very much affected by impurities, defects of the substrate (screw dislocations, steps, etc.), which may lead to a crossover to a slower growth because the domain walls can no longer diffuse freely as in the ideal case of fig. 42. Again computer simulations (Grest and Srolovitz, 1985; Albano et al., 1992) have contributed significantly to clarify these problems, see e.g. fig. 45.

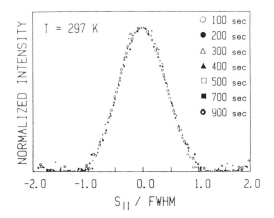

Fig. 44. Dynamic scaling in the growth of Bragg peaks for LEED scattering from oxygen monolayers at $\theta = 1/2$ and $T = 297$ K for adsorption in the p(2×1) structure on W(110). Different symbols denote different times t (in seconds) after the adsorption has taken place. The halfwidth (used for normalization exactly as in fig. 43) is denoted by FWHM ("full width at half maximum"). From Wu et al. (1989).

3. Surface effects on bulk phase transitions

In sect. 2, we have summarized the general theory of phase transitions with an emphasis on low-dimensional phenomena, which are relevant in surface physics, where a surface acts as a substrate on which a two-dimensional adsorbed layer may undergo phase transitions. In the present section, we consider a different class of surface phase transitions: we assume e.g. a semi-infinite system which may undergo a phase transition in the bulk and ask how the phenomena near the transition are locally modified near the surface. sect. 3.1 considers a bulk transition of second order, while sections 3.2 and 3.4–3.6 consider bulk transitions of first order. In this context, a closer look at the roughening transitions of interfaces is necessary (sect. 3.3). Since all these phenomena have been extensively reviewed recently, we shall be very brief and only try to put the phenomena in perspective.

3.1. Surface effects on bulk critical phenomena

We return here to mean field theory with a scalar order parameter $\phi(z)$ and consider now a thick film geometry, assuming hard walls (or surface against vacuum, respectively) at $z = 0$ and $z = L$. Starting again from eq. (14), we may disregard the x and y-coordinates [as in our treatment of the interfacial profile, eqs. (177)–(181)], but now we have to add a perturbation $2F_s^{(bare)}$ to

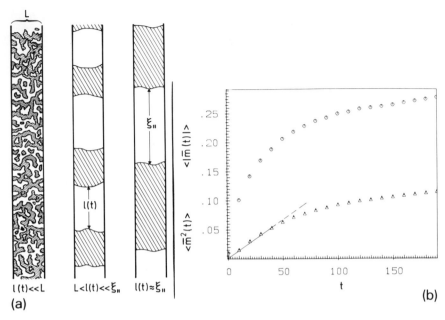

Fig. 45. (a) Schematic description of the domain growth in the c(2×2) structure on the terraces of a regularly stepped surface (as shown in fig. 3), assuming that adatoms in different terraces do not interact with each other, and each terrace can hence be considered independently of all the others. Two types of domains occur, one is shown shaded, the other is left white, while domain walls are indicated by thin solid lines. The thick solid lines show the boundary of the terraces. Three stages of domain growth are indicated: in the first stage (left part), the typical linear dimension of domains $\ell(t)$ is much smaller than L. In the second stage $\ell(t)$ is larger than L but much smaller than the correlation length in thermal equilibrium $\xi_\parallel$, given by eq. (76), middle part. In the last stage (right part) the domain size $\ell(t)$ in the direction parallel to the steps saturates at its equilibrium value $\xi_\parallel$. (b) Time evolution of the average absolute value of the order parameter $\langle|\bar{m}(t)|\rangle$ of the c(2×2) structure (octagons) and of the mean square order parameter $\langle\bar{m}^2(t)\rangle$ (triangles) plotted vs. time, for the nearest neighbor lattice gas model at $T/T_c = 0.85$, $L = 24$, $M = 288$ lattice spacings (cf. fig. 3 for a definition of the geometry). Straight line indicates the law $\langle\bar{m}^2(t)\rangle \propto t$ which is related to $\ell(t) \propto t^{1/2}$ [eq. (217)]. From Albano et al. (1992).

the free energy which describes additional forces due to the two hard walls, change of local interactions (e.g. due to missing neighbors), etc. Then the free energy becomes

$$
\frac{\Delta\mathcal{F}\{\phi(z)\}}{k_B T 3} = \int_o^L dz \left\{ \frac{1}{2}r\phi^2(z) + \frac{1}{4}u\phi^4(z) - \frac{H}{k_B T}\phi(z) \right.
$$
$$
\left. + \frac{1}{2}\frac{R^2}{d}\left(\frac{d\phi}{dz}\right)^2 \right\} + \frac{2F_s^{(bare)}}{k_B T}. \tag{219}
$$

Disregarding any long range forces due to the walls (Dietrich, 1988), we may assume that $F_s^{(\text{bare})}$ depends on the local order parameter $\phi_1 \equiv \phi(z = 0)$ $\{= \phi(z = L)\}$ only, and again we expand $F_s^{(\text{bare})}$ in powers of ϕ_1, keeping only the lowest-order terms (Binder and Hohenberg, 1972)

$$\frac{F_s^{(\text{bare})}}{k_B T} = -\frac{H_1}{k_B T}\phi_1 + \frac{R^2}{2d}\lambda^{-1}\phi_1^2. \tag{220}$$

Here we have omitted constant terms in the free energy and anticipated that a molecular field treatment of an Ising Hamiltonian that describes exactly the situation considered here, namely eq. (1), is consistent with eq. (220). The coefficient of the linear term in eq. (220) thus describes the action of a local "field" (i.e., the variable conjugate to the local order parameter) right at the hard walls (or free surfaces, respectively). If ϕ is the order parameter of gas–liquid condensation, the field H_1 can be interpreted as the binding potential of particles at the hard wall.

The quadratic term in eq. (220) is simply the counterpart of the term $\frac{1}{2}r\phi^2$ in the bulk. Since the latter changes sign at T_c, the need arises to include a term $\frac{1}{4}u\phi^4(z)$ in the bulk, but since there is in general no reason to assume that the coefficient of the quadratic term changes sign at the same temperature, one may stop at the quadratic order of the expansion in eq. (220). The constant λ has the dimension of a length and is called the extrapolation length, see fig. 46. As in the problem describing the interfacial profile [eq. (177)] we obtain the Euler–Lagrange equation

$$r\phi(z) + u\phi^3(z) - \frac{R^2}{d}\frac{d^2\phi}{dz^2} = \frac{H}{k_B T}, \quad 0 < z < L, \tag{221}$$

from minimization of the free energy functional, eq. (219). The bare surface contribution, eq. (220), yields boundary conditions

$$\left.\frac{d\phi}{dz}\right|_{z=0} - \frac{\phi_1}{\lambda} = -\frac{H_1 d}{R^2}, \quad z = 0; \quad \left.\frac{d\phi}{dz}\right|_{z=0} + \frac{\phi_1}{\lambda} = \frac{dH_1}{R^2}, \quad z = L. \tag{222}$$

Let us first specialize to the case $L \to \infty$, which means that the boundary condition at $z = L$ [eq. (222)] can be replaced by $\phi(z \to \infty) = \phi_b$ (in a ferromagnet, ϕ_b is the bulk magnetization m_b, fig. 46). For $T > T_c$, the term $u\phi^3(z)$ in eq. (221) can be neglected, and using $\phi_b = H/(k_B T r)$ the solution of eq. (221) is

$$\phi(z) = A \exp\left(-\frac{z}{\xi}\right) + \frac{H}{k_B T r} = A \exp\left(\frac{z}{\xi}\right) + \phi_b, \tag{223}$$

where $\xi = R/\sqrt{rd}$ [eq. (37)] and the amplitude A is fixed by the boundary condition at $z = 0$,

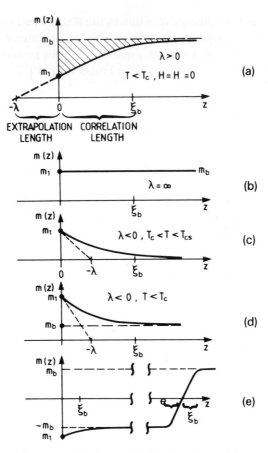

Fig. 46. Schematic order parameter (magnetization) profiles $m(z)$ near a free surface, according to mean field theory. Various cases are shown: (a) Extrapolation length λ positive. The transition of the surface from the disordered state to the ordered state is driven by the transition in the bulk ("ordinary transition"). The shaded area indicates the definition of the surface magnetization m_s. (b) Extrapolation length $\lambda = \infty$. The transition of the surface is called "special transition" ("surface-bulk-multicritical point"). (c), (d) Extrapolation length $\lambda < 0$, temperature above the bulk critical temperature (c) or below it (d). The transition between states (c) and (d) is called the "extraordinary transition". (e) Surface magnetic field H_1 competes with bulk order ($m_b > 0$, $H_1 < H_1^*$ such that $m_1 < -m_b$). In this case a domain of oppositely oriented magnetization with macroscopic thickness ("wetting layer") separated by an interface from the bulk would form at the surface, if the system is at the coexistence curve ($T < T_c$, $H = 0$). From Binder (1983).

$$A \equiv \frac{\lambda H_1 d/(k_B T R^2) - H/(k_B T r)}{1 + (\lambda/\xi)},$$

$$\phi_1 = \frac{\lambda H_1 d/(k_B T R^2)}{1 + \lambda/\xi} + \frac{H}{k_B T r} \frac{1}{\xi/\lambda + 1}$$

(224)

Now local susceptibilities can be defined as (Binder and Hohenberg, 1972; Binder, 1983), for $\lambda > 0$,

$$\chi_1 = \left(\frac{\partial \phi_1}{\partial H}\right)_{H_1,T} = \frac{1}{k_B T r} \frac{1}{(\xi/\lambda) + 1} \propto \left(\frac{T}{T_c} - 1\right)^{-1/2},$$
$$T \to T_c^+, \qquad (225)$$

$$\chi_{11} = \left(\frac{\partial \phi_1}{\partial H_1}\right)_{H,T} = \frac{\lambda d}{(k_B T R^2)(1 + \lambda/\xi)} \propto \text{const} - \left(\frac{T}{T_c} - 1\right)^{1/2},$$
$$T \to T_c^+. \qquad (226)$$

Apart from this local order parameter ϕ_1 and its derivatives there is also interest in the surface excess order parameter ϕ_s defined from the shaded area underneath the profile in fig. 46a as

$$\phi_s = \int_0^\infty dz[\phi_b - \phi(z)]; \qquad (227)$$

using eq. (223) one finds

$$\phi_s = -A\xi \to \chi_T \xi H \propto \left(\frac{T}{T_c} - 1\right)^{-3/2} H, \qquad T \to T_c^+, \qquad (228)$$

and hence the "surface susceptibility" $\chi_s = (\partial \phi_s / \partial H)_{H_1,T} \propto (T/T_c - 1)^{-3/2}$. All these quantities can also be defined as derivatives of the surface excess free energy $f_s(T, H, H_1)$ defined already in eq. (2): one can show that

$$\phi_s = -\left(\frac{\partial f_s}{\partial H}\right)_{T,H_1}, \quad \chi_s = -\left(\frac{\partial^2 f_s}{\partial H^2}\right)_{T,H_1}, \qquad (229)$$

$$\phi_1 = -\left(\frac{\partial f_s}{\partial H_1}\right)_{T,H}, \quad \chi_1 = -\left(\frac{\partial f_s}{\partial H \partial H_1}\right), \quad \chi_{11} = -\left(\frac{\partial^2 f_s}{\partial H_1^2}\right)_{T,H} \qquad (230)$$

One now can define critical exponents by analogy with bulk critical behavior as ($t = T/T_c - 1$)

$$f_s^{\text{sing}} \propto |t|^{2-\alpha_s}, \qquad \phi_s \propto (-t)^{\beta_s}, \qquad \phi_1 \propto (-t)^{\beta_1}, \qquad (231)$$

$$\chi_s \propto |t|^{\gamma_s}, \qquad \chi_1 \propto |t|^{\gamma_1}, \qquad \chi_{11} \propto |t|^{\gamma_{11}}, \qquad (232)$$

and further exponents can be introduced to describe the response at $T = T_c$ to the fields H and H_1, as well as the behavior of order parameter correlation functions. In particular, due to the broken translational invariance introduced by the surface directions parallel and perpendicular to the surface are no longer equivalent, and one has for the correlation function $G(x) \equiv G(\rho, z) = \langle \phi(0, 0)\phi(\rho, z)\rangle$ where one site is at the surface, and ρ is a coordinate parallel to it

$$G_\parallel(\rho) \equiv G(\rho, 0) \propto \rho^{-(d-2+\eta_\parallel)}, \quad G_\perp(z) \equiv G(0, z) \propto z^{-(d-2+\eta_\perp)},$$
$$T = T_c, \qquad (233)$$

where exponents $\eta_\parallel$, $\eta_\perp$ different from η [defined in eq. (43)] are introduced. In Landau's theory, the exponents defined so far have the values

$$\alpha_s = \tfrac{1}{2}, \quad \beta_s = 0, \quad \beta_1 = 1, \quad \gamma_s = \tfrac{3}{2}, \quad \gamma_1 = \tfrac{1}{2}, \quad \gamma_{1,1} = -\tfrac{1}{2},$$
$$\eta_\perp = 1, \quad \eta_\parallel = 2; \qquad (234)$$

some of these exponents have been calculated already in eqs. (225), (226) and (228). As for bulk critical phenomena, one can introduce a scaling hypothesis for the surface free energy, which reads

$$f_s^{\text{sing}}(T, H, H_1) = |t|^{2-\alpha-\nu} \tilde{f}_s(\tilde{H}, \tilde{H}_1) \qquad (235)$$

where $\tilde{H}$ is defined in eq. (84), and $\tilde{H}_1 \propto H_1 |t|^{-(\gamma_{1,1}+\beta_1)}$. This shows that the surface introduces only a single new exponent at the ordinary transition, all others can be found from scaling relations, some of which we quote here

$$\alpha_s = \alpha + \nu, \quad \beta_s = \beta - \nu, \quad \gamma_s = \gamma + \nu = 2\gamma_1 - \gamma_{11},$$
$$\beta_1 + \gamma_1 = \beta + \gamma, \quad \gamma_1 = \nu(2 - \eta_\perp), \quad \gamma_{11} = \nu(1 - \eta_\parallel), \qquad (236)$$
$$\eta_\parallel = 2\eta_\perp - \eta_b,$$

referring to Binder (1983) or Diehl (1986) for detailed derivations. For this ordinary transition, these exponents are rather accurately known from renormalization group expansions (Diehl, 1986) and Monte Carlo calculations (Binder and Landau, 1984; Landau and Binder, 1990). As an example, fig. 47 gives a plot of the surface layer magnetization m_1 vs. $(1 - T/T_c)$, for the Ising model of eq. (1) (Binder and Landau, 1984). Here no magnetic fields are included ($H = H_1 = 0$), but the exchange constant J_s in the surface plane is varied. For small enough values of the ratio J_s/J the slope of the straight lines on this log–log plot is independent of J_s/J, indicating the exponent $\beta_1 \approx 0.78$ which implies (via eqs. (236), using also the best numerical values (Le Guillou and Zinn-Justin, 1980) for the bulk Ising exponents for $d = 3$, $\alpha \approx 0.11$, $\nu \approx 0.63$, $\eta \approx 0.03$, $\gamma \approx 1.24$, $\beta \approx 0.325$) the following set of exponents describing the surface critical behavior of Ising models:

$$\alpha_s \approx 0.74, \quad \beta_s \approx -0.305, \quad \gamma_s \approx 1.87, \quad \beta_1 \approx 0.78,$$
$$\gamma_1 \approx 0.78, \quad \gamma_{11} \approx -0.31, \quad \eta_\perp \approx 0.76, \quad \eta_\parallel \approx 1.49 \qquad (237)$$

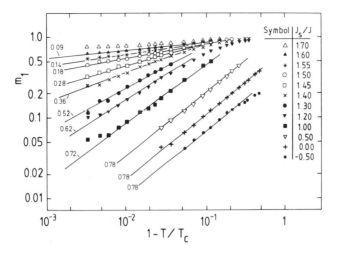

Fig. 47. Log–log plot of the surface layer magnetization m_1 vs. reduced temperature, for various ratios of the exchange J_s in the surface planes to the exchange J in the bulk. Slopes of the straight lines yield effective exponents β_1^{eff} (indicated by the number). Data are from Monte Carlo simulation of $50 \times 50 \times 40$ lattices with two free 50×50 surfaces and otherwise periodic boundary conditions. From Binder and Landau (1984).

Similar studies have been carried out for XY (Landau et al., 1989) and Heisenberg ferromagnets with free surfaces (Binder and Hohenberg, 1974), yielding a somewhat larger value of $\beta_1 (\beta_1 \approx 0.84)$. This result is compatible with an experimental determination of β_1 for the isotropic ferromagnet Ni, $\beta_1 \approx 0.82$ (Alvarado et al., 1982).

Another study of the critical exponent β_1 for the second-order transition of the alloy Fe$_3$Al which undergoes an order–disorder transition at about 500°C from the $D0_3$ phase to the B_2 phase has been carried out by means of evanescent X-ray scattering (Mailänder et al., 1991), yielding $\beta_1 = 0.77 \pm 0.02$, in good agreement with eq. (237). From an analysis of the diffuse scattering, using detailed theories (Dietrich and Wagner, 1984; Gompper, 1986) the exponent $\eta_\parallel = 1.52 \pm 0.04$ could also be extracted (Mailänder et al., 1990), again in agreement with eq. (237). The behavior of alloy surface ordering is somewhat more intricate than that of ferromagnets, however, since usually one component of the alloy is enriched at the surface (Johnson and Blakely, 1979), and this variation of a "non-ordering field" near the surface may induce a surface field H_1 acting on the order parameter (Schmid, 1993). Here we shall not discuss these problems with alloys further, but refer the interested reader to a recent thorough review (Dosch, 1991).

We now return to the decrease of the effective exponent β_1^{eff} seen in fig. 47 when J_s/J increases. This behavior is interpreted (Binder and Hohenberg, 1974; Binder and Landau, 1984; Landau and Binder, 1990) in terms of a

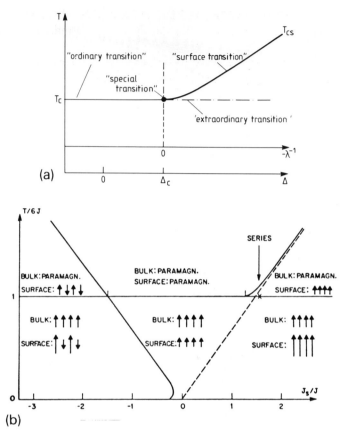

Fig. 48. (a) Schematic phase diagram of the surface of a semi-infinite anisotropic ferromagnet, showing the plane of variables temperature T and enhancement $\Delta = J_s/J - 1$ of the surface exchange constant [in mean field theory, this is related to the inverse of the extrapolation length λ introduced in eqs. (220) and (222)]. At the surface transition line $T_{cs}(\Delta)$, the surface layer undergoes a two-dimensional ferromagnetic ordering while the bulk still stays disordered. The surface free energy then exhibits for $T = T_c$ where the bulk orders also some (weak) singularities (this is called the "extraordinary transition"). From Binder (1983). (b) Transition temperatures for the surface of an Ising simple cubic lattice with exchange J_s in the surface different from the exchange J in the bulk, according to a layer-wise molecular field approximation. For negative J_s an antiferromagnetic ordering of the surface layer occurs. The point marked "series" is the first estimation of the special transition ($J_{sc}/J \approx 1.5$) due to the extrapolation of high temperature series expansions. The ordering of surface and bulk is schematically indicated by arrows. From Binder and Hohenberg (1974).

crossover towards the surface bulk multicritical point, which is estimated to occur at $J_s/J = 1.52 \pm 0.02$ (Landau and Binder, 1990). Figure 48 shows both a schematic phase diagram near this multicritical point, explaining the notation already used in fig. 46, and a numerical phase diagram that results from eq. (1) when treated in molecular field approximation (Binder and Hohenberg, 1974).

Of particular interest is the behavior near the multicritical point, where again a crossover scaling description applies (Binder and Landau, 1984, 1990)

$$f_s(T, H, H_1, \Delta - \Delta_c) = a_s^m |t|^{2-\alpha-\nu} \tilde{f}_{sm}^{\pm} \left\{ bH|t|^{-(\beta+\gamma)}, \right.$$
$$c_m H_1 |t|^{-(\beta_1^m + \gamma_1^m)},$$
$$\left. d_m(\Delta - \Delta_c)|t|^{-\varphi_m} \right\} \quad (238)$$

where a_s^m, b, c_m, d_m are (non-universal) scale factors, $\tilde{f}_{sm}^{\pm}$ is a scaling function, and now two exponents $\beta_1^m + \gamma_1^m$, φ_m are needed to describe the surface critical behavior at this "special transition". The precise values of these exponents are still somewhat controversial (Landau and Binder, 1990; Ruge and Wagner, 1992). An important consequence of eq. (237) is that the shape of the phase diagram near Δ_c in fig. 48a is also controlled by the crossover exponent, namely

$$\frac{T_{cs}}{T_c} - 1 \propto (\Delta - \Delta_c)^{1/\varphi_m}. \quad (239)$$

At this point, we emphasize that the "surface transition" at T_{cs} is a purely two-dimensional ordering phenomenon and hence it is simply described by the two-dimensional Ising exponents as listed in table 1. Of course, fig. 48 holds therefore only for one-component (Ising-like) ordering: for an XY-model, the surface transition still exists but has the character of a Kosterlitz–Thouless transition, and similar modifications are also predicted for the phase diagrams describing the antiferromagnetic surface ordering of ferromagnets, fig. 49 (Binder and Landau, 1985). These phase diagrams of "magnetic surface reconstruction" (Trullinger and Mills, 1973) are rather speculative, we are not aware of any explicit calculations, apart from a study of the ferromagnetic special transition of XY magnets (Peczak and Landau, 1991). In fig. 49, it is argued that the ferromagnetic ordering of the bulk acts like a symmetry-breaking magnetic field on the antiferromagnetic order of the surface, and thus even for $n = 3$ a Kosterlitz–Thouless (1973) transition to a "spin-flop" kind of arrangement (but without true long range order [LRO]) is still possible. Similarly, the Kosterlitz–Thouless transition for $n = 2$ is turned into an Ising-like transition for $T < T_c$ for the spin components which are perpendicular to the direction of the bulk magnetization.

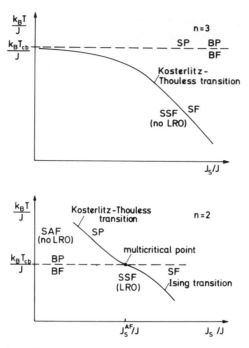

Fig. 49. Schematic phase diagrams predicted for the antiferromagnetic order at the surfaces of isotropic Heisenberg magnets ($n = 3$, upper part) and of XY magnets ($n = 2$), lower part. Phases occurring are bulk paramagnetic (BP), bulk ferromagnetic (BF), surface paramagnetic (SP), surface ferromagnetic (SF), surface antiferromagnetic (SAF) and surface spin flop (SSF). From Binder and Landau (1985).

Evidence for a "surface transition" (to a ferromagnetic state of the surface) has been found for ferromagnetic Gd (Weller and Alvarado, 1988; Rau and Robert, 1987). The interpretation of such experiments, however, is complicated by the fact that surface anisotropies may be much stronger than the magnetic anisotropy in the bulk, and hence complicated crossover phenomena may occur.

As a last point of this section, we draw attention to the fact that for Ising-type orderings the disturbance of the ordering near the surface decays exponentially fast with the distance z from the surface, $m(z) \approx m_b [1 - \text{const} \exp(-z/\xi)]$, cf. fig. 50, while for a Heisenberg ferromagnet a power law decay is observed, $m(z) \approx m_b [1 - \text{const}(\xi/z)]$, in accord with spin wave theory (Binder and Hohenberg, 1974). While the local interactions at the surface [modelled by J_s in eq. (1)] determine the value of m_1 and hence the amplitude of the deviation $m_b - m(z)$, the range of this deviation is controlled solely by the correlation length ξ of the bulk. This is a feature correctly predicted by Landau's theory, eq. (223), although this theory does

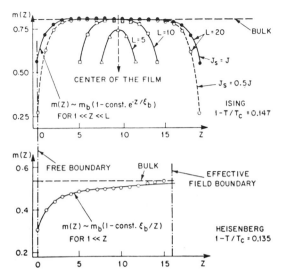

Fig. 50. Magnetization profiles across thin Ising films [eq. (1) with $H = H_1 = 0$], upper part, and near the surface of semi-infinite Heisenberg ferromagnets, lower part (where bulk behavior in the Monte Carlo simulation is enforced by an effective field boundary condition at $z = 16$). Note that in the Ising case (where three film thicknesses $L = 5$, 10, and 20 are shown) the surface layer magnetization $m_1 = m(z = 0)$ is independent of L, and for $L \geq 10$ already the bulk value of the order parameter is reached in the center of the film. For the Heisenberg model, on the other hand, at a comparable temperature distance from T_c the free surface produces a long-range perturbation of the local magnetization $m(z)$. From Binder and Hohenberg (1974).

not yield the power-law form of the magnetization deviation for isotropic magnets.

3.2. Wetting phenomena

We consider now the adsorption of fluids from the gas phase on hard walls. From a macroscopic point of view, one can consider the formation of fluid droplets of spherical cap-like shape and thin fluid layers spread out over the whole substrate surface as competing possibilities and ask which geometry leads to a minimum of the free energy. Three interfacial free energies play a role — the gas–liquid interfacial tension f_{int}, as well as the surface free energy density f_s^ℓ of liquid in contact with the wall, and of gas f_s^g in contact with the wall. A "sessile" (i.e., stable) droplet with contact angle θ given by (Young 1805)

$$\cos \theta = \frac{f_s^g - f_s^\ell}{f_{int}} \tag{240}$$

occurs as long as $f_s^g < f_s^\ell + f_{int}$: a surface that satisfies this condition is called non-wet. The density profile of the gas atomically close to the wall then is qualitatively described by fig. 5, upper part. In contrast, if $f_s^g > f_s^\ell + f_{int}$ it is energetically more favorable for the droplets to completely spread out and form a (thick) film of fluid phase, as sketched in the lower part of fig. 5, describing a wet surface. Approximating the description of the gas–liquid condensation transition in terms of a lattice gas model, and invoking the analogy between Ising magnets and lattice gases (sect. 2.3), it is clear that the wetting behavior can be (qualitatively) described by Ising models again, and to this fact we have already alluded in fig. 46e (which is nothing but the same situation as the lower part of fig. 5, but in "magnetic notation"). Let us examine (Cahn, 1977) the behavior of eq. (238) near the critical point T_c: Since $f_{int} \propto (-t)^\mu$ with $\mu = (d - 1)v$, while $f_s^g - f_s^\ell$ should be proportional to the density difference at the surface, $f_s^g - f_s^\ell \propto \rho_1^\ell - \rho_1^g \propto (-t)^{\beta_1}$, where β_1 is nothing but the critical exponent of the surface layer order parameter, considered in the previous subsection. Since $\beta_1 < \mu$, one always expects that $\cos \theta$ increases as $T \to T_c$ up to a temperature T_w such that for $T_w < T < T_c$ the surface is wet (Cahn, 1977; see also Ebner and Saam, 1977).

The theory of wetting phenomena has been extensively reviewed recently (Sullivan and Telo da Gama, 1986; Dietrich, 1988); we present here a very brief introduction only. We return to the free energy of a semi-infinite system [eqs. (219), (220)] and rescale the parameters such (Schmidt and Binder, 1987) that all the parameters of the bulk free energy density are absorbed in the rescaled bulk field h, order parameter $\mu(\mathcal{Z})$ and rescaled distance $\mathcal{Z}$,

$$\frac{\Delta\mathcal{F}\{\mu(\mathcal{Z})\}}{k_B T_c} = \int_0^\infty d\mathcal{Z} \left\{ \frac{1}{2}(\frac{\partial}{\partial\mathcal{Z}}\mu)^2 - \mu^2 + \frac{1}{2}\mu^4 - \mu h \right\}$$
$$- \frac{h_1}{\gamma}\mu_1 - \frac{1}{2}\frac{g}{\gamma}\mu_1^2, \tag{241}$$

$\mu_1 = \mu(\mathcal{Z} = 0)$, and h_1/γ is then the rescaled surface field H_1, g/γ the rescaled coefficient $-R^2\lambda^{-1}/d$. Now the Euler–Lagrange equations describing the solution that minimizes eq. (241) read

$$\frac{1}{2}\frac{\partial^2\mu}{\partial\mathcal{Z}^2} + \mu - \mu^3 + h = 0, \quad \gamma\frac{\partial\mu}{\partial\mathcal{Z}}\bigg|_{\mathcal{Z}=0} + h + h_1 + g\mu(\mathcal{Z} = 0) = 0.$$
$$\tag{242}$$

Multiplying the first of these equations by $\partial\mu(\mathcal{Z})/\partial\mathcal{Z}$ and integrating from $\mathcal{Z}' = \mathcal{Z}$ to $\mathcal{Z}' = \infty$ one obtains for $h = 0$

$$\left(\frac{\partial\mu(\mathcal{Z})}{\partial\mathcal{Z}}\right)^2 = [\mu^2(\mathcal{Z}) - 1]^2, \quad \text{i.e.} \quad \frac{\partial\mu(\mathcal{Z})}{\partial\mathcal{Z}} = -|\mu^2(\mathcal{Z}) - 1| \tag{243}$$

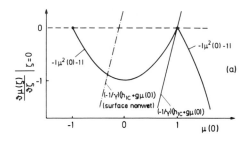

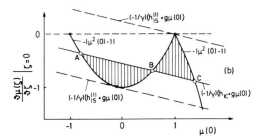

Fig. 51. Plot of $-\partial \mu(\mathcal{Z})/\partial \mathcal{Z}|_{\mathcal{Z}=0}$ versus $\mu(0)$ for the cases of a second-order wetting transition (a) and a first-order wetting transition (b). The solution consistent with the boundary condition always is found by intersection of the curve $|\mu^2(0) - 1|$ with the straight line $[h_1 + g\mu(0)]/\gamma$. In case (a) this solution is unique for all choices of h_1/γ (keeping the order parameter g/γ fixed). Critical wetting occurs for the case where the solution (denoted by a dot) occurs for $\mu(0) = +1$, where then $\partial\mu(\mathcal{Z})/\partial \mathcal{Z}|_{\mathcal{Z}=0} = 0$ and hence the interface is an infinite distance away from the surface. For $h_1 > h_{1c}$ the surface is non-wet while for $h_1 > h_{1c}$ the surface is wet. In case (b) the solution is unique for $h_1 < h_{1s}^{(1)}$ (only a non-wet state of the surface occurs) and for $h > h_{1s}^{(2)}$ (only a wet state of the surface occurs). For $h_{1s}^{(2)} > h_1 > h_{1s}^{(1)}$ three intersections (denoted by A, B, C in the figure) occur, B being always unstable, while A is stable and C metastable for $h_{1c} > h_1 > h_{1s}^{(1)}$, and A is metastable and C stable for $h_{1s}^{(2)} > h > h_{1c}$. At h_{1c} where the exchange of stability between A and C occurs (i.e., the first-order wetting transition) the shaded areas in fig. 51b are equal. This construction is the surface counterpart of the Maxwell-type construction of the first-order transition in the bulk Ising model (cf. fig. 37). From Schmidt and Binder (1987).

using $\mu(\mathcal{Z} \to \infty) \to -1$. Equation (243) for $\mathcal{Z} = 0$ must also satisfy the boundary condition in eq. (242). Figure 51 illustrates the graphical solution of these equations, both for the case of critical wetting which occurs for $\mu(0) = +1$ and at

$$h_{1c} = -g, \qquad g < -2\gamma \tag{244}$$

and for the case of first-order wetting which occurs for $g > -2\gamma$ (fig. 52). In this case, $h_{1s}^{(1)} = -g$ is a stability limit, a "surface spinodal" (Nakanishi and Pincus, 1983) of the metastable wet phase while the other stability limit occurs when the straight line $\partial\mu(Z)/\partial Z|_{Z=0} = -h_1/\gamma - g\mu(0)/\gamma$ is tangential to the curve $\partial\mu(Z)/\partial Z|_{Z=0} = -|\mu^2(0)-1|$, see fig. 51b. One finds (Schmidt and Binder, 1987) that this happens for $h_{1s}^{(2)}/\gamma = 1 + (g/\gamma)^2/4$.

One can show that the second-order wetting transition is characterized by a divergence of the susceptibility χ_1, in this mean field theory of critical wetting

$$\chi_1 \propto (h_1 - h_{1c})^{-1}, \tag{245}$$

while the layer susceptibility $(\partial\phi_1/\partial H_1)_{T,H=0}$ stays finite and exhibits a jump singularity there. A divergence of χ_{11} does occur, however, at the wetting tricritical point in fig. 52. Also it is interesting to note that at the non-wet side of the wetting transition the order parameter profile [eq. (5a)] is always just a piece of the interfacial profile, as obtained in eq. (179), but

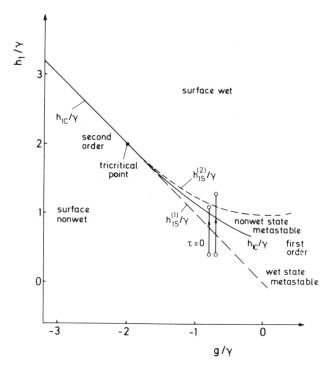

Fig. 52. Phase diagram of the surface plotted in terms of the scaled variables h_1/γ and g/γ. For $g/\gamma < -2$ one observes critical wetting and for $g/\gamma > -2$ one observes first-order wetting. In the latter regime, mean field theory predicts metastable wet and non-wet regions limited by the two surface spinodal lines $h_{1s}^{(1)}$ and $h_{1s}^{(2)}$, respectively. Also two "quenching experiments" are indicated where starting at a rescaled time $\tau = 0$ from a stable state in the non-wet region one suddenly brings the system by a change of h_1 into the metastable wet or unstable non-wet region, respectively. From Schmidt and Binder (1987).

shifted such that the inflection point is not at $z = 0$ but rather at $z = z_o$, $\phi(z) = +\phi_b \tanh[(z - z_o)/2\xi]$, z_o being defined such that the boundary condition is fulfilled, $\phi(z = 0) = \phi_1$. As the wetting transition (in critical wetting) is approached, one has $\phi_1 \to -\phi_b$. (In fig. 46e, it was assumed that $H_1 < 0$, $\phi_b > 0$, while in figs. 51 and 52 the inverse situation $H_1 > 0$, $\phi_b < 0$ is considered: of course, due to spin reversal symmetry of an Ising magnet, or the equivalent particle-hole symmetry of a lattice gas, these situations are completely analogous). From the hyperbolic tangent profile it then follows that the thickness of the wetting layer (z_o) diverges as $z_o \propto |\ln|\phi_1 + \phi_b|| \propto |\ln|h_1 - h_{1c}||$ as one approaches the wetting transition from the non-wet side.

It also is of great interest to study the situation where a non-zero bulk field h is present, for the case where the surface would be wet at $h = 0$. In gas–fluid condensation (fig. 5), this translates into the situation that the chemical potential μ of the gas is initially chosen such that one stays in the one-phase region ($\mu < \mu_{coex}$, undersaturated gas) and then one increases the gas pressure and hence μ such that $\mu \to \mu_{coex}$. Then the coverage (or adsorbate surface excess density ρ_s per substrate area) diverges (fig. 53c). Note that here we have tacitly assumed that the gas–liquid interface even close to the wall can be considered as a smooth, delocalized object (fig. 6), so that ρ_s vs. μ yields a smooth curve, with at most one transition ("prewetting") and no sequence of layering transitions (fig. 4) occurs. The conditions when we have layer-by-layer growth (multilayer adsorption) and when we have wetting will be discussed in the next subsection.

Again we emphasize that wetting phenomena are not restricted to the gas–liquid transition, but analogous phenomena occur for all transitions that belong to the same "universality class". A particularly, practically important, example are binary (fluid or solid) mixtures that undergo phase separation in the bulk (fig. 55).

The mean field theory of wetting phenomena with short range forces due to the wall, as outlined in eqs. (241)–(245) and described in figs. 51–55, is obviously closely related to the theory of surface critical phenomena, where basically the same description [eqs. (219), (220)] was used as a starting point. This holds true also on a more microscopic level: the Ising-lattice gas Hamiltonian, eq. (1), can be used to study both surface critical phenomena and wetting! These relations are clearly understood if one considers global phase diagrams in the space of variables T, J_s, H_1 and H; fig. 56a, b (Nakanishi and Fisher, 1982; Binder and Landau, 1988; Binder et al., 1989). It thus turns out that the surface–bulk-multicritical point (considered in sect. 3.1) is also the endpoint of the line of tricritical wetting transitions (figs. 56, 57). By extensive numerical Monte Carlo work (e.g. fig. 57; Binder and Landau, 1988; Binder et al. 1989) of the model eq. (1) both second-order and first-order wetting transitions as well as the wetting tricritical

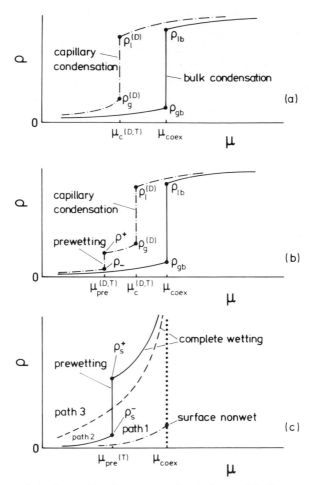

Fig. 53. Schematic isotherms (density ρ versus chemical potential μ) corresponding to the gas–liquid condensation in capillaries of thickness D, for the case without (a) and with (b) prewetting, and adsorption isotherm (c) for a semi-infinite system, where the surface excess density ρ_s is plotted vs. μ. Full curves in (a) and (b) plot the density ρ vs. μ for a bulk system, phase coexistence occurs there between ρ_{gb} (bulk gas) and $\rho_{\ell b}$ (bulk liquid), while in the capillary due to the adsorption of fluid at the walls the transition is shifted from μ_{coex} to a smaller value $\mu_c(D, T)$ (with $\mu_{\text{coex}} - \mu_c(D, T) \propto 1/D$, the "Kelvin equation"), and the density jump (from $\rho_g(D)$ to $\rho_\ell(D)$) is reduced. Note also that in the case where a semi-infinite system exhibits a first-order wetting transition T_w, for $T > T_w$ one may cross a line of (first-order) prewetting transitions (fig. 54) where the density in the capillary jumps from ρ_- to ρ_+ [or in the semi-infinite geometry, the surface excess density jumps from ρ_s^- to ρ_s^+, cf. (c)], which means that a transition occurs from a thin adsorbed liquid film to a thick adsorbed film. As $\mu \to \mu_{\text{coex}}$, the thickness of the adsorbed liquid film in the semi-infinite geometry then diverges to infinity in a smooth manner, $\rho_s \propto |\ln(\mu - \mu_{\text{coex}})|$, which is called "complete wetting", if $T > T_w$, while ρ_s goes towards a finite non-zero value at $\mu = \mu_{\text{coex}}$ if

points could be located (fig. 57), while these calculations still remained inconclusive with respect to the nature of critical wetting with short range surface forces. In fact, in order to consider fluctuations beyond the mean field theory of critical wetting, one uses the "drumhead model" of a smooth interface [fig. 6, eq. (185)] and describes the effect of the surface by an effective potential $V_{\text{eff}}(h)$ acting on the local position of the interface $h(x, y)$. A renormalization group treatment (Brézin et al., 1983; Lipowsky et al., 1983; Fisher and Huse, 1985; Lipowsky and Fisher, 1987) shows that due to capillary-wave excitations $d = 3$ dimensions is a marginal case for the validity of mean field theory of critical wetting, and one expects a non-universal behavior with exponents which depend on the interfacial stiffness, κ, which is not yet known very accurately for the Ising model (Fisher and Wen, 1992).

However, we shall not go into detail about this problem, in particular since it is not clear whether this problem is relevant to experiment (Dietrich, 1988). In particular, for fluids and fluid binary mixtures the long range of the van der Waals attractions between the atoms changes the behavior significantly. While for bulk critical phenomena the potential between two atoms behaves as $V(r) \propto r^{-6}$ for large distances, and this is a strong enough fall-off so the critical behavior is the same as that for short range systems, for an atom close to a wall the situation is different, since one has to integrate over all pair potentials between atoms making up the hard wall and the considered adsorbate atom. As a result, a potential results that decreases with distance from the wall rather slowly, $V(z) \propto z^{-3}$, and this translates into a long range interface potential $V_{\text{eff}}(h) \propto h^{-3}$. For such long range surface potentials, the capillary wave type fluctuations of the interface are irrelevant, in a renormalization group sense, and mean field theory can be used. Since the conditions for observing critical wetting are rather restrictive (Dietrich, 1988), it is no surprise that most experimental reports find first-order wetting transitions (see Dietrich, 1988, for a recent review). However, it is also evident, that experiments on wetting phenomena have their own problems — slow equilibration of thick adsorbed layers is a problem, adsorbed impurities at the surface may be a severe problem, and sometimes subtle finite size effects need to be considered. A discussion of these phenomena is outside of our scope here. See Franck (1992) and Beysens (1990) for recent reviews of wetting experiments.

---

the surface is non-wet. In the case where one has critical wetting (or when one has first-order wetting but chooses a path in the (T, μ)-plane that does not cross the prewetting line, see fig. 54) one has a smooth adsorption isotherm [broken curve marked *path 3* in (c)]. From Binder and Landau (1992a).

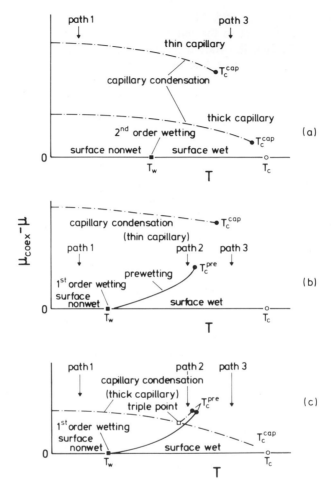

Fig. 54. Schematic phase diagrams for wetting and capillary condensation in the plane of variables temperature and chemical potential difference. (a) Refers to a case in which the semi-infinite system at gas–liquid condensation ($\mu_{coex} - \mu = 0$) undergoes a second-order wetting transition at $T = T_w$. The dash-dotted curves show the first-order (gas–liquid) capillary condensation at $\mu = \mu_c(D, T)$ which ends at a capillary critical point T_c^{cap}, for two choices of the thickness D. For all finite D the wetting transition then is rounded off. (b), (c) refer to a case where a first-order wetting transition exists, which means that ρ_s remains finite as $T \to T_w^-$ and there jumps discontinuously towards infinity. Then for $\mu_{coex} - \mu > 0$ a transition may occur during which the thickness of the layer condensed at the wall(s) jumps from a small value to a larger value ("prewetting"). For thick capillaries, this transition also exists (c) but not for thin capillaries because then $\mu_{coex} - \mu_c$ (D,T) simply is too large. Full dots in this figure denote two-dimensional critical points, full squares denote wetting transitions, open circles show bulk three-dimensional criticality, and the open square denotes a capillary triple point. In (b) and (c) three paths 1, 2 and 3 are shown which refer to the three adsorption isotherms of fig. 53c. From Binder and Landau (1992a).

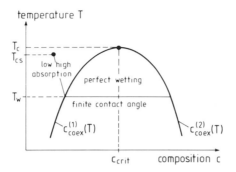

Fig. 55. Schematic phase diagram of a binary mixture with an unmixing transition in the bulk (miscibility gap from composition $c_{coex}^{(1)}(T)$ to $c_{coex}^{(2)}(T)$ ending in a critical point T_c, c_{crit}) and a first-order wetting transition at T_w at the surface of the mixture and a wall. For $T > T_w$, a (thick) layer of concentration with the other branch of the coexistence curve, $c_{coex}^{(2)}(T)$ is adsorbed at the wall. The prewetting line ending in a surface critical point T_{cs} is also shown. After Cahn (1977).

As a final point of this subsection, we mention that wetting phenomena may also occur in one dimension less: when we consider adsorption on stepped surfaces in the submonolayer range, fig. 3, the change of the binding potential near the boundary of a terrace may give rise to preferential adsorption along this boundary. In this case, the fluctuations of the interface in a wetting transition are much more relevant, as is already obvious from the much larger width produced by these capillary wave fluctuations for an one-dimensional interface [eq. (190)] in comparison with a two-dimensional one [eq. (189)], and the same conclusion is evident from computer-simulated pictures of a two-dimensional version of eq. (1), fig. 58 (Albano et al., 1989a), where one can see that an interface that is not bound tightly to the terrace boundary indeed undergoes large fluctuations in its local position. In this model a critical wetting transition can be located exactly at (Abraham, 1980)

$$\exp\left[\frac{2J}{k_BT}\right]\left\{\cosh\left(\frac{2J}{k_BT}\right) - \cosh\left(\frac{2H_1}{k_BT}\right)\right\} = \sinh\left(\frac{2J}{k_BT}\right), \qquad (246)$$

see fig. 59a. In this case the singular part of the boundary free energy can be written as ($\tilde{t}$ denotes here the distance from the critical wetting line)

$$\frac{\mathcal{F}_s^{(sing)}}{k_BT} = \tilde{t}^{\nu_\parallel}\tilde{F}_s\left\{\tilde{t}^{-(\nu_\parallel+\nu_\perp)}H, L\tilde{t}^{\nu_\perp}\right\}, \qquad (247)$$

where we have anticipated a finite width L of the terrace, and the exponents of the correlation lengths $\xi_\parallel \propto \tilde{t}^{-\nu_\parallel}$, $\xi_\perp \propto \tilde{t}^{\nu_\perp}$ are known exactly (Abraham and Smith, 1986; Abraham, 1988) as $\nu_\parallel = 2$, $\nu_\perp = 1$. (Note that a similar

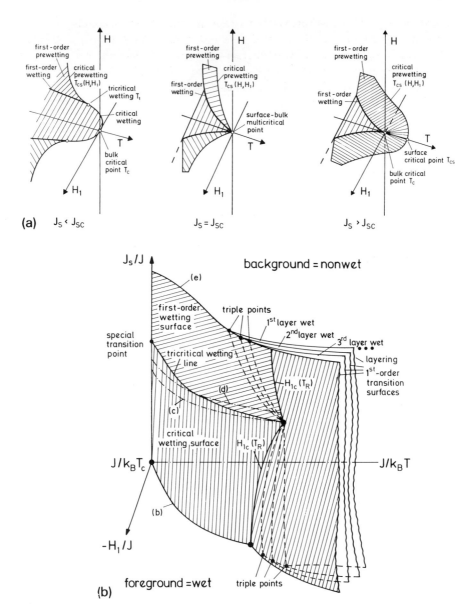

Fig. 56. (a) Schematic phase diagrams of a semi-infinite Ising magnet in the vicinity of the bulk critical point T_c as a function of temperature T, bulk field H, and surface field H_1. In the shaded part of the plane $H = 0$ the system (for $T < T_c$) is non-wet, while outside of it (for $T < T_c$) it is wet. The wetting transition is shown by a thin line where it is second order and by a thick line where it is first order. First-order prewetting surfaces terminate in the plane $H = 0$ at the first-order wetting line. Critical and multicritical points are indicated

ansatz as eq. (247) with prefactor written in general dimensionality $\tilde{t}^{(d-1)\nu_\parallel}$ applies in the higher dimensions d, too, but in mean field theory as treated above $\nu_\parallel = 1$, $\nu_\perp = 0$, implying logarithmic growth laws of the wetting layer as the transition is approached.) Equation (247) implies that the surface excess magnetization (i.e., the thickness of the adsorbed boundary layer) diverges as $(L \to \infty, \tilde{t} = 0)$

$$m_s = -\left(\frac{\partial \mathcal{F}_s^{\text{sing}}}{\partial H}\right)_{T,H_1} \propto H^{-\nu_\perp/(\nu_\parallel+\nu_\perp)} = H^{-1/3}, \qquad H \to 0, \qquad (248)$$

and similarly $\Delta m_1 = m_1(H) - m_1(0) \propto H^{(\nu_\parallel-1)/(\nu_\parallel+\nu_\perp)} = H^{1/3}$. While the latter result is in rough agreement with corresponding simulations, fig. 59b, eq. (248) has not yet been verified numerically.

In models with several ground states (3 state Potts model, clock models etc.) a further wetting phenomenon may occur at interfaces between coexisting domains: e.g., in a model with an interface between domains in states 1 and 2 the third phase may intrude in the interface (Selke, 1984; Sega et al., 1985; Dietrich, 1988).

3.3. Multilayer adsorption

While in figs. 53 and 54 it was assumed that there exists a wetting transition temperature T_w, such that the surface excess density ρ_s for $\mu \to \mu_{\text{coex}}$ reaches a finite limit for $T < T_w$, and diverges smoothly for $T > T_w$, already in fig. 4 it has been emphasized that the growth of the adsorbed film can also proceed via a sequence of layering transitions. This behavior can also be understood in terms of a mean field approximation based on the Ising-lattice gas Hamiltonian, eq. (1): however, it is then necessary to avoid the continuum approximation, eqs. (219)–(222), because one must take into account that the order parameter changes rapidly on the scale of a lattice spacing. Therefore one has to work on the basis of a layerwise mean field approximation (Pandit and Wortis, 1982; Pandit et al., 1982)

$$m_1 = \tanh\left[H + H_1 + \frac{q_{//}J_s}{k_B T}(m_1) + \frac{J}{k_B T}(m_2)\right], \qquad n \geq 2, \qquad (249)$$

in the figure. After Nakanishi and Fisher (1982). (b) Conjectured phase diagram of eq. (1) for $H = 0$ and semi-infinite geometry in the space of variables J_s/J, H_1, $J/k_B T$. For T less than the roughening transition temperature T_R the "surface" of wetting transitions which separates the wet state of the surface (in the foreground) and non-wet state (in the background) splits into several surfaces, describing the individual layering transitions. Note that the topology how these layering transition surfaces end (assumed here in lines of surface triple points) is still speculative, while the topology near the bulk critical point is established by detailed numerical calculations. From Binder and Landau (1988).

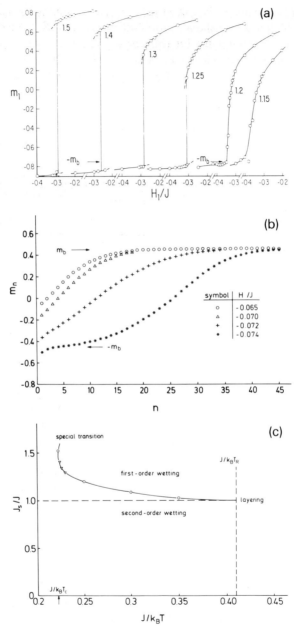

Fig. 57. (a) Location of first-order, second-order and tricritical wetting transitions in the Ising model via Monte Carlo studies of a $50 \times 50 \times 40$ system (with two free 50×50 surfaces which have exchange J_s between nearest neighbors, while all other exchange constants are equal to J, and where a negative surface magnetic field H_1 acts) by a study of the surface layer

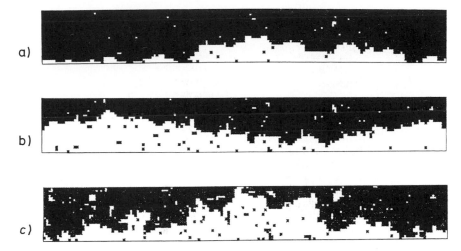

Fig. 58. Snapshot pictures of a lattice gas model of adsorption on an $L \times M$ terrace (fig. 3) choosing $L = 24$, $M = 288$ and the two-dimensional version of eq. (1) with $H = 0$, $J_s = J$, and antiparallel boundary fields $H_1 = -3J$, $H_L = +3J$, at the time step $t = 24000$ MCS/site of a Monte Carlo simulation, and three temperatures: $T = 0.68T_c$ (a), $T = 0.78(T_c)$ (b), and $T = 0.88T_c$ (c). Sites taken by adsorbed atoms are shown in black, empty sites are left white. From Albano et al. (1989a).

$$m_n = \tanh\left[H + \frac{q_{//}J}{k_B T}(m_n) + \frac{J}{k_B T}(m_{n-1} + m_{n+1})\right], \qquad n \geq 2, \quad (250)$$

$q_{//}$ being the coordination number in the lattice plane parallel to the wall. This set of equations indeed leads to an infinite sequence of transitions: however, all these layering transitions extend right up towards the bulk critical temperature. It is believed, however, that the latter feature is an artefact of the mean field approximation and what happens in a correct theory is that the sequence of layering transitions terminates near the roughening transition temperature T_R (Pandit et al., 1982). Figure 60 shows

---

magnetization m_1 vs. H_1/J at $H = 0$ $J/k_B T = 0.25$, for several choices of J_s/J as indicated. The system is simulated in a (metastable) state with positive magnetization m_b in the bulk. For first-order wetting transitions, m_1 on the wet side of the transition is more negative than $-m_b$ (arrows). From such data, $J_s/J \equiv 1.2$ is found at the tricritical wetting transition. From Binder and Landau (1988). (b) Layer magnetization m_n plotted vs. layer number n for $J/k_B T = 0.226$, $J_s/J = 1.33$, $H = 0$, a $128 \times 128 \times 160$ system and two free 128×128 surfaces, and four values of the surface field. Note that within the achieved accuracy $H_1/J = -0.074$ is the tricritical value. From Binder et al. (1989). (c) Line of tricritical wetting transitions separating first-order wetting (above) from second-order wetting (below), in the plane of variables J_s/J and $J/k_B T$. The dashed vertical line $J/k_B T_R \approx 0.41$ shows the roughening transition (Mon et al., 1989). From Binder et al. (1989).

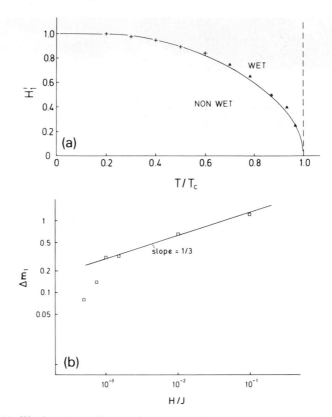

Fig. 59. (a) Wetting phase diagram for a semi-infinite square Ising model with nearest neighbor exchange J and a free surface where a boundary field H_1 acts ($H_1' = H_1/J$). The solid curve represents eq. (246) while the points result from various extrapolations of Monte Carlo data obtained from $L \times M$ strips as in fig. 58, with $L \leq 24$. From Albano et al. (1989a). (b) Log–log plot of $\Delta m_1 = m_1(H) - m_1(0)$ vs. magnetic field H, for $H_1' = H_1/J = 0.5$ and $T = T_w 0.863 T_c$. The straight line shows the theoretical exponent 1/3. From Albano et al. (1990).

schematically the predicted phase diagrams. The numerical study of such phenomena by Monte Carlo simulation of eq. (1) turns out to be rather difficult (Binder and Landau, 1988, 1992b; see fig. 61). These data are thus not suitable for a test of the theoretical prediction (Nightingale et al., 1984)

$$T_R - T_c(n) \propto (\ln n)^{-2}, \qquad n \to \infty \tag{251}$$

In fig. 60 for the case of strong substrate attraction all layering transition lines accumulate at the point $T = 0$, $H = 0$ (i.e., $\mu = \mu_{coex}$). However, this is only true due to the specific assumption of a short range force arising from the surface, which in the Ising lattice gas framework leads to a surface

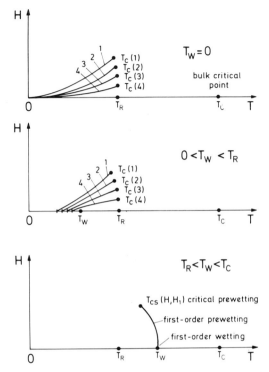

Fig. 60. Schematic phase diagrams of a semi-infinite Ising magnet as a function of bulk field H and temperature T. Three possible "scenarios" are shown (which of them is realized depends on the ratios between the surface and bulk interactions, H_1/J and J_s/J; note that many additional "scenarios" can be thought of and it is not yet clear under which conditions these phase diagrams actually occur). In an adsorption problem, the upper part corresponds to a "strong substrate", the surface being wet at all temperatures, the middle and lower part correspond to "intermediate substrate systems". The surface is only wet if T exceeds a certain temperature T_w. If T_w exceeds the roughening temperature T_R, one just has one first-order prewetting line ending in a prewetting critical point $T_{cs}(H, H_1)$; this is the situation discussed in Sec. 3.2. On the other hand, if $T_w < T_R$, one has an infinite sequence of first-order layering transitions (labeled by the number $n = 1, 2, 3, 4, \ldots$ of the layer in the figure.) These layering transitions end in layering critical points $T_c(n)$, with $\lim_{n\to\infty} T_c(n) = T_R$. After Pandit et al. (1982).

magnetic field H_1 acting in the first layer only, while there is no direct influence of the surface in all higher layers, $n \geq 2$ [eq. (250)]. However, in the framework of layerwise mean field approximations (De Oliveira and Griffiths, 1978; Ebner, 1980; Tarazona and Evans, 1983; Patrykiejew et al., 1990) or of Monte Carlo simulations (Kim and Landau, 1981; Ebner, 1981; Patrykiejew et al., 1990) it is easy to study other assumptions as well, such as a surface potential $V_n = -A/n^3$ acting on an adatom in the nth layer.

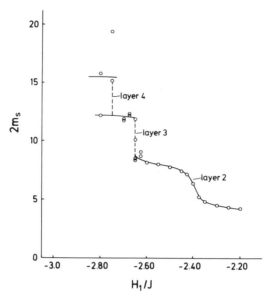

Fig. 61. Surface excess magnetization plotted vs. surface magnetic field H_1/J at $J/k_B T = 0.42$ (note $J/k_B T_R \approx 0.41$) and $J_s/J = 0.5$, for a $L \times L \times D$ system with $L = 128$, $D = 40$. At this temperature, one has exceeded the layering critical points of layers $n = 1$ and $n = 2$, but layering transitions for larger n ($n = 3$ and $n = 4$) still occur. From Binder and Landau (1992b).

Figures 62 and 63 show some typical phase diagrams resulting from a lattice gas model with nearest neighbor exchange but long range surface potential for two choices of the substrate potential strength parameter A. While for strong substrate binding potentials one has a simple sequence of layering transitions at $\mu_{coex}^n(T)$ for the condensation of all layers ($n = 1,2,3,\ldots$), for weaker substrate potentials it happens that the first $2,3,\ldots$ layers may condense together, and also surface triple points occur where layering transitions of different layers start to occur together (e.g., in fig. 63b such a triple point occurs at about $T^* \approx 2.0$ where the layering transitions of layer 3 and of layers $1+2$ coexist).

At this point we emphasize that all our discussion of multilayer adsorption has so far been in the framework of lattice gas models, and thus has left out a very important parameter: this is the misfit between the lattice spacing preferred by the substrate and the lattice spacing of the adsorbate. The strain energy building up in thick layers commensurate with the substrate is expected to prevent wetting in many materials (Wortis, 1985; Huse, 1984; Gittes and Schick, 1984). It is also possible that in the first layer (or first few layers) adjacent to the substrate a structure of the adlayer forms which does not match that of bulk adsorbate material [at least this possibility is

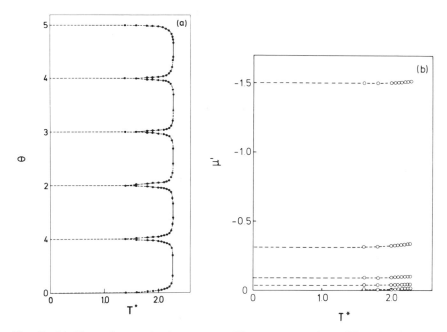

Fig. 62. (a) Phase diagram in the coverage (θ)-temperature plane ($T^* = k_B T/J$ where J is the exchange constant between nearest neighbors) of a nearest-neighbor lattice gas model on the simple cubic lattice with a free surface, and a potential $V(n) = -A/n^3$ with $A = 2.5J$ acting in the nth layer. (b) The corresponding phase diagram in the grand-canonical ensemble ($\mu' = (\mu - \mu_{\text{coex}})/J$). Note that $V(n)$ was cut off for $n > 4$. Thus the curve for the layering transition with $n = 5$ merges at the bulk coexistence curve. From Patrykiejew et al. (1990).

suggested by various model calculations, see Ebner et al. (1983); Wagner and Binder (1986); Georgiev et al. (1990)].

An interesting aspect, of course, that also would deserve detailed discussion are dynamic phenomena associated with wetting and multilayer adsorption, such as the spreading of droplets on walls that should wet (De Gennes, 1985), the growth of wetting layers after quenching experiments (Lipowsky, 1985a; Lipowsky and Huse, 1986; Grant et al., 1987; Grant, 1988; Mon et al., 1987; Schmidt and Binder, 1987; Binder, 1990; Patrykiejew and Binder, 1992; Mannebach et al., 1991; Steiner et al., 1992). While for the growth of wetting layers with short range forces various theories (Lipowsky, 1985a; Schmidt and Binder, 1987) predict a logarithmic growth of the thickness of the layer with time and this seems to be observed in simulations (Mon et al., 1987), for long range forces Lipowsky and Huse (1986) predict a faster growth (power laws $\theta(t) \propto t^x$ with $x = 1/4$ or $1/5$ (non-conserved case) or $x = 1/8$ or $1/10$ (conserved order parameter); the two values in

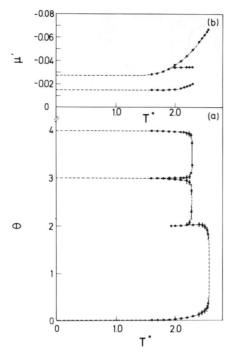

Fig. 63. Phase diagram of multilayer adsorption in the coverage (θ)-temperature plane (a), and corresponding grand-canonical phase diagram (b) for the same model as fig. 62 but a weaker substrate potential ($A = 0.93$ J). From Patrykiejew et al. (1990).

each case refer to non-retarded or retarded van der Waals forces, respectively). We are not aware of an experimental verification of this prediction yet — experiments on the dynamics of the build-up of a wetting layer in polymer mixtures seem to be more consistent with a logarithmic growth (Steiner et al., 1992). In the case of multilayer adsorption, each layer is adsorbed via (two-dimensional) nucleation and growth on top of the previous one, and then the dynamics of growth reflects the details of the phase diagram. As an example, fig. 64 presents some recent computer simulations (Patrykiejew and Binder, 1992) that were motivated by related experiments (Mannebach et al., 1991). Finally we emphasize once again that all these considerations refer to growth phenomena at the surface in a case where the coexisting gas phase is still undersaturated or at most one has saturated gas right at the coexistence curve. We do not discuss here thin film growth from supersaturated gases via heterogeneous nucleation at surfaces, although this would be a topic of great practical interest (Venables et al., 1984; Zinke-Allmang et al., 1992). We also do not discuss here the kinetics of surface enrichment in mixtures (Binder and Frisch, 1991) or surface-induced spin-

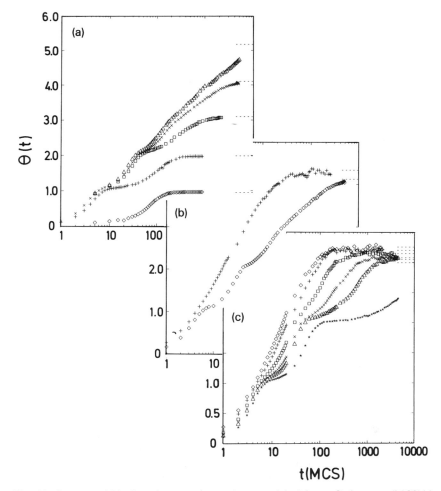

Fig. 64. Coverage $\theta(t)$ plotted versus time t (measured in Monte Carlo steps (MCS)/site, where in one time unit on the average one evaporation or condensation event per lattice site is attempted, for the same model as shown in fig. 62. Case (a) refers to $T^\star = 2.25$ and five choices of the chemical potential difference $\mu'[\mu' = -1.4$ ($\diamond$), -0.25 ($+$), -0.06 ($\square$), -0.02 ($\times$) and -0.004 ($\triangle$). Case (b) refers to $\mu' = -0.02$ and two temperatures, $T^\star = 2.25$ ($\diamond$) and 3.5 ($+$). Case (c) refers to $\mu^\star = -0.06$ and temperatures $T^\star = 2.0$ ($\star$), 2.2 ($\triangle$), 2.3 ($\times$), 2.5 ($\square$), 3.0 ($+$) and 3.5 ($\diamond$). Thin dashed lines at the right-hand side mark equilibrium values. Note that the ordinate scale is linear while the abscissa scale is logarithmic, implying that straight lines on the plot give evidence for $\theta(t) \propto \ln t$. From Patrykiejew and Binder (1992).

odal decomposition (Wiltzius and Cumming, 1991; Bruder and Brenn, 1992; Jones et al., 1991; Puri and Binder, 1992; Shi et al., 1993; Tanaka, 1993; Ball and Essery, 1990).

3.4. The roughening transition

In fig. 7, we have already alluded to the possibility that an interface between a crystal and the gas (or even vacuum) which is atomistically sharp and well-localized even on the scale of lattice spacings at low temperatures may get diffuse and delocalized at higher temperatures due to spontaneous thermal fluctuations. This roughening transition was first directly observed for interfaces between He^4 crystals coexisting with its own superfluid (Avon et al., 1980; Balibar and Castaing, 1980). The existence of either rough or non-rough crystal surfaces has important implications on crystal growth (Müller-Krumbhaar, 1978) and in this context the existence of this transition had already been postulated by Burton et al. in 1951. We have seen (sect. 2.5) that the roughness of the surface is a basic condition for a continuum-type description of interfacial fluctuations in terms of the capillary wave Hamiltonian [eq. (185)], and thus provided understanding that growth of wetting layers can only happen above the roughening transition temperature T_R of the gas–solid temperature of the adsorbate material (sect. 3.2), while the layer-by-layer adsorption (sect. 3.3) can happen only for $T < T_R$: T_R is an accumulation point of layering critical points (eq. (251), figs. 56, 60).

In this subsection we shall add further aspects to this transition, emphasizing in particular that the step free energy $k_B T s(T)$ vanishes at T_R (Weeks, 1980; van Beijeren and Nolden, 1987). This fact has particular implications for "vicinal" (i.e., high index-) surfaces, e.g. Cu(11ℓ), with ℓ odd: at low temperatures such a surface can be viewed as a dense regular array of steps relative to a (100) surface.

Now a roughening of such a vicinal surface means that kinks may occur on the steps and these steps may shift relative to the neighboring ones. There is now ample evidence that roughening transitions of surfaces such as Cu(113) do occur (e.g., Liang et al., 1987; Salanon et al., 1988; Lapujoulade et al., 1990) and that these observations can be understood in terms of terrace–step–kink models (Selke and Szpilka, 1986) and related continuum models, where an effective step–step interaction is taken into account (Villain et al., 1985).

The step-free energy $k_B T s(T)$ can be conveniently introduced by considering the interfacial tension of interfaces which are tilted through an angle θ relative to a low-index lattice plane (fig. 65). For small θ the angular-dependent surface tension takes the form

$$f_{\text{int}}(\theta) = f_{\text{int}}(0) + \frac{s}{a}|\theta| + o(\theta^2), \tag{252}$$

where $f_{\text{int}}(0)$ is the interfacial free energy per unit area for a flat interface oriented perpendicular to a lattice axis, a is the lattice spacing, and s the free energy cost per step. While $f_{\text{int}}(\theta)$ is analytic in θ at $\theta = 0$ for $T > T_R$ where the interface is rough, $f_{int}(\theta)$ has a quadratic expansion in θ there, this is not so in the regime where the interface is rigid, since the density of steps for an interface inclined through an angle θ is proportional to $|\theta|$.

While the picture of the tilted interface developed so far is pretty obvious for $T \rightarrow 0$, at non-zero temperature one has to worry about thermal fluctuations. There are two types of contributions (Privman, 1992). Those on scales of the correlation lengths of the two coexisting phases can be adsorbed in the definition of $s(T)$. However, the steps when viewed from above will not be just straight lines (as fig. 3 suggests), they can have kink-type shifts either to the right or to the left and will therefore behave random-walk like. Even though "kinks" cost energy, entropy causes large excursions of the steps [the problem is fully analogous to fluctuations of one-dimensional interfaces, eq. (190)], and thus the step-wandering due to these "soft mode"-like step fluctuations creates long range step–step interactions.

Chui and Weeks (1976) argued that the solid-on-solid (SOS) model, commonly used in simulations of crystal growth and roughening (fig. 7),

$$\mathcal{H}_{SOS} = J \sum_{\langle i,j \rangle} |h_j - h_i|, \tag{253}$$

where h_i is the (discrete) height variable that results as the lattice analog of the continuum interface $z = h(x, y)$ considered in sect. 2.5 (the plane (x,y) is then represented by a set of lattice sites i), can be replaced by a discrete gaussian model. Following Weeks (1980) we write

$$\mathcal{H}_{DG} = J \sum_{\langle i,j \rangle} (h_i - h_j)^2 = \frac{J}{2} \sum_q |h_q|^2 G^{-1}(q), \tag{254}$$

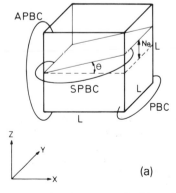

Fig. 65. (a) Boundary conditions used to impose at tilted interface in an Ising ferromagnet: antiperiodic (APBC) in the z-direction, periodic in the y-direction (PBC), and screw periodic boundary conditions (SPBC) in the x-direction.

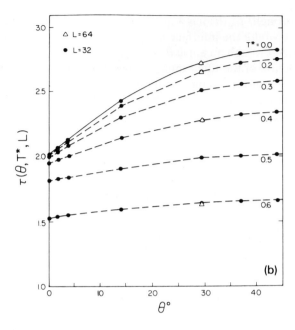

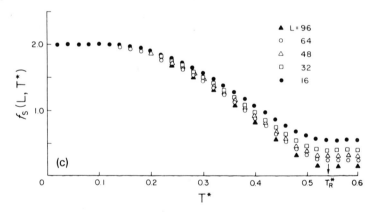

Fig. 65 (contd.). (b) Angle dependence of the (anisotropic) interfacial tension in units of J. The reduced temperature $T^* = T/T_c$. The solid curve is the theoretical variation, and the dashed lines are added as a guide to the eye. From Mon et al. (1989). (c) Temperature dependence of the step free energy (denoted here as $f_s(L, T)$ for systems of size L^3) in units of the Ising exchange constant J plotted vs. the lattice spacing. The roughening temperature (which can be extracted from these data by a finite size scaling analysis) is indicated by an arrow. From Mon et al. (1989).

where for a square lattice with lattice spacing unity $G^{-1}(q) = 4 - 2\,(\cos q_x + \cos q_y) \approx q^2$ (for small q), h_q being the Fourier transform of the height variable h_i. Thus for small q eq. (254) differs from the capillary wave Hamiltonian, eq. (186), basically by the discreteness of the h_i's. In the partition function

$$Z_{DG} = \int \mathcal{D}\{h_i\}\Pi_j\, W(h_j) \exp\left[-\frac{\mathcal{H}_{DG}}{k_B T}\right] \tag{255}$$

the discreteness of the $\{h_i\}$ shows up via a weighting function $W(h_j)$ that ensures that only integer heights contribute,

$$W(h_j) = \sum_{n_j=-\infty}^{+\infty} \delta(h_j - n_j) = \sum_{k_j=-\infty}^{+\infty} \exp\left[rk_j h_j\right], \tag{256}$$

using the representation of the delta function in terms of the Poisson summation formula. Here $k_j = 2\pi n$ for integer n. From eqs. (255) and (256) we find

$$Z_C \equiv \frac{Z_{DG}}{Z_0} = \sum_{k_j=-\infty}^{+\infty} \left\langle \exp\left(i\sum_j k_j h_j\right)\right\rangle_o, \tag{257}$$

where Z_o is the unweighted partition function of the Gaussian model (eq. (255) with $W \equiv 1$), and $\langle\ldots\rangle_o$ a corresponding average in the un-weighted Gaussian ensemble. Equation (257) is the characteristic function for the Gaussian distribution. One can show that the $\{k_j\}$ also have a Gaussian distribution which involves the inverse matrix to G_1^{-1}, namely ($G(jj')$ is the lattice Green's function)

$$Z_C = \sum_{k_j=-\infty}^{+\infty} \exp\left[-\frac{k_B T}{2J}\sum_{jj'} k_j\, G(jj')k_{j'}\right],$$
$$G(j, j') = \frac{1}{2N}\sum_q \frac{\exp[iq\cdot(r_i - r_j)]}{G^{-1}(q)} \tag{258}$$

One can reinterpret the partition function Z_C as that of a neutral two-dimensional Coulomb gas in which the k_j represent the charges (note the q^{-2} dependence at small q in eq. (258) which characterizes the Coulomb interaction). The reduced temperature $k_B T/J$ has been inverted in going from the discrete gaussian model in eq. (254) to the Coulomb gas in eq. (258). Thus the insulating dielectric phase (with tightly bound pairs of opposite charges), where the correlations decay with a power law, appears now at high temperatures ($T > T_R$), while the "conducting" phase where the "free" charges give rise to usual Debye screening occurs at $T < T_R$, the screening length corresponding to the finite correlation length of the height–height correlation function. Thus by this treatment one has mapped the roughening transition on the Kosterlitz–Thouless transition encountered for the planar spin model in sect. 2.4. From this one can show that for $T < T_R$ the correlation length ξ of height–height correlations and the step-free energy behave as

$$\xi^{-1}(T) \propto s(T) \propto \exp\left[-\frac{c}{(T_R - T)^{1/2}}\right], \qquad T \to T_R^-, \tag{259}$$

where c is a constant. Equation (259) was tested in recent Monte Carlo simulations (Mon et al., 1989; see fig. 65). Another test of the theory of roughening was carried out for a model of an "antiphase domain boundary" of ordered alloys (fig. 66). These simulations (Schmid and Binder, 1992a, b) also provided evidence for the usefulness of the capillary wave theory to describe the interface in the rough phase.

As a final point of this subsection, we mention that the properties of the angle-dependent interfacial tension $f_{int}(\hat{n})$ in crystals ($\hat{n}$ being a unit vector perpendicular to the interface) control also the equilibrium crystal shape (this is known as the Wulff (1901) construction, see Rottmann and Wortis (1984) for a review). A non-rough interface then corresponds to a *facet* of the crystal, i.e. a planar region perpendicular to some (low indexed!) crystallographic direction. The equilibrium crystal shapes are now composed of such facets and of smoothly curved regions (at $T > 0$). The edges at which distinct regions meet (facet/facet, facet/curved or curved/curved) are non-analyticies of the crystal shape which occur along particular directions $\hat{r}(T)$

Fig. 66. (a) Geometry of an "antiphase domain boundary" in (100) direction for the ordered B2 phase of body-centered cubic (bcc) alloys: in this phase, two interpenetrating simple cubic sublattices I, II are occupied preferentially by A, B atoms in a binary alloy (such as FeAl, for instance). The order parameter ϕ can then be defined as a concentration difference between the sublattices. $\phi = c_A^I - c_A^{II}$. At the interface (broken straight lines), where two B2 domains displaced by a vector $(1/2)a_0(1, 1, 1)$ meet, ϕ changes sign. (b) Monte Carlo results for the temperature dependence of the interfacial width $W(T)$ for an Ising model of FeAl alloys with nearest (V_1), next-nearest (V_2) and third nearest neighbor (V_3) crystallographic interactions ($V_2/|V_1| = -0.167$, $V_3/|V_1| = 0.208$, and a nearest neighbor magnetic exchange $J_1/|V_1| = 1.65$ between Fe atoms (magnetic moments being described as classical unit vectors there). The chemical potential difference is chosen such that the B2 order–disorder transition occurs at $T_c/|V_1| = 7.1$, much higher than T_R which is estimated from a plot of $1/W^4$ vs. T by linear extrapolation (dotted curve) as $T_R/|V_1| = 2.7 \pm 0.1$, since the Kosterlitz–Thouless theory implies $W^2(T) \propto (T_R - T)^{-1/2}$ for $T < T_R$. Three different linear dimensions $L_\parallel$ are shown. From Schmid and Binder (1992a). (c) Plot of the constant a characterizing the logarithmic divergence of the interfacial width in the rough phase, $W^2 = a^2 \ln(L_\parallel/\xi)$, where $L_\parallel$ is the linear dimension of the system parallel to the interface and ξ the effective correlation length, in the form $(a^2 - 1/\pi^2)^2$ vs. T, to test the Kosterlitz–Thouless theory of roughening which implies that $a^2 \approx \pi^{-2} + c'(T - T_R)^{1/2}$ near T_R, with c' a non-universal constant. From Schmid and Binder (1992a).

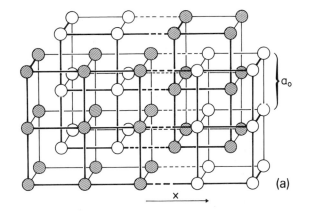

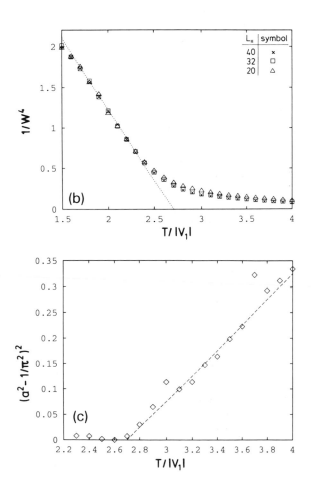

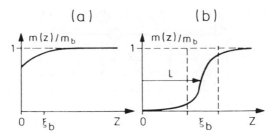

Fig. 67. Order parameter profiles $m(z)/m_b$ associated with surface-induced disorder. The coordinate z measures the distance from the surface ($z = 0$). ξ_b is the bulk correlation length and m_b the bulk order parameter. If case (a) persists up to the first-order transition temperature T_c, this means the surface stays ordered up to T_c, while case (b) shows surface induced disordering; a layer of thickness L gets disordered already at $T < T_c$, and as $T \to T_c^-$ the (delocalized) interface at mean position $z = L$ from the surface advances into the bulk, $L(T) \to \infty$ as $T \to T_c^-$, and the surface order parameter $m_1 = m(z = 0)$ then vanishes continuously, $m_1 \propto (1 - T/T_c)^{\beta_1}$. From Dosch et al. (1988).

from the center of gravity of the crystal to its surface. As the temperature is raised, each facet of orientation $\hat{n}$ disappears at a particular temperature $T_R(\hat{n})$, which is nothing else than the roughening temperature of that crystal surface. Since in practice equilibrium crystal shapes and their facetting transitions can only be observed for He^4 crystals at the solid/superfluid phase boundary (Avon et al., 1980; Balibar and Castaing, 1980), while the shapes of other crystals result from the dynamics of crystal growth and stay in their shape in metastable equilibrium, since an adjustment of their shape to the equilibrium shape would require long range transport of large fractions of the crystal volume, we will not be going into details here but rather refer the reader to the literature (Rottmann and Wortis, 1984; Wortis, 1985; van Beijeren and Nolden, 1987).

3.5. Surface-induced ordering and disordering; surface melting

We now consider the interface between a vacuum and a system that undergoes a first-order (i.e., discontinuous) order–disorder transition in the bulk at a temperature T_c. Due to "missing neighbors" at a surface, we expect that the order parameter at temperatures $T < T_c$ is slightly reduced in comparison with its bulk value (fig. 67). If this situation persists up to $T \to T_c^-$, such that both the bulk order parameter $\phi(z \to \infty)$ and the surface order parameter $\phi_1 \equiv \phi(z = 0)$ vanish discontinuously, the surface stays ordered up to T_c, a situation that is not of very general interest. However, it may happen (Lipowsky, 1982, 1983, 1984, 1987; Lipowsky and Speth, 1983) that the surface region disorders somewhat already at $T < T_c$, and this disordered layer grows as $T \to T_c^-$ and leads to a continuous vanishing of the surface layer order parameter, $\phi_1 \propto (1 - T/T_c)^{\beta_1}$ although the bulk order parameter $\phi_b(T) = \phi(z \to \infty)$ vanishes discontinuously at T_c. This "surface-induced disordering" (SID) can be considered as a wetting phenomenon: the disordered phase wets the interface between the vacuum and the ordered phase upon approaching T_c, where the disordered phase becomes thermodynamically stable (cf. fig. 13b, c). The approach $T \to T_c$ means complete wetting therefore, the interface between the disordered surface layer and the ordered bulk being completely unbound from the surface. We treat this again in terms of the one-component Landau theory, similar as in eqs. (219), (220), but now one has to choose the bulk free energy density $f(\phi)$ appropriate for a first-order transition (to represent fig. 13b and c instead of fig. 13a),

$$\frac{\mathcal{F}\{\phi(z)\}}{k_B T} = \int_0^\infty dz \left\{ f(\phi) + \frac{1}{2} \frac{R^2}{d} \left(\frac{d\phi}{dz} \right)^2 \right\} - \frac{H_1}{k_B T} \phi_1 + \frac{R^2}{2d} \lambda^{-1} \phi_1^2,$$

(260)

with

$$f(\phi) = -\frac{H}{k_B T} \phi + \frac{r}{2} \phi^2 - \frac{b}{n} \phi^n + \frac{c}{m} \phi^m,$$

(261)

where the coefficients $r,b,c > 0$ and $n,m = (4, 6)$ for the situation of fig. 13b, while $n,m = (3, 4)$ for systems which allow a cubic invariant such as the Potts model (see sect. 2.1). This problem can be studied with an approach fully analogous to that of eqs. (241)–(245); for $H = H_1 = 0$ the behavior is controlled by a comparison of the parameter $r_1 \equiv R^2 \lambda^{-1}/d$ and $r(T_c)$: If $r_1 \geq \sqrt{r(T_c)}$ the disordered phase wets the interface (Lipowsky and Speth, 1983). In the case $-\bar{x}\sqrt{r(T_c)} < r_1 < \sqrt{r(T_c)}$ there is no wetting, the constant $\bar{x}$ being $\bar{x} = 2^{1/3} - 1$ for $(n, m) = (3, 4)$ and $\bar{x} = \sqrt{2} - 1$ for $(n, m) = (4, 6)$. If $r_1 < -\bar{x}\sqrt{r(T_c)}$, the surface coupling is so strong that the ordered phase wets the interface between the vacuum and the disordered phase, for $T \to T_c^+$, and one has surface-induced order instead of disorder. While reports of surface-induced disordering exist for alloys such as Cu_3Au (Sundaram et al., 1973; McRae and Malic, 1984; Alvarado et al., 1987; Dosch et al., 1988), ferroelectrics such as $NaNO_2$ (Marquardt and Gleiter, 1982), etc., we are not aware of any experimental observation of surface induced order at a first-order transition in the bulk. However, the possibility of this phenomenon (beyond Landau's theory!) has been demonstrated by Monte Carlo simulations for a free (010) surface of a model of a face-centered cubic AB alloy, where the bulk ordering temperature T_c is relatively low due to "frustration" effects, cf. fig. 68 (Schweika et al., 1990). In all cases, the theory predicts the thickness of the wetting layer $\ell(T)$ to grow logarithmically, $\ell(T) \propto |\ln|T - T_c||$. The exponent β_1 in this one-component Landau theory

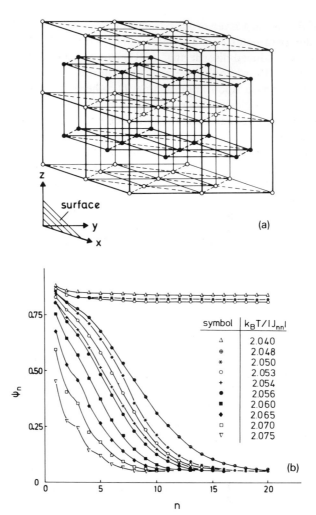

Fig. 68. (a) Section of the ordered lattice of the AB alloy in $L1_o$ (Cu AuI) structure. This structure results in an Ising model with nearest neighbor interactions $J_{nn} < 0$ and next nearest neighbor interactions $J_{nnn} > 0$ (thick lines). A and B atoms are shown as open and full circles, respectively, and the orientation of the coordinate axes is indicated. For this structure, nearest neighbor interactions between atoms of the same kind are unfavorable ("frustrated"), and are shown by broken bonds, while favorable nearest neighbor interactions (between different planes) are also present but not shown. Note that no "frustrated" interactions occur in the surface plane. (b) Absolute value of the local order parameter ψ_n in the nth lattice plane, $n = 1$ being the free surface of part (a), for various temperatures as indicated. The three upper curves refer to ordered states in the bulk. Note that data for $k_B T/|J_{nn}| = 2.050$ and 2.053 are plotted for both ordered and disordered starts. All data refer to $J_{nnn}/|J_{nn}| = 0.05$, and a $61 \times 60 \times 60$ lattice with two free 60×60 surfaces. Curves are guide to the eye only. Here it was estimated that $k_B T_c/|J_{nnn}| = 2.0525 \pm 0.0010$, while the surface order sets in at $k_B T_{cs}/|J_{nnn}| = 2.107 \pm 0.002$. From Schweika et al. (1990).

is found as $\beta_1 = 1/2$, for both choices of (n, m) mentioned above, except for the case where $r_1 = [r(T_c)]^{1/2}$, which corresponds to a wetting tricritical point, where $\beta_1 = 1/3$ for $(n,m) = (3,4)$ while $\beta_1 = 1/4$ for $(n,m) = (4,6)$, see Lipowsky and Speth (1983).

One can make the connection between surface induced disordering and wetting even more explicit by mapping (Kroll and Lipowsky, 1983) eqs. (260) and (261) onto eqs. (219) and (220). For $(n, m) = (3, 4)$ one can make the substitution $\phi(z) = \psi(z) + b/(3c)$ in eqs. (260) and (261) to obtain a new bulk free energy $f(\psi) = -h\psi - t\psi^2/2 + g\psi^4/4$, with $t = -r + b^2/3c$, $g = c$, $h = H/k_BT + [b/(3c)][2b^2/(9c) - r]$, and a bare surface free energy $f(\psi_1) = -h_1\psi_1 + t_1\psi_1^2/2$ where $h_1 = H_1/k_BT - (br_1)/(3c)$ while $t_1 = r_1$. Thus the phase diagram for SID in the parameter space (H_1,H,r) is obtained by simply tilting the wetting phase diagram (h_1, t,h) out of the plane $h = 0$. E.g., let us study the path $H_1 = 0$, $H = 0$, $\delta = r(T) - r(T_c) \to 0$, which is physically relevant for SID, in the transcription to wetting: with $r(T_c) = 2b^2/9c$ one finds that $T = T_w - T_c\delta r$ and $\Delta\mu = -b\delta r/(3c) = -[b/(3c)](T_w - T)/T_c$. Thus SID corresponds to complete wetting at T_w along a particular path for the chemical potential difference $\Delta\mu \to 0$. Since complete wetting and critical wetting satisfy scaling hypotheses as bulk scaling phenomena do, namely (Nakanishi and Fisher, 1982), in d dimensions with $d \leq 3$,

$$f_s^{sing} = \tilde{t}^{(d-1)\nu_\parallel} \tilde{f}\left(\Delta\mu\, \tilde{t}^{-\Delta}\right), \qquad \xi_\parallel = \tilde{t}^{-\nu_\parallel} \tilde{\xi}\left(\Delta\mu\, \tilde{t}^{-\Delta}\right), \qquad (262)$$

$\xi_\parallel$ being the correlation length along the wall, and the exponent $\Delta = (d + 1)\nu_\parallel/2$ $(= 2\nu_\parallel$ for $d = 3)$. For critical wetting in mean field theory one has $\nu_\parallel = 1$, while for complete wetting one has (Lipowsky, 1985b) $\nu_\parallel^{co} = 2/(d+1)$ $(= 1/2$ for $d = 3)$. This fact also implies $\xi_\parallel \propto (\Delta\mu)^{-1/2}$ for the path applicable to SID, $\xi_\parallel \propto (T_c - T)^{1/2}$. Using $\phi(z = 0) = \exp[-\ell(T)/\xi_b]$, since $\phi(z = 0)$ is determined by the "tail" of the order parameter profile, fig. 67, $\ell(T) \approx \frac{1}{2}\xi_b \ln|1 - T/T_c|$ yields then $\phi(z) \propto (1 - T/T_c)^{1/2}$ as well. Figure 69 shows an example of simulations verifying this behavior (Helbing et al., 1990).

At this point, we now follow Lipowsky (1987) to discuss the effect of short range versus long range forces. In the spirit of fig. 6, we describe the problem in terms of an effective potential for the interface height h above the surface. For the short range case we have $(t = 1 - T/T_c)$

$$V_{eff}(h) = c\, f_{int} \exp\left(-\frac{2h}{\xi_b}\right) + T_c(S_{dis} - S_{ord})\, th \qquad (263)$$

where c is a constant of order unity, f_{int} the interfacial tension between the ordered (ord) and disordered (dis) phases, S_{ord}, S_{dis} are their bulk entropies, and ξ_b is the correlation length in the disordered phase. From Landau's theory, eqs. (260), (261) one can show that $c = \xi_b(\xi_b - \lambda)/[\lambda(\xi_b + \xi_b^o)]$,

ξ_b^o being the correlation length in the ordered phase, and λ is again the extrapolation length. In this excess free energy per unit area $V_{\text{eff}}(h)$ the exponential h-dependence of the repulsive term is due to the exponential tails of the order parameter profile $\phi(z)$ discussed above. Now the equilibrium thickness h can be obtained from minimizing $V(h)$,

$$\frac{\partial V_{\text{eff}}(h)}{\partial h} = 0 \Rightarrow \frac{h}{\xi_b} = -\frac{1}{2}\ln t + \ln \text{const}, \tag{264}$$

which is the logarithmic law mentioned above. If we now take the long range van der Waals forces (Dzyaloshinskii et al., 1961) into account and neglect retardation effects, one obtain an additional contribution, H being the "Hamaker constant" (Lipowsky, 1985b; Israelachvili, 1985)

$$V_{\text{LR}}(h) = Hh^{-2}, \qquad H = \frac{\pi}{12}\epsilon\sigma^6(\rho_{\text{ord}} - \rho_{\text{dis}})\rho_{\text{dis}}, \tag{265}$$

where σ, ϵ are the range and strength parameters of the basic Lennard–Jones type attractive part of the pair potential ($V_{\text{attr}}(r) = -\epsilon(r/\sigma)^{-6}$), and ρ_{ord}, ρ_{dis} are the particle number densities of the ordered and disordered phases, respectively. From eq. (265) one can see that the long range forces are important if there is a large density difference between the phases. In the regime close enough to T_c where they dominate a minimization of $V_{\text{eff}}(h)$ with respect to h now leads to a faster growth, $h \propto t^{-1/3}$. [The same law applies for complete wetting with non-retarded van der Waals forces,

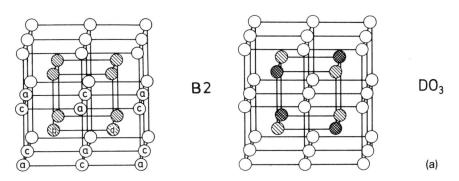

Fig. 69. (a) Part of the body-centered cubic lattice ordered in the B2 structure (left part) and in the DO$_3$ structure (right part). Left part shows assignment of four sublattices a, b, c and d. In the B2 structure (cf. also fig. 66a), the concentrations of A atoms are the same at the a and c sublattices, but differ from the concentrations of the b, d sublattices, while in the DO$_3$ structure the concentration of the b sublattice differs from that of the d sublattice, but both differ from those of the a, c sublattices (which are still the same). In terms of an Ising spin model, these sublattice concentrations translate into sublattice "magnetizations" m_a, m_b, m_c, m_d, which allow to define three order parameter components $\psi_1 = m_a + m_c - m_b - m_d$, $\psi_2 = m_a - m_c + m_b - m_d$, and $\psi_3 = -m_a + m_c + m_b - m_d$.

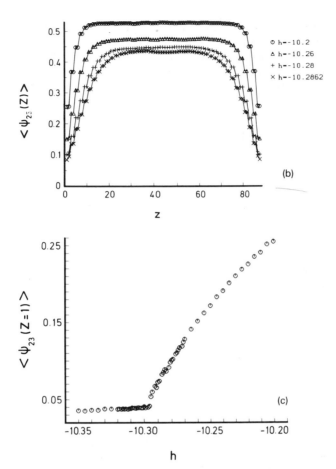

Fig. 69 (contd.). (b) Plot of the order parameter profile $\psi_{23}(z) \equiv (\psi_2^2(z) + \psi_3^2(z))^{1/2}$. vs. the coordinate z across a system of size $L_\parallel \times L_\parallel \times L_\perp$, with $L_\parallel = 20$, $L_\perp = 89$, and two free $L_\parallel \times L_\parallel$ surfaces, choosing nearest and next nearest neighbor antiferromagnetic exchange, $J_{nnn}/J_{nn} = 1/2$. Temperature chosen is $k_B T/|J_{nn}| - 1$ and four different fields $h = H/|J_{nn}|$ probing the transition from the DO$_3$ phase to the disordered phase. The growth of a layer of the disordered phase at both surfaces is clearly seen. From Helbing et al. (1990). (c) Variation of the local order parameter at the surface near the first-order transition of the model described in (b). Note that due to the finite cross section ($L_\parallel \times L_\parallel$) the root mean square order parameter must be non-vanishing even in the disordered phase. From Helbing et al. (1990).

where the coverage varies with the chemical potential difference $\Delta\mu$ as $\theta \propto (\Delta\mu)^{-1/3}$, see Dietrich (1988), which is also established experimentally (Krim et al., 1984).]

Another long range force that must be considered for surface-induced disordering of solids are elastic forces (Wagner, 1978), which tend to truncate the divergence of h as $t \to 0$ (Lajzerowicz, 1981; Speth, 1985). An analogous truncation of the growth of the coverage θ in the case of wetting or multilayer adsorption due to elastic effects (Huse, 1984; Gittes and Schick, 1984) has already been noted. Although such elastic effects might be present in the surface-induced disordering of alloys such as Cu_3Au, the data of Dosch et al. (1988) confirm the simple logarithmic growth law, eq. (264). But the behavior of the surface layer order parameter exponent β_1 is still controversial: while McRae and Malic (1984) found $\beta_1 = 0.31 \pm 0.05$ Alvarado et al. (1987) found $\beta_1 = 0.82 \pm 0.07$ (for the (100) surface). Since the order parameter of Cu_3Au has three components (it belongs to the class of the 4-state Potts model), and there is an interplay with surface enrichment in this alloy, even on the level of the mean field approximation the behavior is rather complicated (Morán-López et al., 1985; Kroll and Gompper, 1987), β_1 is non-universal and not given by $\beta_1 = 1/2$ even at the level of mean field theory. Capillary wave effects are expected to change this exponent, as for critical wetting (Lipowsky, 1987). But always one would predict $\beta_1 \geq 1/2$ and thus the results of McRae and Malic (1984) are not understood.

A particularly interesting first-order transition is the melting of crystals, and application of the "surface-induced disordering" concept (fig. 67) to this transition immediately suggests the possibility of "*surface melting*". In fact, the idea that the melting of crystals starts at their surface is very old (Tammann, 1910). On a phenomenological level, the treatment analogous to eqs. (263)–(265) should apply to surface melting as well.

We simply have to take into account that three interfacial tensions compete: f_{int}^{sv} for the solid–vapor interface, f_{int}^{lv} for the liquid–vapor interface, and f_{int}^{sl} for the solid–liquid interface. We have surface melting if $f_{int}^{sv} - (f_{int}^{lv} + f_{int}^{sl}) \equiv \Delta f > 0$. The effective free energy replacing eq. (263) is, for short range forces, T_m being the melting temperature

$$F(h) = f_{int}^{lv} + f_{int}^{sl} + (T_m - T)h(S_{liq} - S_{solid}) + \Delta f \exp\left(-\frac{2h}{\xi_b}\right). \quad (266)$$

Due to the density difference between the crystal and the melt, there is also a long range part as given by eq. (265). Since this density difference is rather small, one would expect to see logarithmic growth of the fluid layer up to a crossover temperature T^* slightly below the melting temperature, where then the power law growth $h \propto t^{-1/3}$ takes over.

Unfortunately, the analytical theory of surface melting has been so far not developed in a more explicit way, simply because one is lacking reliable quantitative theories of the melting transition in the bulk. Thus theories that locate the onset of surface melting from lattice-dynamical phonon instabilities (Pietronero and Tosatti, 1979; Jayanthi et al., 1985a, b; Tosatti, 1988; Trayanov and Tosatti, 1988) need to be viewed with the caution that analogous treatments of melting in the bulk do not work. While computer simulations (e.g. Broughton and Woodcock, 1978; Broughton and Gilmer, 1983, 1986; Stoltze et al., 1988) in principle are very attractive, the study of slow long-wavelength phenomena (as are involved in such interface unpinning transitions) for microscopic models with realistic potentials is very difficult, and thus the results (in our opinion) are not yet very conclusive. (For the same reason, we have not discussed any simulations of wetting phenomena with realistic potentials, e.g. Henderson and van Swol (1984, 1985), and confined the discussion to the simpler lattice gas model in sect. 3.2.) Density functional theories of freezing (see Haymet, 1992, for a recent review and further references) may be a more promising starting point for a theory of surface melting, in view of the fact that density functional theories of wetting for gas–liquid transitions (Evans, 1990, 1992) are fairly successful. For first steps in this direction see Löwen et al. (1989) and Löwen and Beier (1990).

Experimental evidence for surface melting comes primarily from ion scattering studies (van der Veen et al., 1990) and evanescent scattering of X-rays (Dosch, 1991). For Pb(110) a $h(t) \propto \ln t$ law has been seen up to $T_m - T^\star \approx 0.3$ K while closer to T_m a behavior $h(t) \propto t^{-0.315}$ was found, in agreement with the expected behavior outlined above. For Al(110) only the logarithmic regime was seen (van der Veen et al., 1990; Dosch, 1991), while the close-packed Al(111) surface does not show surface melting. Also for ice (0001) surfaces the logarithmic growth law was seen (Golecki and Jaccard, 1978). The advantage of the X-ray technique is that it also gives detailed information on the enhancement of the Debye–Waller factor (i.e., the mean-square displacement of atoms $\langle u^2 \rangle$) at the surface.

4. Discussion

In the present chapter, we have attempted to give an introductory review of the theory of phase transitions, with a special emphasis on surface physics: both surface effects on bulk phase transitions were discussed (local critical phenomena on surfaces that differ in their character from critical phenomena in the bulk, surface-induced ordering and disordering in conjunction with first-order transitions, e.g. surface melting, and other wetting phenomena) and phase transitions in strict two-dimensional geometry, as they occur in adsorbed layers. In fact, monolayers at submonolayer coverage present rich and unique examples of types of phase transitions and "universality classes" of critical phenomena, which cannot be studied otherwise. Using simple Landau-type theories as an unifying tool, we have sketched how one can classify the various types of phase transitions in two dimensions, and understand surface effects on both bulk second-order and first-order transitions in terms of the order parameter profile resulting from the gradient term of the free energy functional under the action of appropriate boundary conditions. But we have also emphasized the fact that due to restricted dimensionality statistical fluctuations are extremely important, and thus one must not rely on Landau's theory too much. In fact, the statistical fluctuations lead to qualitatively new phenomena which cannot be understood by Landau's theory at all: spontaneous domain formation at the lower critical dimension (e.g. quasi-one-dimensional Ising systems such as may occur for adsorption on stepped surfaces (figs. 3, 45a, etc.), spontaneous interface delocalization due to capillary wave excitations, spontaneous destruction of long range order in two dimensional systems, if the order would break a continuous symmetry such as the spontaneous magnetization would do in isotropic ferro- or antiferromagnets, by the so-called "Goldstone modes" (i.e., long-wavelength excitations which cost no excitation energy for $\lambda \to \infty$, i.e., magnons in isotropic magnets, acoustic phonons in crystals that are incommensurate with the substrate periodicity, etc.). A particularly interesting phenomenon is the destruction of the low temperature phase of isotropic XY-magnets, which exhibits an algebraic (i.e., power-law) decay of spin correlations at large distances, by the spontaneous unbinding of topological excitations (vortex–antivortex pairs). This Kosterlitz–Thouless transition would also be the mechanism for a metal–insulator transition of a two dimensional Coulomb gas, and what is most important in the present context, it also describes the roughening transition of crystal surfaces (or of interfaces in alloys, anisotropic magnets, etc.). This roughening transition also plays a particular role for multilayer adsorption being observed rather than wetting. This fact again underlines the close relation between all the phenomena treated in the present article.

On the other hand, it is clear that a detailed exposition of the theoretical knowledge that has been accumulated on these subjects is far beyond the scope of our treatment, which rather is intended as a kind of tutorial providing physical insight and a guide to the more specialized literature. Thus, technical aspects of the more advanced theoretical methods (renormalization group, transfer matrix calculations, Coulomb gas methods and conformal invariance considerations and — last but not least — Monte Carlo computer simulation) have not been discussed at all here. Also, the emphasis has been on the discussion of the simplest models — Ising/lattice gas models, classical spin models, Potts, clock and ANNNI models, Frenkel–Kontorova model, etc. — and the developed concepts were mostly illustrated with Monte Carlo computer simulation results from the author's group, mentioning corresponding experimental results only rather occasionally. The reason for this choice is that the Monte Carlo study of models can pinpoint the phenomena under consideration more easily and stringently than experiments can usually do: the latter are often affected by a simultaneous interplay of many different effects, which usually are hard to disentangle, the interactions and thus the appropriate model description often is not known very precisely, and various non-ideal effects come into play which are interesting in their own right but may obscure the issues under discussion (lattice defects of the substrate, finite size over which the substrate periodicity (fig. 1) applies, chemisorbed immobile impurities, adsorbate-induced relaxation or reconstruction of the substrate, etc.). In fact, randomly distributed quenched impurities are known to have dramatic effects on phase transitions in two dimensions: if they lead to a linear coupling to the order parameter, like in the random-field Ising model, they already destroy long range order at arbitrary small concentrations (Nattermann and Villain, 1988; Imry and Ma, 1975) due to spontaneous break-up in an irregular domain pattern (fig. 16); if the random defects couple to the square of the local order parameter only (in the framework of Landau's theory, this means that the coefficient $r(T)$ of the quadratic term $(1/2)r(T)\phi^2$ gets a random component, see Stinchcombe, 1983) a different critical behavior results for models where in the pure case the specific heat diverges (Harris, 1974). Also for this phenomenon evidence from computer simulations exists (Selke, 1993; Matthews-Morgan et al., 1981, 1984), while it is clearly much harder to establish this change of critical behavior experimentally. In contrast, the rounding of two-dimensional phase transitions by "random fields" has been seen experimentally both in quasi-two-dimensional Ising antiferromagnets (Ferreira et al., 1983) and in $CO_{1-x}(N_2)_x$ mixtures physisorbed on graphite (Wiechert and Arlt, 1993) as well as in corresponding simulations (Morgenstern et al., 1991; Binder, 1984c; Pereyra et al., 1993). In any case, experimental studies of phase transitions at surfaces must watch out carefully for any (unwanted) effects due to quenched disorder, which are often hard to control.

A very important limitation of experimental studies of phase transitions in adsorbed monolayers is the size over which the substrate is flat and more or less ideal (Marx, 1985). The typical linear dimension L controlling size effects in adsorption experiments on graphite is of the order of 100 Å, but it can be varied only by choosing different types of grafoil (e.g. Bretz 1977). While the resulting rounding and shifting of adsorbed He^4 at the order–disorder transition to the $(\sqrt{3} \times \sqrt{3})$ structure is qualitatively consistent with the theoretical ideas on finite size effects on phase transitions (Fisher, 1971; Barber, 1983; Privman, 1990; Binder, 1992b, c), a quantitative comparison is hardly possible due to an (unknown) variation in the size and shape of the regions over which grafoil is homogeneous. Only for the first-order transitions of Ne and O_2 adsorbed on grafoil has a quantitative interpretation of the rounding of the delta-function singularity of the specific heat (reflecting the latent heat at the transition) due to finite size in terms of the corresponding theory (Challa et al., 1986) been possible (Marx, 1989). Monte Carlo simulations are also affected by finite size (Privman, 1990; Binder 1992a–c), but there the size and shape of the system can be precisely controlled and varied over a reasonable range, and in conjunction with finite size scaling theory (Fisher, 1971; Barber, 1983) these size effects have become a powerful tool for the study of critical phenomena. Since these finite size effects have been reviewed recently by the author (Binder, 1992b, c), no further discussion of these problems is given here.

Phase transitions in adsorbed layers obviously are very sensitive to the interplay between the binding forces to the substrate and the lateral interactions among the adatoms (figs. 1, 2). Often the phase diagrams in the sub-monolayer range reflect these interactions in detail (for very simple examples see fig. 28). Precise knowledge of such interactions for particular substrate/adsorbate systems is desirable for many reasons. Since on the phase diagrams of such systems there is a very rich experimental information available, an attractive line of research is concerned with the theoretical explanation of such information in terms of corresponding detailed atomistic models with appropriately chosen interactions. An example of this approach (Binder and Landau, 1981) was shown already in fig. 25, referring to H on Pd(100), attempting to explain the LEED data of Behm et al. (1980), see fig. 70, in terms of a lattice gas model which is adjusted such that the Monte Carlo data (fig. 71) corresponding to the experiment "mimick" the data as closely as possible. The analysis of the computer simulation "data"

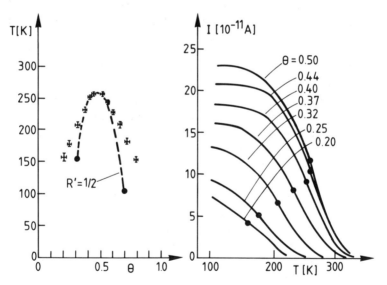

Fig. 70. Experimental phase diagram for H adsorbed on Pd(100), left part, as extracted from the temperature variation of LEED intensities at various coverages θ (right part). Crosses denote the points $T_{1/2}$ where the LEED intensities have dropped to one half of their low temperature values (denoted by full dots in the right part). Dashed curve is a theoretical phase diagram obtained by Binder and Landau (1981) for $R' = \varphi_t/\varphi_{nnn} = 1/2$ (only the regime of second-order transition ending in tricritical points [dots] are shown). Experimental data are taken from Behm et al. (1980). From Binder and Landau (1981).

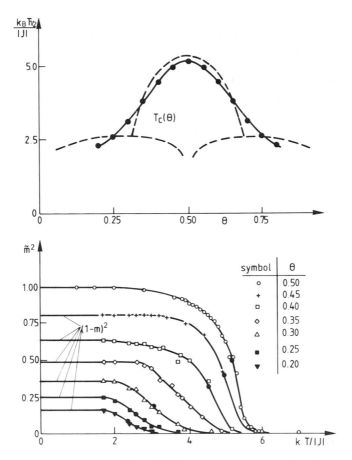

Fig. 71. Order parameter square of the c(2×2) structure (lower part) plotted vs. temperature at constant coverage, as obtained from interpolation of data taken at constant chemical potential. Full dots denote temperatures $T_{1/2}$ where the order parameter square has dropped to 50 % of its low temperature value. These data are for $R = -1$, $R' = \varphi_t/\varphi_{nn} = 0$. Upper part shows phase diagram as derived from $T_{1/2}$ (dots and full curve) in comparison with the correct phase diagram (broken curves). From Binder and Landau (1981).

can be carried out in close analogy to the analysis of experimental data. In this way, one could not only show (Binder and Landau, 1981) that a lattice gas model with a reasonable choice of interaction parameters can describe most of the experimental results for H on Pd(100), but also that the estimation of the phase diagram from the inflection points in intensity vs. temperature curves (fig. 70) overestimates the ordering temperature at off-stoichiometric coverages and is insensitive to the location of the tricritical point (fig. 71). Although the set of lateral interaction parameters resulting from such an analysis certainly is not very precise (and maybe not even unique), the method seems to be competitive with predictions of such inter-action parameters from electronic structure work (Muscat, 1985, 1986b), and this type of analysis has become very popular and has been applied to many different systems (e.g. H on Fe(110), see Kinzel et al. (1982); H on Ni(111), see Roelofs (1982) and Roelofs et al. (1986); O on W(110), see Rikvold et al. (1984); O on Mo(110), see Dünweg et al. (1991); Si on W(110), see Amar et al. (1985); O on Ni(111), see Roelofs et al. (1981); Se on Ni(100), see Bak et al. (1985); N_2 on graphite, see Marx et al. (1993); etc.). A particularly useful methodic development is the simulation of structure factors observed with finite resolution in reciprocal space (Bartelt et al., 1985a–c). Some of these calculations have been reviewed by Roelofs (1982) and by Binder and Landau (1989) and hence this subject is not followed up again here.

Another very important phase transition phenomenon at surfaces that is not treated in the present article is *surface reconstruction* (for a review, see e.g. Willis, 1985). The reason for omitting this topic — as well as for other shortcomings of the present article — simply is lack of expertise of the present author. Also he has to apologize to many colleagues where work pertinent to aspects of subjects that have been discussed here was only briefly mentioned or not at all: a complete coverage of all material would have been very difficult to achieve, and it would have destroyed the introductory character of this chapter. Thus also examples taken from computer simulations in order to illustrate more general points have often been taken from the work of the author's group just for simplicity, and never should it be taken as implication that other contributions are not similarly valuable. Despite all these caveats, we do hope that this article will stimulate increasing interest in the subject of phase transitions at surfaces — there are still many interesting open problems which pose opportunity for further work and present particularly rewarding challenges.

Acknowledgements

The author likes to thank many collaborators whose valuable research has contributed to the author's developing understanding of the subject, and have led to examples of results that have been presented here. In particular, he is grateful to E.V. Albano, D.W. Heermann, W. Helbing, P.C. Hohenberg, D.P. Landau, K.K. Mon, A. Patrykiejew, W. Paul, A. Sadiq, F. Schmid, I. Schmidt, W. Schweika and S. Wansleben. He also has greatly benefitted from stimulating interactions with B. Dünweg, T.L. Einstein, G. Ertl, J.D. Gunton, K. Kaski, W. Kinzel, K. Knorr, D. Marx, G. Marx, A. Milchev, P. Nielaba, M. Rovere, W. Selke and H. Wiechert.

References

Abraham, D.B., 1980, Phys. Rev. Lett. **44**, 1165.

Abraham, D.B., 1988, J. Phys. **A 21**, 1741.

Abraham, D.B., and E.R. Smith, 1986, J. Stat. Phys. **43**, 621.

Albano, E.V., K. Heermann, D.W. Binder and W. Paul, 1989a, Surface Sci. **223**, 151.

Albano, E.V., K. Heermann, D.W. Binder and W. Paul, 1989b, Z. Physik **B 77**, 445.

Albano, E.V., K. Heermann, D.W. Binder and W. Paul, 1990, J. Stat. Phys. **61**, 161.

Albano, E.V., K. Heermann, D.W. Binder and W. Paul, 1992, Physica **A 183**, 130.

Alexander, S., 1975, Phys. Lett. **A 54**, 353.

Allen, S.M., and J.W. Cahn, 1979, Acta Metall. **27**, 1085.

Als-Nielsen, J., and R.J. Birgeneau, 1977, American J. Phys. **45**, 554.

Alvarado, S.F., A. Campagna, F. Ciccacci and M. Hopper, 1982, J. Appl. Phys. **53**, 7920.

Alvarado, S.F., M. Campagna, A. Fattah and W. Uelhoff, 1987, Z. Physik **B 66**, 103.

Amar, J., S. Katz and J.D. Gunton, 1985, Surface Sci. **155**, 667

Amit, D.J., 1984, Field Theory, The Renormalization Group and Critical Phenomena (World Scientific, Singapore).

Anisimov, M.A., S.B. Kiselev, J.V. Sengers and S. Tang, 1992, Physica **A 188**, 487.

Ashkin, J., and E. Teller, 1943, Phys. Rev. **64**, 178.

Avon, J.E., L.S. Balfour, C.G. Kuper, J. Landau, S.G. Lipson and S. Schulman, 1980, Phys. Rev. Lett. **45**, 814.

Bak, P., 1984, in: Chemistry and Physics of Solid Surfaces VII, eds. R. Vanselow and R.F. Howe, (Springer, Berlin), p. 317.

Bak, P., and V.J. Emery, 1976, Phys. Rev. Lett. **36**, 978.

Bak, P., P. Kleban, W.N. Unertl, J. Ochab, G. Akinci, N.C. Bartelt and T.L. Einstein, 1985, Phys. Rev. Lett. **54**, 1539.

Balibar, S., and B. Castaing, 1980, J. Phys. (Paris) Lett. **41**, L 329.

Ball, R.C., and R.L.H. Essery, 1990, J. Phys.: Condensed Matter **2**, 10303.

Barber, M.N., 1983, in: Phase Transition and Critical Phenomena eds. C. Domb and J.L. Lebowitz (Academic Press, London), p. 145.

Bartelt, N.C., T.L. Einstein and L.D. Roelofs, 1985a, Surface Sci. **149**, L 47.

Bartelt, N.C., T.L. Einstein and L.D. Roelofs, 1985b, Phys. Rev. **B 32**, 2993.

Bartelt, N.C., T.L. Einstein and L.D. Roelofs, 1985c, J. Vac. Sci. Technol. **A 3**, 1568.

Baxter, R.J., 1971, Phys. Rev. Lett. **26**, 832.

Baxter, R.J., 1972, Ann. Phys. **70**, 193.

Baxter, R.J., 1973, J. Phys. **C 6**, L 445.

Baxter, R.J., 1980, J. Phys. **A 13**, L 61.

Baxter, R.J., 1982, Exactly Solved Models in Statistical Mechanics (Academic Press, London).

Beale, P.D., P.M. Duxbury and J.M. Yeomans, 1985, Phys. Rev. **B 31**, 7166.

Behm, R.J., K. Christmann and G. Ertl, 1980, Surface Sci. **99**, 320.

Berlin, T.H., and M. Kac, 1952, Phys. Rev. **86**, 821.

Berezinskii, V.L., 1971, Soviet Phys. JETP **32**, 493.

Berezinskii, V.L., 1972, Soviet Phys. JETP **34**, 610.

Beysens, D., 1990, in: Liquids at Interfaces, eds. J. Charvolin, J.F. Joanny and J. Zinn-Justin (Elsevier, Amsterdam), p. 499.

Binder, K., 1976, in: Phase Transitions and Critical Phenomena, Vol. 5b, eds. C. Domb and M. S. Green (Academic Press, London) p. 1.

Binder, K., 1977, Phys. Rev. **B 15**, 4425.

Binder, K., 1979a, in: Trends in Physics (Proceedings of the 4th EPS General Conference, York 1978) ed. Institute of Physics (A. Hilger, Bristol), p. 164.

Binder, K. (ed.), 1979b, The Monte Carlo Method in Statistical Physics (Springer, Berlin).

Binder, K., 1981a, Z. Physik B 43, 119.

Binder, K., 1981b, in: Stochastic Nonlinear Systems in Physics, Chemistry and Biology, eds. L. Arnold and R. Lefever (Springer, Berlin), p. 62.

Binder, K., 1983, in: Phase Transitions and Critical Phenomena, Vol. 8, eds. C. Domb and J.L. Lebowitz (Academic Press, London), p. 1.

Binder, K., 1984a, Applications of the Monte Carlo Method in Statistical Physics (Springer, Berlin).

Binder, K., 1984b, Phys. Rev. A 29, 341.

Binder, K., 1984c, Phys. Rev. B 29, 5184.

Binder, K., 1985, Z. Physik B 61, 13.

Binder, K., 1987, Rep. Prog. Phys. 50, 783.

Binder, K., 1990, in: Kinetics of Ordering and Growth at Surfaces, ed. M.G. Lagally (Plenum Press, New York, NY), p. 31.

Binder, K., 1991, in: Materials Science and Technology, Vol. 5: Phase Transformations of Materials, ed. P. Haasen (VCH, Weinheim), p. 405.

Binder, K. (ed.), 1992a, The Monte Carlo Method in Condensed Matter Physics (Springer, Berlin).

Binder, K., 1992b, Annu. Rev. Phys. Chem. 43, 33.

Binder, K., 1992c, in: Computational Methods in Field Theory, eds. H. Gausterer and C.B. Lang (Springer, Berlin), p. 59.

Binder, K., and H.L. Frisch, 1991, Z. Physik B 84, 403.

Binder, K., and D.W. Heermann, 1985, in: Scaling Phenomena in Disordered Systems, eds. R. Pynn and A. Skjeltorp (Plenum, New York, NY), p. 207.

Binder, K., and D.W. Heermann, 1988, Monte Carlo Simulation in Statistical Physics: An Introduction (Springer, Berlin).

Binder, K., and P.C. Hohenberg, 1972, Phys. Rev. B 6, 3461.

Binder, K., and P.C. Hohenberg, 1974, Phys. Rev. B 9, 2194.

Binder, K., and D.P. Landau, 1980, Phys. Rev. B 21, 1941.

Binder, K., and D.P. Landau, 1981, Surface Sci. 108, 503.

Binder, K., and D.P. Landau, 1984, Phys. Rev. Lett. 52, 318.

Binder, K., and D.P. Landau, 1985, Surface Sci. 151, 409.

Binder, K., and D.P. Landau, 1988, Phys. Rev. B 37, 1745.

Binder, K., and D.P. Landau, 1989, in: Molecule-Surface Interactions. Advances in Chemical Physics, Vol. 76 (K.P. Lawley, ed., (J. Wiley & Sons, New York) p. 91.

Binder, K., and D.P. Landau, 1990, Physica A 163, 17.

Binder, K., and D.P. Landau, 1992a, J. Chem. Phys. 96, 1444.

Binder, K., and D.P. Landau, 1992b, Phys. Rev. B 46, 4844.

Binder, K., and H. Müller-Krumbhaar, 1974, Phys. Rev. B 9, 2328.

Binder, K., and D. Stauffer, 1974, Phys. Rev. Lett. 33, 1006.

Binder, K., and D. Stauffer, 1976a, Advances Phys. 25, 343.

Binder, K., and D. Stauffer, 1976b, Z. Physik B 24, 407.

Binder, K., W. Kinzel and D.P. Landau, 1982, Surface Sci. 117, 232.

Binder, K., D.P. Landau and S. Wansleben, 1989, Phys. Rev. B 40, 6971.

Bishop, D.J., and J.D. Reppy, 1978, Phys. Rev. Lett. 40, 1727.

Blinc, R., and B. Zeks, 1974, Soft Modes in Ferroelectrics and Antiferroelectrics (North-Holland, Amsterdam).

Brazovskii, S.A., I.E. Dzyaloshinskii and S.P. Obukhov, 1977, Zh. Eksp. Teor. Fiz. 72, 1550.

Bretz, M., 1977, Phys. Rev. Lett. 38, 501.

Brézin, E., B.I. Halperin and S. Leibler, 1983, Phys. Rev. Lett. 50, 1387.

Broughton, J.Q., and J.H. Gilmer, 1983, Acta Metall. 31, 845.

Broughton, J.Q., and J.H. Gilmer, 1986, J. Chem. Phys. 84, 5741.

Broughton, J.Q., and L.V. Woodcock, 1978, J. Phys. **C 11**, 2743.

Bruce, A.D., and R.A. Cowley, 1981, Structural Phase Transitions (Taylor and Francis, London).

Bruder, F., and R. Brenn, 1992, Phys. Rev. Lett. **69**, 624.

Burkov, S.E., and A.L. Talapov, 1980, J. Phys. (Paris) **41**, L387.

Burton, W.K., N. Cabrera and F.C. Frank, 1951, Philos. Trans. Roy. Soc. London, Ser. **A 243**, 299.

Cahn, J.W., 1977, J. Chem. Phys. **66**, 3667.

Cahn, J.W., and J.E. Hilliard, 1958, J. Chem. Phys. **28**, 258.

Cahn, J.W., and J.E. Hilliard, 1959, J. Chem. Phys. **31**, 688.

Cardy, J.L., 1987, in: Phase Transitions and Critical Phenomena, Vol. 11, eds. C. Domb and J.L. Lebowitz (Academic Press, London), p. 55.

Challa, M.S.S., D.P. Landau and K. Binder, 1986, Phys. Rev. **B 34**, 1841.

Chinn, M.D., and S.C. Fain, 1977, Phys. Rev. Lett. **39**, 146.

Chui, S.T., and J.D. Weeks, 1976, Phys. Rev. **B 14**, 4978.

Dammann, B., and J.D. Reger, 1993, Europhys. Lett. **21**, 157.

Dash, J.G., 1978, Phys. Rep. **38C**, 177.

De Gennes, P.G., 1968, Solid State Comm. **6**, 163.

De Gennes, P.G., 1985, Rev. Mod. Phys. **57**, 827.

De Groot, S.R., and P. Mazur, 1962, Non-Equilibrium Thermodynamics (North-Holland, Amsterdam).

Den Nijs, M.P.M., 1979, J. Phys. **A 12**, 1857.

De Oliveira, M.J., and R.B. Griffiths, 1978, Surface Sci. **238**, 317.

Diehl, H.W., 1986, in: Phase Transitions and Critical Phenomena, Vol. 10, eds. C. Domb and J.L. Lebowitz (Academic Press, London), p. 76.

Dietrich, S., 1988, in: Phase Transitions and Critical Phenomena, Vol. 12, eds. C. Domb and J.L. Lebowitz (Academic Press, London), p. 1.

Dietrich, S., and H. Wagner, 1984, Z. Physik **B 56**, 207.

Domany, E., M. Schick, J.S. Walker and R.B. Griffiths, 1978, Phys. Rev. **B 18**, 2209.

Domb, C., and M.S. Green, (eds.), 1974, Phase Transitions and Critical Phenomena, Vol. 3 (Academic Press, London).

Domb, C., and M.S. Green, (eds.), 1976, Phase Transitions and Critical Phenomena, Vol. 6 (Academic Press, London).

Dosch, H., 1991, Evanescent Scattering and Phase Transitions in Semi-Infinite Matter (Springer, Berlin).

Dosch, H., L. Mailänder, A. Lied, J. Peisl, F. Grey, R.L. Johnson and S. Krummacher, 1988, Phys. Rev. Lett. **60**, 2382.

Dünweg, B., A. Milchev and P.A. Rikvold, 1991, J. Chem. Phys. **94**, 3958.

Dzyaloshinskii, I.E., 1964, Zh. Ehsp. Teor. Fiz. **35**, 1518.

Dzyaloshinskii, I.E., E.M. Lifshitz and L.P. Pitaevskii, 1961, Adv. Phys. **10**, 165.

Ebner, C., 1980, Phys. Rev. **A 22**, 2776.

Ebner, C., 1981, Phys. Rev. **A 23**, 1925.

Ebner, C., and W.F. Saam, 1977, Phys. Rev. **38**, 1486.

Ebner, C., C. Rottmann and M. Wortis, 1983, Phys. Rev. **B 28**, 4186.

Eckert, J., W.D. Ellenson, J.B. Hastings and L. Passell, 1979, Phys. Rev. Lett. **43**, 1329.

Einstein, T.L., 1982, in: Chemistry and Physics of Solid Surfaces IV, eds. R. Vanselow and R. Howe (Springer, Berlin), p. 251.

Elitzur, S., R. Pearson and J. Shigemitsu, 1979, Phys. Rev. **D 19**, 3698.

Evans, R., 1990, in: Liquids at Interfaces, eds. J. Charvolin, J.F. Joanny and Zinn-Justin (Elsevier, Amsterdam), p. 1.

Evans, R., 1992, in: Fundamentals of Inhomogeneous Fluids, ed. D. Henderson (M. Dekker, New York, NY), p. 85.

Ferreira, I.B., A.R. King, V. Jaccarino, J.L. Cardy and A.J. Guggenheim, 1983, Phys. Rev. B 28, 5192.

Fisher, D.S., and D.A. Huse, 1985, Phys. Rev. B 32, 247.

Fisher, M.E., 1968, Phys. Rev. 176, 257.

Fisher, M.E., 1969, J. Phys. Soc. Japan, Suppl. 26, 87.

Fisher, M.E., 1971, in Critical Phenomena, Proc., 1970 E. Fermi Int. School Phys., ed. M.S. Green (Academic Press, London) p. 1.

Fisher, M.E., 1974, Rev. Mod. Phys. 46, 587.

Fisher, M.E., 1975, Phys. Rev. Lett. 34, 1634.

Fisher, M.E., 1984, J. Stat. Phys. 34, 667.

Fisher, M.E., and R.J. Burford, 1967, Phys. Rev. 156, 583.

Fisher, M.E., and D.R. Nelson, 1974, Phys. Rev. lett. 32, 1634.

Fisher, M.E., and H. Wen, 1992, Phys. Rev. Lett. 68, 3654.

Fisk, S., and B. Widom, 1969, J. Chem. Phys. 50, 3219.

Franck, C., 1992, in: Fundamentals of Inhomogeneous Fluids, ed. D. Henderson (M. Dekker, New York, NY), p. 277.

Frank, F.C., and J.H. van der Merwe, 1949a, Proc. Roy. Soc. (London) A 198, 205.

Frank, F.C., and J.H. van der Merwe, 1949b, Proc. Roy. Soc. (London) A 198, 216.

Frenkel, J., and T. Kontorova, 1938, Phys. Z. Sowjet. 13, 1.

Furukawa, H., 1985, Adv. Phys. 34, 703.

Georgiev, N., A. Milchev, M. Paunov and B. Dünweg, 1992, Surface Sci. 264, 455.

Ginzburg, V.L., 1960, Soviet Phys. Solid State 2, 1824.

Gittes, F.T., and M. Schick, 1984, Phys. Rev. B 30, 209.

Glauber, R.J., 1963, J. Math. Phys. 4, 293.

Golecki, I., and J. Jaccard, 1978, J. Phys. C 11, 4229.

Gompper, G., 1986, Z. Phys. B 62, 357.

Grant, M., 1988, Phys. Rev. B 37, 5705.

Grant, M., K. Kaski and K. Kankaala, 1987, J. Phys. A 20, L571.

Grest, G.S., and J. Sak, 1978, Phys. Rev. B 17, 3607.

Grest, G.S., and D.J. Srolovitz, 1985, Phys. Rev. B 32, 3014.

Griffiths, R.B., 1970, Phys. Rev. Lett. 24, 715.

Grinstein, G., and S.-K. Ma, 1982, Phys. Rev. Lett. 49, 685.

Gunton, J.D., M. San Miguel and P.S. Sahni, 1983, in: Phase Transitions and Critical Phenomena, Vol. 8, eds. C. Domb and J.L. Lebowitz (Academic Press, London), p. 267.

Haldane, F.D.M., P. Bak and T. Bohr, 1983, Phys. Rev. B 28, 2743.

Halperin, B.I., and D.R. Nelson, 1978, Phys. Rev. Lett. 41, 121.

Halperin, B.I., P.C. Hohenberg and S.-K. Ma, 1974, Phys. Rev. B 10, 139.

Harris, A.B., 1974, J. Phys. C 7, 1671.

Harris, A.B., O.G. Mouritsen and A.J. Berlinsky, 1984, Can. J. Phys. 62, 915.

Haymet, A.D.J., 1992, in: Fundamentals of Inhomogeneous Fluids, ed. D. Henderson (M. Dekker, New York, NY) p. 363.

Helbing, W., B. Dünweg, K. Binder and D.P. Landau, 1990, Z. Phys. B – Condensed Matter 80, 401.

Henderson, J.R., and F. van Swol, 1984, Mol. Phys. 51, 991.

Henderson, J.R., and F. van Swol, 1985, Mol. Phys. 56, 1313.

Henzler, M., and H. Busch, 1990, Phys. Rev. B 41, 4891.

Hohenberg, P.C., 1967, Phys. Rev. 158, 383.

Hohenberg, P.C., and B.I. Halperin, 1977, Rev. Mod. Phys. 49, 435.

Hornreich, R.M., M. Luban and S. Shtrikman, 1975, Phys. Rev. Lett. 35, 1678.

Huse, D.A., 1981, Phys. Rev. **B 24**, 5180.
Huse, D.A., 1982, Phys. Rev. Lett. **49**, 1121.
Huse, D.A., 1984, Phys. Rev. **B 29**, 6985.
Imry, Y., and S.-K. Ma, 1975, Phys. Rev. Lett. **35**, 1399.
Israelachvili, J.N., 1985, Intermolecular and Surface Forces (Academic Press, London).
Jancovici, B., 1967, Phys. Rev. Lett. **19**, 20.
Jasnow, D., 1984, Rep. Progr. Phys. **47**, 1059.
Jayanthi, C.S., E. Tosatti and A. Fasolino, 1985a, Surface Sci. **152/153**, 155.
Jayanthi, C.S., E. Tosatti and L. Pietronero, 1985b, Phys. Rev. **B 31**, 3456.
Johnson, W.C., and J.M. Blakely, (eds.), 1979, Interfacial Segregation (American Society for Metals, Cleveland, OH).
Jones, R.A.L., L.J. Norton, E.J. Kramer, F.S. Bates and P. Wiltzius, 1991, Phys. Rev. Lett. **66**, 1326.
José, J.V., L.P. Kadanoff, S. Kirkpatrick and D.R. Nelson, 1977, Phys. Rev. **B 16**, 1217.
Kadanoff, L.P., and F.J. Wegner, 1971, Phys. Rev. **B 4**, 3989.
Kaski, K., K. Binder and J.D. Gunton, 1984, Phys. Rev. **B 29**, 3996.
Kawabata, C., and K. Binder, 1977, Solid State Comm. **22**, 705.
Kawasaki, K., 1972, in: Phase Transitions and Critical Phenomena, Vol. 2, eds. C. Domb and M.S. Green (Academic Press, London), p. 2.
Kihara, T., Y. Midzuno and T. Shizume, 1954, J. Phys. Soc. Japan **9**, 681.
Kim, I.M., and D.P. Landau, 1981, Surface Sci. **110**, 415.
Kinzel, W., W. Selke and K. Binder, 1982, Surface Sci. **121**, 13.
Klein, W., and C. Unger, 1983, Phys. Rev. **B 28**, 445.
Knops, H.J.F., 1980, Ann. Phys. **128**, 448.
Komura, S., and H. Furukawa (eds.), 1988, Dynamics of Ordering Processes in Condensed Matter (Plenum Press, New York, NY).
Kosterlitz, J.M., 1974, J. Phys. **C 7**, 1046.
Kosterlitz, J.M., and D.J. Thouless, 1973, J. Phys. **C 6**, 1181.
Krim, J., J.G. Dash and J. Suzanne, 1984, Phys. Rev. Lett. **52**, 640.
Krinsky, S., and D. Mukamel, 1977, Phys. Rev. **B 16**, 2313.
Kroll, D.M., and G. Gompper, 1987, Phys. Rev. **B 36**, 7078.
Kroll, D.M., and R. Lipowsky, 1983, Phys. Rev. **B 28**, 6435.
Lajzerowicz, J., 1981, Ferroelecics **35**, 219.
Landau, D.P., 1983, Phys. Rev. **B 27**, 5604.
Landau, D.P., and K. Binder, 1981, Phys. Rev. **B 24**, 1391.
Landau, D.P., and K. Binder, 1985, Phys. Rev. **B 31**, 5946.
Landau, D.P., and K. Binder, 1990, Phys. Rev. **B 41**, 4633.
Landau, L.D., and E.M. Lifshitz, 1958, Statistical Physics (Pergamon Press, Oxford)
Landau, D.P., S. Tang and S. Wansleben, 1988, J. Phys. (Paris) **C-8**, 1525.
Landau, D.P., R. Pandey and K. Binder, 1989, Phys. Rev. **B 39**, 12302.
Lapujoulade, J., B. Salanon, F. Fabre and B. Loisel, 1990, in: Kinetics of Ordering and Growth at Surfaces, ed. M.G. Lagally (Plenum Press, New York, NY) p. 355.
Larher, Y., 1978, J. Chem. Phys. **68**, 257.
Le Guillou, J.C., and J. Zinn-Justin, 1980, Phys. Rev. **B 21**, 3976.
Liang, K.S., E.B. Sirota, K.L. D'Amico, G.J. Hughes and S.K., Sinha, 1987, Phys. Rev. Lett. **59**, 2447.
Lifshitz, I.M., 1962, Soviet Physics JETP **15**, 939.
Lifshitz, I.M. and V.V. Slyozov, 1961, J. Phys. Chem. Solids **19**, 35.
Lipowsky, R., 1982, Phys. Rev. Lett. **49**, 1575.
Lipowsky, R., 1983, Z. Physik **B 51**, 165.
Lipowsky, R., 1984, J. Appl. Phys. **55**, 2485.

Lipowsky, R., 1985a, J. Phys. **A 18**, L585.
Lipowsky, R., 1985b, Phys. Rev. **B 32**, 1731.
Lipowsky, R., 1987, Ferroelectrics **73**, 69.
Lipowsky, R., and D.A. Huse, 1986, Phys. Rev. Lett. **52**, 353.
Lipowsky, R., and M.E. Fisher, 1987, Phys. Rev. **B 36**, 2126.
Lipowsky, R., and W. Speth, 1983, Phys. Rev. **B 28**, 3983.
Lipowsky, R., D.M. Kroll and R.K.P. Zia, 1983, Phys. Rev. **B 27**, 4499.
Löwen, H., and T. Beier, 1990, Phys. Rev. **B 41**, 4435.
Löwen, H., T. Beier and H. Wagner, 1989, Europhys. Lett. **9**, 791.
Ma, S.-K., 1976, Modern Theory of Phase Transitions (Benjamin Press, Reading).
Mailänder, L., H. Dosch, J. Peisl and R.L. Johnson, 1990, Phys. Rev. Lett. **64**, 2527.
Mailänder, L., H. Dosch, J. Peisl and R.L. Johnson, 1991, in: Advances in Surface and Thin Film Diffraction Symposium, Boston, MA, USA, 27-29 November 1990 (Material Research Society, Pittsburgh, PA) p. 87.
Mannebach, H., V.G. Volkmann, J. Faul and K. Knorr, 1991, Phys. Rev. Lett. **67**, 1566.
Marquardt, P., and H. Gleiter, 1982, Phys. Rev. Lett. **48**, 1423.
Marx, D., O. Opitz, P. Nielaba and K. Binder, 1993, Phys. Rev. Lett. **70**, 2908.
Marx, R., 1985, Phys. Rep. **125**, 1.
Marx, R., 1989, Phys. Rev. **B 40**, 2585.
Matthews-Morgan, D., D.P. Landau and R.H. Swendsen, 1981, Phys. Rev. **B 24**, 1468.
Matthews-Morgan, D., D.P. Landau and R.H. Swendsen, 1984, Phys. Rev. Lett. **53**, 679.
McRae, E.G., and R.A. Malic, 1984, Surface Sci. **148**, 551.
McTague, J.P., and M. Nielsen, 1976, Phys. Rev. Lett. **37**, 596.
Mermin, N.D., 1967, J. Math. Phys. **8**, 1061.
Mermin, N.D., 1968, Phys. Rev. **176**, 250.
Mermin, N.D., and H. Wagner, 1966, Phys. Rev. Lett. **17**, 1133.
Mon, K.K., K. Binder and D.P. Landau, 1987, Phys. Rev. **B 35**, 3683.
Mon, K.K., D.P. Landau, K. Binder and S. Wansleben, 1989, Phys. Rev. **B 39**, 7089.
Moran-Lopez, J.L., F. Mejia-Lira and K.H. Bennemann, 1985, Phys. Rev. Lett. **54**, 1936.
Morgenstern, I., K. Binder and R.M. Hornreich, 1981, Phys. Rev. **B 23**, 287.
Mouritsen, O.G., 1985, Phys. Rev. **B 32**, 1632.
Mouritsen, O.G., 1990, in: Kinetics of Ordering and Growth at Surfaces, ed. M.G. Lagally (Plenum Press, New York, NY) p. 1.
Mouritsen, O.G., and A.J. Berlinsky, 1982, Phys. Rev. Lett. **48**, 181.
Müller-Krumbhaar, H., 1978, in: Current Topics in Materials Science, Vol. 1, ed. H. Kaldis (North-Holland, Amsterdam) p. 1.
Mukamel, D., M.E. Fisher and E. Domany, 1976, Phys. Rev. Lett. **37**, 565.
Muscat, J.P., 1985, Progr. Surface Sci. **18**, 59.
Muscat, J.P., 1986, Phys. Rev. **B 33**, 8136.
Nakanishi, H., and M.E. Fisher, 1982, Phys. Rev. Lett. **49**, 1565.
Nakanishi, H., and P. Pincus, 1983, J. Chem. Phys. **79**, 997.
Nattermann, T., and J. Villain, 1988, Phase Transitions **11**, 5.
Nauenberg, M., and B. Nienhuis, 1974, Phys. Rev. Lett. **33**, 941.
Nelson, D.R., and B.I. Halperin, 1979, Phys. Rev. **B 19**, 2457.
Nelson, D.R., and J.M. Kosterlitz, 1977, Phys. Rev. Lett. **39**, 1201.
Nienhuis, B., 1982, J. Phys. A.: Math. Gen. **15**, 199.
Nienhuis, B., 1987, in: Phase Transitions and Critical Phenomena, Vol. 11, eds. C. Domb and J. L. Lebowitz (Academic Press, London), p. 1.
Nightingale, M.P., 1977, Phys. Lett. **59 A**, 486.
Nightingale, M.P., W.F. Saam and M. Schick, 1984, Phys. Rev. **B 30**, 3830.
Novaco, A.D., and J.P. McTague, 1977, Phys. Rev. Lett. **38**, 1286.

Ohta, T., D. Jasnow and K. Kawasaki, 1982, Phys. Rev. Lett. **49**, 1223.

Oitmaa, J., 1981, J. Phys. **A 14**, 1159.

Onsager, L., 1944, Phys. Rev. **65**, 117.

Ostlund, S., 1981, Phys. Rev. **B 24**, 398.

Pandit, R., and M. Wortis, 1982, Phys. Rev. **B 25**, 3226.

Pandit, R., M. Schick and M. Wortis, 1982, Phys. Rev. **B 25**, 5112.

Patrykiejew, A., and K. Binder, 1992, Surface Sci. **273**, 413.

Patrykiejew, A., K. Binder and D.P. Landau, 1990, Surface Sci. **238**, 317.

Pearson, R.B., 1980, Phys. Rev. **B 22**, 2579.

Penrose, O., and J.L. Lebowitz, 1971, J. Stat. Phys. **3**, 211.

Pereyra, V., P. Nielaba and K. Binder, 1993, J. Phys.: Condensed Matter **5**, 6631.

Peczak, P., and D.P. Landau, 1991, Phys. Rev. **B 43**, 1048.

Piercy, P., and H. Pfnür, 1987, Phys. Rev. Lett. **59**, 1124.

Pietronero, L., and E. Tosatti, 1979, Solid St. Comm. **32**, 255.

Pokrovskii, V.L., and A.L. Talapov, 1979, Zh. Eksp. Teor. Fiz. **75**, 1151.

Potts, R.B., 1952, Proc. Camb. Philos. Soc. **48**, 106.

Privman, V. (ed.), 1990, Finite Size Scaling and Numerical Simulation of Statistical Systems (World Scientific, Singapore).

Privman, V., 1992, Int. J. Mod. Phys. **C 3**, 857.

Privman, V., A. Aharony and P.C. Hohenberg, 1991, in: Phase Transitions and Critical Phenomena, Vol. 14, eds. C. Domb and J.L. Lebowitz (Academic Press, London), p. 1.

Puri, S., and K. Binder, 1992, Phys. Rev. **A 46**, R4487.

Rau, C., and M. Robert, 1987, Phys. Rev. Lett. **58**, 2714.

Riedel, E., and F.J. Wegner, 1969, Z. Physik **225**, 195.

Rikvold, P.A., K. Kaski, J.D. Gunton and M.C. Yalabik, 1984, Phys. Rev. **B 29**, 6285.

Roelofs, L.D., 1982, in: Chemistry and Physics of Solid Surfaces IV, eds. R. Vanselow and R. Howe (Springer, Berlin), p. 219.

Roelofs, L.D., N.C. Bartelt and T.L. Einstein, 1981, Phys. Rev. Lett. **47**, 1348.

Roelofs, L.D., T.L. Einstein, N.C. Bartelt and J.D. Shore, 1986, Surface Sci. **176**, 295.

Rottmann, C., and M. Wortis, 1984, Phys. Rep. **103**, 59.

Rous, P., 1995, in: Cohesion and Structure of Surfaces, ed. D.G. Pettifor (Elsevier, Amsterdam), ch. 1.

Ruge, C., S. Dunkelmann and F. Wagner, 1992, Phys. Rev. Lett. **69**, 2465.

Sadiq, A., and K. Binder, 1983, Surface Sci. **128**, 350.

Sadiq, A., and K. Binder, 1984, J. Stat. Phys. **35**, 517.

Salanon, B., F. Fabre, J. Lapujoulade and W. Selke, 1988, Phys. Rev. **B 38**, 7385.

Sarbach, S., and I.D. Lawrie, 1984, in: Phase Transitions and Critical Phenomena, Vol. 9, eds. C. Domb and J.L. Lebowitz (Academic Press, London) p. 1.

Schick, M., 1981, Progr. Surface Sci. **11**, 245.

Schmid, F., 1993, Z. Phys. B – Condensed Matter **91**, 77.

Schmid, F., and K. Binder, 1992a, Phys. Rev. **B 46**, 13565.

Schmid, F., and K. Binder, 1992b, Phys. Rev. **B 46**, 13553.

Schmidt, I., and K. Binder, 1987, Z. Physik **B 67**, 369.

Schulz, H.J., 1983, Phys. Rev. **B 28**, 2746.

Schweika, W., K. Binder and D.P. Landau, Phys. Rev. Lett. **65**, 3321.

Sega, I., W. Selke and K. Binder, 1985, Surface Sci. **154**, 331.

Selke, W., 1984, Surface Sci. **144**, 176.

Selke, W., 1988, Phys. Rept **170**, 213.

Selke, W., 1992, in: Phase Transitions and Critical Phenomena, Vol. 15, eds. C. Domb and J.L. Lebowitz (Academic Press, London), p. 1.

Selke, W., 1993, in: Computer Simulation Studies in Condensed Matter IV, eds. D.P. Landau, K.K. Mon and H.B. Schüttler (Springer, Berlin), p. 18.

Selke, W., and A.M. Szpilka, 1986, Z. Physik B 62, 381.

Shi, B.Q., C. Harrison and A. Cumming, 1993, Phys. Rev. Lett. 70, 206.

Speth, W., 1985, Z. Physik B 61, 325.

Stanley, H.E., 1968, Phys. Rev. 176, 718.

Stanley, H.E., 1971, An Introduction to Phase Transitions and Critical Phenomena (Oxford University Press, Oxford).

Stauffer, D., 1992, Physica A 184, 201.

Steiner, U., J. Klein, E. Eiser, A. Budkowski and L.J. Fetters, 1992, Science 258, 1126.

Stephanov, M.A., and M.M. Tsypin, 1991, Nucl. Phys. B 366, 420.

Stinchcombe, R.B., 1983, in: Phase Transitions and Critical Phenomena, Vol. 7, eds. C. Domb and J.L. Lebowitz (Academic Press, London), p. 150.

Stoltze, P., J.K. Nørskov and U. Landman, 1988, Phys. Rev. Lett. 61, 440.

Straley, J.P., and M.E. Fisher, 1973, J. Phys. C 6, 1310.

Sullivan, D.E., and M.M. Telo da Gama, 1986, in: Fluid Interfacial Phenomena, ed. C.A. Croxton (Wiley, New York, NY), p. 45.

Sundaram, V.S., B. Farrell, R.S. Alben and W.D. Robertson, 1973, Phys. Rev. Lett. 31, 1136.

Swendsen, R.H., and S. Krinsky, 1979, Phys. Rev. Lett. 43, 177.

Tammann, G., 1910, Z. Phys. Chem. 68, 205.

Tanaka, H., 1993, Phys. Rev. Lett. 70, 53.

Tarazona, P., and R. Evans, 1983, Mol. Phys. 48, 799.

Tolédano, J.C., and P. Tolédano, 1987, The Landau Theory of Phase Transitions (World Scientific, Singapore).

Tosatti, E., 1988, in: The Structure of Surfaces II, eds. J.F. van der Veen and M.A. van Hove (Springer, Berlin 1988), p. 535.

Trayanov, A., and E. Tosatti, 1988, Phys. Rev. B 38, 6961.

Trullinger, S.E., and D.L. Mills, 1973, Solid State Commun. 12, 819.

Van Beijeren, H., and I. Nolden, 1987, in: Structure and Dynamics of Surfaces II, eds. W. Schommers and P. Blanckenhagen (Springer, Berlin), p. 259.

Van der Veen, J.F., and J.W.H. Frenken, 1986, Surface Sci. 178, 382.

Van der Veen, J.F., B. Pluis and A.G. Denier van der Gon, 1990, in: Kinetics of Ordering and Growth at Surfaces, ed. M.G. Lagally (Plenum Press, New York, NY), p. 343.

Van Hove, L., 1954, Phys. Rev. 93, 1374.

Venables, J.A., G.D.T. Spiller and M. Hanbrücken, 1984, Rep. Progr. Phys. 47, 399.

Villain, J., 1975, J. Phys. (Paris) 36, 581.

Villain, J., 1980, in: Ordering in Strongly Fluctuating Condensed Matter Systems, ed. T. Riste (Plenum, New York, NY), p. 222.

Villain, J., 1982, J. Phys. (Paris) 43, L551.

Villain, J., D.R. Grempel and J. Lapujoulade, 1985, J. Phys. F 15, 809.

Volkmann, U.G., and K. Knorr, 1989, Surface Sci. 221, 379.

Wagner, H., 1978, in: Hydrogen in Metals, eds. G. Alefeld and H. Völkl (Springer, Berlin), p. 5.

Wagner, P., and K. Binder, 1986, Surface Sci. 175, 421.

Weeks, J.D., 1980, in: Ordering in Strongly Fluctuating Condensed Matter Systems, ed. T. Riste (Plenum Press, New York, NY), p. 293.

Weeks, J.D., G.H. Gilmer and H.J. Leamy, 1973, Phys. Rev. Lett. 31, 549.

Wegner, F.J., 1967, Z. Physik 206, 465.

Weller, D., and D. Alvarado, 1988, Phys. Rev. B 37, 9911.

Widom, B., 1972, in: Phase Transitions and Critical Phenomena, Vol. 2, eds. C. Domb and M.S. Green (Academic Press, London), p. 79.

Wiechert, H., and S.-A. Arlt, 1993, Phys. Rev. Lett. **71**, 2090.

Willis, R.F., 1985, in: Dynamical Phenomena at Surfaces, Interfaces, and Superlattices, eds. F. Nizzoli, K.H. Rieder and R.F. Willis (Elsevier, Amsterdam), p. 126.

Wilson, K.G., and M.E. Fisher, 1972, Phys. Rev. Lett. **28**, 548.

Wilson, K.G., and J. Kogut, 1974, Phys. Rep. **12 C**, 75.

Wiltzius, P., and A. Cumming, 1991, Phys. Rev. Lett. **66**, 3000.

Wortis, M., 1985, in: Fundamental Problems in Statistical Mechanics VI, ed. E.G.D. Cohen (North-Holland, Amsterdam), p. 87.

Wu, F.Y., 1982, Rev. Mod. Phys. **54**, 235.

Wu, P.K., M.C. Tringides and M.G. Lagally, 1989, Phys. Rev. **B 39**, 7595.

Wulff, G., 1901, Z. Kristallogr. Mineral **34**, 449.

Yeomans, J., 1992, Statistical Mechanics of Phase Transitions (Oxford University Press, Oxford).

Young, A.P., 1978, J. Phys. **C 11**, L 453.

Young, A.P., 1980a, in: Strongly Fluctuating Condensed Matter Systems, ed. T. Riste (Plenum Press, New York, NY) p. 11.

Young, A.P., 1980b, in: Strongly Fluctuating Condensed Matter Systems, ed. T. Riste (Plenum Press, New York, NY) p. 271.

Young, T., 1805, Philos. Trans. Roy. Soc. London **95**, 65.

Zangwill, A., 1988, Physics at Surfaces (Cambridge University Press, Cambridge).

Zettlemoyer, A.C. (ed.), 1969, Nucleation (Marcel Dekker, New York, NY).

Zinke-Allmang, M., L.C. Feldman and M.H. Grabow, 1992, Surf. Sci. Rep. **16**, 377.

Chapter IV

SURFACE STRUCTURE AND REACTIVITY

MICHAEL BOWKER

Department of Chemistry, University of Reading, Whiteknights Park, P.O. Box 224, Reading,
Berkshire RG6 6AD, United Kingdom
I.R.C in Surface Science, University of Liverpool, Oxford Road, P.O. Box 147,
Liverpool L69 3BX, United Kingdom

Contents

Cohesion and Structure of Surfaces
Edited by D.G. Pettifor
© *1995 Elsevier Science B.V. All rights reserved.*

Abstract

The surface of a material can be considered as an effect chemical and it is now widely recognised to be such in industry. It is an effect chemical because a small amount of material at the surface can significantly alter its properties and improve it in terms of efficiency, whether this be as a catalyst, or a protective coating for instance.

The structure of the surface involved can be all important in determining optimised properties and this is part of the reason why studies using single crystals of well-defined structure, and ultra high vacuum conditions, have come into vogue in the last twenty years. In this article the relationship of the reactivity of a surface to its morphology and composition is discussed, particularly in light of the thermodynamic driving force for adsorption and catalysis, namely, surface free energy. Examples of the application of modern methods in surface science to the study of adsorption and reaction systems are given in abundance.

1. Introduction

In a real sense it can be said that the whole of chemistry is concerned with surfaces. In general, chemistry is involved with the transformation of one molecular system into another by the interaction and rearrangement of valence electrons, and this occurs in the surface, outermost, most weakly bound valence electrons. Bond distances so formed are very roughly equivalent to the sum of the radii of the atoms involved.

The solid surfaces we are concerned with here are similar to those described above, and are represented by regions of high valence electron density, at the surface of which reactions may take place (fig. 1). Solid surfaces are not regions of anomaly, they are part of the mainstream of chemical reactivity, albeit a rather special part.

The surface represents a region of great opportunity for the technologist, since the properties of the interface can be completely modified by the addition of only small amounts of material. Thus, the surface can be considered as an "effect" chemical and suitably modified surfaces represent "high added value" materials. The properties can be changed, for instance, from high energy surfaces to low energy, low wettability ones by coating with fluorinated polymers (such as PTFE), can be hardened by plasma treatment in reactive gases to produce longer lasting tools, can be made biocompatible by coating with appropriate polymers, or can have their corrosion resistance improved with protective coatings.

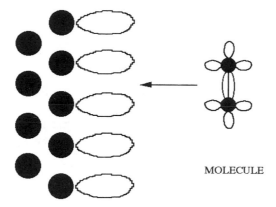

MOLECULE

SOLID SURFACE

Fig. 1. A simplistic picture of solid–gas interaction.

Surface science techniques have expanded their application in industry in the last few years for just these reasons — the nature of the surface has been recognised as an important factor in the performance of many materials. This is especially the case in the electronic device industry where low levels of surface contamination can severely deteriorate electronic conductivity and barrier properties.

However, in most of these areas the surface modification is carried out on a bulk scale. Surface coatings are usually applied at the level of micron thickness. Although small amounts of impurities can severely affect device performance, commercial devices are built of macroscopic layers whose bulk conductivity is the important parameter.

One of the few areas where the properties of the top monolayer is crucial to economic success is catalysis. In such technology, reactions are performed at the outermost atomic layer, by direct interaction with the incoming molecules, and submonolayer amounts of impurities can severely affect the rates of reaction either positively or negatively. A schematic illustration of the catalytic event is presented in fig. 2, which shows a catalytically active component (a metal particle in this case) supported on a relatively inert material, such as alumina, which is used to thermally stabilise the metal with respect to loss of surface area ("sintering" — which results in an increase in particle size and decrease in particle number density per unit area). The events which occur during the catalytic cycle are illustrated.

Some catalytic processes are limited by gas phase diffusion, usually those with very high surface area catalysts which in turn means very small diameter, restricted pores within the material. After the gas has diffused to the active

PARAMETERS AFFECTING CATALYTIC EFFICIENCY

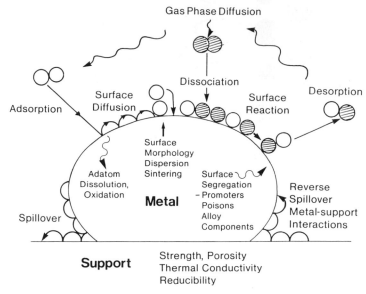

Fig. 2. A schematic diagram of the range of processes involved in a catalytic conversion, showing the steps involved in a reaction of two diatomics to produce a hybrid molecule.

surface it has to adsorb, usually first via a weakly held physisorbed species which can be highly mobile in two dimensions, then into a more strongly held chemisorbed state. Within this state the adsorbed species may also diffuse and encounter another reactive entity and, with sufficient energy, react to form the adsorbed product. It is the case that for many catalytic processes this is the rate determining step, and modification of the nature of the surface complex involved in this transition *will* affect the rate of product formation. Once the adsorbed product is formed it then desorbs from the surface, usually over a relatively small activation barrier. In some cases, especially at high conversions where the product is present in the gas phase at high concentrations, then this final step of desorption may be rate limiting and can result in product build-up on the surface, with consequent retardation of the synthesis rate. Indeed it is thought by many authors that the classical catalytic reaction of ammonia synthesis on Fe is just such a case with NH_3 self-poisoning the reaction at high pressures and high conversions. The illustration in fig. 2 acts as a basis for the division of this article as follows.

1. *Adsorption* (sect. 3). The role and nature of adsorption, for a variety of the types of molecules and functional groups involved in catalysis, is

described in detail in this section, including activated and non-activated adsorption.

2. *Desorption* (sect. 4). Desorption is discussed in less detail than adsorption, simply because it is generally the reverse of the adsorption process. Nevertheless, desorption is used widely in surface science and catalysis as a diagnostic of the adsorbed state of atoms and molecules, and these aspects will be dealt with in this section.

3. *Surface diffusion* (sect. 5). It will be shown that surface diffusion is an important aspect of catalysis in several respects. Firstly, diffusion in weakly held "precursor" states can lead to higher reactivity than might be expected on the basis of the simple Langmuirian kinetics. Secondly, the rate of diffusion can completely modify surface reaction models from those expected for phase separated, island models of reaction, to the completely homogeneous concentration situation. Thirdly, it is clear from recent studies, especially with Scanning Tunnelling Microscopy (STM), that we can no longer consider the surface as a rigid "checker-board", but that surface atoms of the solid may be in a continuous state of flux during the course of a catalytic reaction. Indeed, it may be the situation, in some cases, that the surface can adapt itself to provide the configuration necessary for the reaction.

4. *Surface reaction* (sections 6–8). This is the largest section, concerned with the reaction event itself occurring between adsorbed species. It will include a brief description of reaction kinetics at surfaces, together with a classification of the kind of catalytic reactions which are important and a consideration of work mainly carried out on well-defined surfaces where the surface structure is well-characterised. It includes further sections on the effect of atomic number (electronic structure) on reactions on metals in the transition series, on the effects of surface structure on reaction rates, and on important aspects of catalysis, namely poisoning and promoter effects.

In the concluding section, consideration will be given to the relationship between studies on single crystals and the behaviour of particulate catalysts, especially with respect to the relevance of one to the other. A very old concept in catalysis is that of the "active site". This concept recently has come again to the forefront of research in this area and consideration of the possibility of directly observing the "active site" will be given in the final section.

The schematic image of a catalyst surface in fig. 2 can be visualised better in fig. 3, which shows what such a surface might look like at different magnifications, from the microscopic, atomic scale view to a catalyst pellet. The atomic level view is an idealised one and is a reproduction of a rather beautiful field ion microscope (FIM) image of the hemispherical tip of an Ir needle. This is a single crystal and shows the heterogeneity of the surfaces exposed at the curved tip, with a variety of well-defined planes, steps and even missing atoms/defects present in a few places. The SEM picture

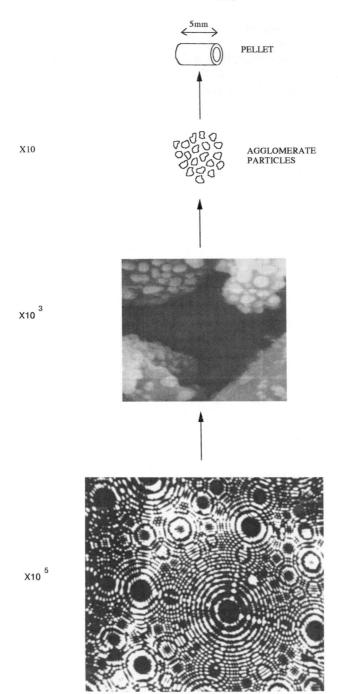

shows a particulate supported catalyst (in this case Ag on α-alumina) with fairly well-defined metal particles which are close to hemispherical. At lower magnification this catalyst is a fine powder and that in turn is made up as the catalyst pellets which are finally used in a full scale catalytic plant.

It is important to note that catalysts are generally high surface area, three dimensional surfaces, quite different in that respect from the macroscopic two dimensional single crystals most commonly used in surface science studies. The reason such high area materials are used can be seen from the consideration of the rate equation for the simplest form of catalytic reaction, an isomerisation under conditions of pre-equilibrium:

$$A + S \Leftrightarrow AS$$

$$AS \rightarrow B + S$$

Here A is isomerised to molecule B by adsorption and conversion at a surface site S. In the scheme there is adsorption/desorption equilibrium for the gas phase reactant and the rate is given by the Langmuir equation (otherwise known as the Michaelis–Menten equation in enzymic catalysis):

$$-\frac{d[A]}{dt} = S_T \cdot \frac{k P_A}{1 + k' P_A} \tag{1}$$

Here k and k' are complex rate constants containing the rate constants associated with all three steps in the isomerisation mechanism above. The important point, though, is that S_T, the total number of "active sites" at the surface is contained in the rate equation, hence the need for high surface area to maximise S_T. In some cases (especially in selective oxidations) it is necessary to limit S_T and surface area to avoid further reaction/decomposition of a desired intermediate product.

Although there may be some debate regarding the difference between single crystal and particulate surfaces, which will be discussed in sect. 9 later, it is clear from eq. (1) why surface science has provided such a good insight into catalytic reactivity at the microscopic level. Simple measurements of the rate of product formation in a microreactor gives information relating to the overall rate constants and for realistic reactions of more complexity than that given above, little useful insight can be gained in that way into the kinetics of individual steps in the reaction, especially regarding the rate determining step. The utility of surface science has been to provide methods to separate these reactions into their elementary steps for individual study and has identified the nature of the intermediates involved in such reactions

Fig. 3. An illustration of the type of make-up of a supported metal catalyst from atomic scale (FIM image of a metal surface) to an SEM image of a $Ag/\alpha Al_2O_3$ catalyst to the catalyst pellet, as might be loaded into an industrial scale reactor.

(AS, for instance, in the scheme above). Examples of such studies will be given in sect. 6.

In the following section, however, consideration will be given to the fundamental properties of surfaces which make them regions of interest for academia and industry. Surfaces are regions of high free energy which acts as the driving force for adsorption and catalysis. Thus the thermodynamic properties of surfaces is the primary subject which needs to be addressed in a consideration of surface reactivity.

2.   Thermodynamic considerations

Surfaces are regions of high energy due to the asymmetry in the interface region and the lowered coordination number of surface atoms. In order to make such interfaces, work has to be done, as shown in fig. 4, to break bonds in the bulk of the material.

The work done, δW, is equivalent to the product of the surface free energy per unit area (G_S) multiplied by the area created:

$$\delta W_{T,P} = G_S \delta A \tag{2}$$

The surface free energy is equivalent to the surface tension. It is this surface free energy which drives adsorption and catalysis and explains why metals

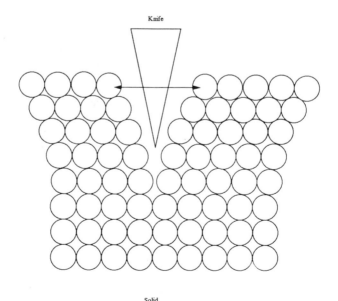

Fig. 4. Making a new surface.

are very active materials for such processes, since they have high cohesive energies in the bulk and so high surface energies when these bonds are broken. Metals generally have high surface energy, while other materials (such as halides, sulphides and oxides) have low surface energy and are relatively inactive in catalytic terms (although they may be used for their particular selectivity to a desired product, and oxide catalysts are often used in such a case).

The structure of a surface which is formed is strongly related to the thermodynamics. Because of the high energy of the interface, surfaces attempt to minimise this energy by increasing the coordination number of surface atoms and this is achieved in several ways, as follows:

I. Surface reconstruction — *all* surfaces reconstruct to reduce the surface energy and to maximise the surface coordination, and these can be classified into two forms.

(i) *Surface relaxation*. Most surfaces show a contraction between layers 1 and 2 (fig. 5), increasing the interaction and binding energy between these two layers, but without a change in packing within the layer. This usually occurs to the slight detriment of layer 2–3 bonding for which an *increased* lattice parameter is found. An example of this kind of relaxation is for Cu(110) where the top layer contracts by ~8% compared with the bulk value, while the second layer is expanded by ~3% (Adams et al., 1983; Copel

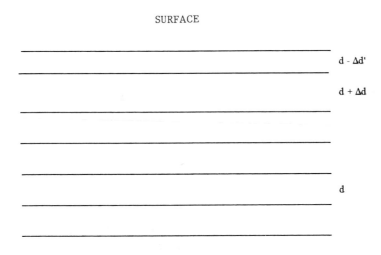

SURFACE

$d - \Delta d'$

$d + \Delta d$

d

BULK

Fig. 5. Surface relaxation. Surfaces generally contract in the outer two layers (a way of minimising surface free energy) and expand slightly in the layer below.

et al., 1986). Thus, as shown in fig. 5, the lattice spacing shows a rapidly damped oscillation down to the bulk distance.

(ii) *Surface rearrangement.* Many transition metal and several oxide surfaces have been shown to undergo gross rearrangement of the surface atom structure to result in an increased or decreased topmost atomic layer density. The most common rearrangements are for fcc (100) and (110) surfaces, perhaps the most well-known and earliest found being that of the reconstruction of Pt(100) to form a pseudo-hexagonal overlayer more like a close packed (111) surface (fig. 6a) (Van Hove et al., 1981). This is easily understandable from the thermodynamic viewpoint given above, since surface atoms with a (100) termination have only C_4 coordination in the top layer, whereas in a (111) termination it is C_6 and the surface layer has a higher density of atoms than the (100) layer. A typical reconstruction of (110) fcc surfaces is shown in fig. 6b — a "missing row" structure with a lower density of top layer

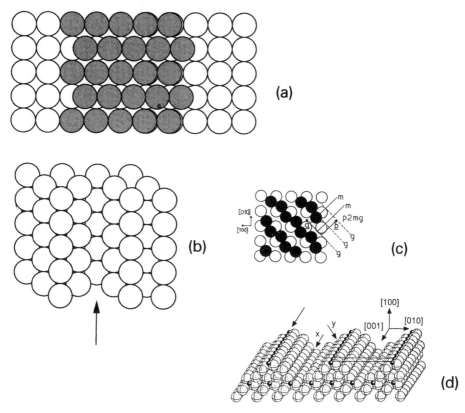

Fig. 6. (a) Reconstruction of Pt(100). (b) A missing row (110) surface, the missing row indicated by the arrow. (c) W(100) reconstruction. (d) TiO_2(100) (1×3) reconstruction.

atoms. However, although the surface appears "rougher", some of the top layer atoms have the same coordination as in the unreconstructed surface, whereas newly exposed atoms (in the second layer) have high coordination and again appear in a (111) like array with a planar coordination closer to C_6. Indeed, such surfaces can be described as microfacetted with (111) planes exposed.

One of the earliest recognised reconstructions was that of W(100) (Debe and King, 1979; Heinz and Muller, 1982), a body-centred cubic lattice, which is of a more subtle nature than those described above, involving lateral displacement of adjacent rows of surface atoms in opposite directions, leaving zig-zag chains of W atoms in the surface layer (fig. 6c). Such effects are not limited to high surface energy metal surfaces, but also occur on lower energy oxides. Figure 6d shows the (1×3) reconstruction of the $TiO_2(100)$ crystal surface which exposes microfacets, and presumably higher average coordination at the interface (Murray et al., 1994).

II. Sintering — the second major way a surface can reduce its surface free energy is by reducing the number of surface atoms it exposes, thus decreasing the surface: bulk ratio. In the situation of a catalyst, the relationship between surface atoms and particle radius is shown in fig. 7 (Bowker, 1983). All catalysts operating under high temperatures and high pressures of reactive gases tend to minimise surface energy in this way, an example being the Ag catalyst shown earlier in fig. 3. After 40 days in a microreactor synthesising ethylene oxide from an ethylene/oxygen reacting gas, the particle number density has reduced and the average particle size has increased significantly. One role of the support in catalysts is to help reduce the rate of such

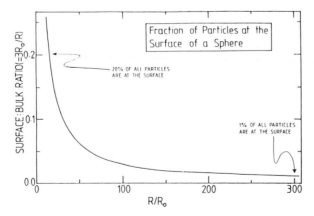

Fig. 7. The relationship between surface:bulk atoms for a spherical particle of radius R in normalised parameters of atomic radius.

processes by anchoring and separating particles. A silver powder would sinter at a much higher rate under such reaction conditions.

III. Chemisorption — Adsorption is found to be generally spontaneous and highly exothermic, due to the thermodynamic relationships involved. It is another way in which the high surface free energy can be reduced, in this case by satisfying the free valencies at the interface by surface compound formation. Indeed, if we consider oxygen adsorption, metals generally have higher surface free energy than their corresponding oxides. Since ΔG then is generally negative and adsorption is accompanied by a decrease in entropy, due to loss of at least one degree of translational freedom, then adsorption is exothermic, although cases of endothermic adsorption are known.

$$\Delta G = \Delta H - \underbrace{T \Delta S}_{\text{positive}} \tag{3}$$

If we consider the generalities of adsorption (the details will be discussed later) of molecules on surfaces, it is important to ponder upon whether adsorption will be molecular or dissociative and the Lennard-Jones type of description of adsorption (Lennard-Jones, 1932) shown in fig. 8 (for the real example of oxygen dissociation on Ag, see Dean and Bowker (1988/89, 1989); Campbell (1985)) makes a good starting point for consideration, the dynamics and kinetics being considered later.

It is likely that three states of adsorption will generally exist, labelled as 1, 2 and 3 in fig. 8. State 1 is the weakly held physisorbed molecule, in a

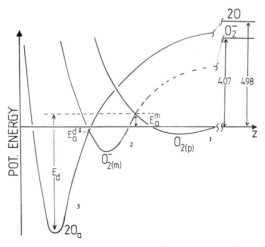

Fig. 8. One-dimensional potential energy diagram for the adsorption and dissociation of oxygen on an Ag catalyst (after Dean and Bowker, 1989).

state similar to condensed vapour, but whose binding depends slightly on the environment (clean metal or covered metal, high coverage or low coverage). State 2 is a chemisorbed molecule, for which there is some degree of charge transfer between the surface and the molecule, often partially filling levels of the molecule which are unoccupied in the gas phase. Nørskov et al. (1981) show that such orbital filling occurs by broadening of these unoccupied levels which fill as they cross (in energy) the highest occupied levels of the solid. The third state is the totally dissociated molecule. It can be seen from fig. 8 that the following relation holds

$$\Delta H_{(a)} = D_{AA} - 2D_{MA} \tag{4}$$

where D_{MA} is the binding energy of the adatom to the surface. In general then, there is a stronger heat of adsorption for tighter atom binding to the surface. This is nicely illustrated by comparison of the general form of the initial heat of adsorption of diatomics across the transition series with such basic quantities as the latent heat of vaporisation, which reflects the cohesive energy of the lattice. The higher the cohesive energy, the higher the surface energy when bonds are broken to form an interface; therefore the stronger is the driving force for surface reactivity, and in particular for adsorption, which acts to reduce this energy by surface compound formation. It must be noted that this is very significant for catalysis, but it is not the case that the most reactive surfaces are the best for catalysis. On the contrary it can be the case that binding of the adsorbate is too strong, which can detract from the bond-breaking which is also a necessary part of the catalytic cycle. Metals which are very reactive, W for instance, tend to self-poison in catalytic reactions, leaving few active sites available for catalytic turnover.

The relation in eq. (4) is nicely reflected in the abilities of metals to dissociate O_2, NO and CO (the dissociation energies being around 500, 630 and 1100 kJ mol^{-1}, respectively). Since the dissociation energy of the oxygen molecular bond is low it is dissociated by all close packed surfaces of metals, including Ag, whereas CO is not easily dissociated on Rh(111). On the other hand, Rh(111) will readily dissociate NO, whereas Pt(111) will not (Root et al., 1983). Of course, there is an interplay between thermodynamics and kinetics here, but nevertheless there is a correlation between ease of dissociation and thermodynamic stability. Generally, the weaker the adsorption heat, the higher the barrier to dissociative adsorption.

3. Adsorption

The mechanism of chemisorption can be divided into two types — *direct activated* and *precursor-mediated* and these are illustrated in the diagram of fig. 9.

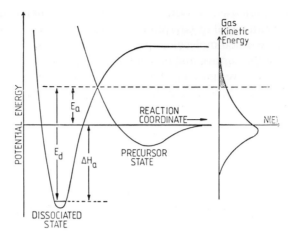

Fig. 9. Simplistic adsorption potential showing the molecules in the Boltzmann distribution of molecular energies which are capable of dissociation.

It must be noted that this is a schematic diagram where the abscissa is *not* a linear distance scale; instead it represents the trajectory pathway of an incoming molecule to a surface. Dissociative adsorption can occur from a weakly held molecular state if the *net* barrier to adsorption is low (precursor mediated) but is of low probability if it is high. Then it is only the "hot" molecules of the Maxwell–Boltzmann distribution of velocities (fig. 9) which can dissociate and they do this by direct passage over the energy barrier (direct activated). The rate of dissociation from a precursor state can be written as follows for the simple case in fig. 9,

$$R_{\text{diss}} = k_1(\theta_A) \cdot \theta_A \tag{5}$$

where $k_1(\theta_A)$ is the rate constant for the dissociation *from the precursor* state (likely to be strongly coverage dependent) and θ_A is the coverage in that state. Thus the rate is a strong function of substrate temperature since the system is equilibrated to that quantity. It is dependent on it in two ways: (i) the rate constant is temperature dependent in the usual way increasing with temperature, while (ii) the coverage has a strong negative order dependence on temperature. This can be represented in more general terms by the following expression, where the terms in eq. (5) are expanded and the coverage of the molecular state is expressed here by the simple coverage independent Langmuir isotherm,

$$R_{\text{diss}} = A \exp\left[-\frac{E_a + \Delta H_m}{RT_s}\right] \cdot \left(\frac{KP}{1 + KP}\right) \tag{6}$$

The net result is that the low coverage rate of formation of dissociative states in this case will show a maximum with temperature, assuming *pre-equilibrium* in the weakly held state is quickly obtained.

For direct activated adsorption, the dominant effect is that of the energy of the gas phase molecules [eq. (7)], though the distribution of this energy into the various degrees of freedom of the molecule can be crucial, as described in sect. 3.1 below.

$$R_{diss} = A \exp\left(-\frac{E_a}{RT_g}\right) \cdot P_A \tag{7}$$

Often, at least for systems where the activation barrier is not *too* large, both of these channels can co-exist, though each tend to dominate in different temperature regimes. In general the precursor mediated channel will dominate at low substrate and gas temperatures, while the direct channel will be dominant at high gas temperatures, examples of this being O_2 adsorption on Cu(110) (Pudney and Bowker, 1990; Hodgson et al., 1993) and N_2 on Fe(111) (Ertl, 1991; Rettner and Stein, 1987), discussed below (sect. 3.3).

3.1.   Direct-activated adsorption

A considerable effort has recently been expended on gaining an understanding of alkane adsorption on metals. In general adsorption is facile for organic molecules which have exposed functional groups available for direct interaction with a surface, for instance the π bonds of alkenes, or the lone pairs of oxygenates. However, alkanes are filled shell molecules with no easily available functional groups and behave almost like inert molecules. In a thermodynamic sense, they are low on the energetic scale and so they are very stable entities. Thus any chemical processes with CH_4 require very high temperatures, an example being the steam reforming catalytic reaction to form synthesis gas over a Ni catalyst (Riddler and Twigg, 1989)

$$CH_4 + H_2O \Leftrightarrow CO, \ CO_2, \ H_2$$

High temperatures are required for both kinetic and thermodynamic reasons. The kinetic limitation is methane dissociation which generally proceeds over a high barrier and there has been significant work in this field in recent years, using a variety of techniques, especially molecular beams, to enhance the measurable probability of dissociation per collision, and using high pressure cells to enhance the number of collisions and hence reaction probability per unit time. Dissociation of the methane molecules requires cleavage of one CH bond to form an adsorbed methyl group and an adsorbed hydrogen atom. The highly activated nature of this dissociation has been demonstrated by a variety of workers (Rettner et al., 1985; Beebe et al., 1987; Brass and Ehrlich, 1987; Winters, 1975; Lee et al., 1987; Ceyer, 1990; Sun and Weinberg, 1990; Luntz and Harris, 1992a, b). On Ni, the dissociation probability is $\sim 10^{-8}$ at 500 K (gas and solid) and is somewhat crystal plane dependent, having the order of reactivity expected ((110) > (100) > (111)) (Beebe et al., 1987). A number of interesting facts about this reaction are demonstrated by recent experiments:

(i) there can be a strong isotope effect for CH_4 vs. CD_4;

(ii) vibrational energy is as important in determining dissociation as translational energy;

(iii) dissociation can be induced by direct collisional activation on the surface.

Regarding (i) it has been shown that there is an enhancement of CH_4 dissociation over CD_4 of around an order of magnitude on Ni(111) (Lee et al., 1987; Ceyer, 1990) much more than can be expected from simple transition state considerations relating to the differences in vibrational partition functions. Such an effect was also found by Beebe et al. (1987) on Ni, though the effect was marked for Ni(100) with a factor of 20 difference in dissociation rate, but with little difference for Ni(110). The explanation for these data is that H atom tunnelling through the activation barrier occurs at close approach of the methane molecule to the surface. The D atom has a much lower probability of tunnelling due to its higher mass. An important point here, however, is that the tunnelling is strongly substrate temperature dependent because the shape of the barrier depends on substrate vibrational motions and greater energy in this mode aids tunnelling (Luntz and Harris, 1992a, b).

Measurements by Rettner et al. (1986) and Lee et al. (1987) show how important vibrational energy is in aiding dissociation, largely by increasing the effective translational energy of the H atoms in the reaction coordinate upon approach to the surface. Becaerle et al. (1987) demonstrated that CH_4, held in a physisorbed state on Ni(111), can be induced to dissociate by being "hammered" by high energy Ar atoms, the translational energy of the incoming atom being transferred to the CH_4 molecule, pushing the molecule into the surface electron distribution and "pushing" the hydrogen atom through the energy barrier to dissociation.

The effect of vibrational energy is particularly important for the dissociation of hydrogen on Cu(110), a system which has been studied in some detail of late (Hayden and Lamont, 1989; Berger et al., 1991; Hodgson et al., 1991; Rettner et al., 1992; Halstead and Holloway, 1990; Darling and Holloway, 1992). Figure 10a shows that as the translational energy of a H_2 beam is increased, above a certain energy the sticking increases from very low values

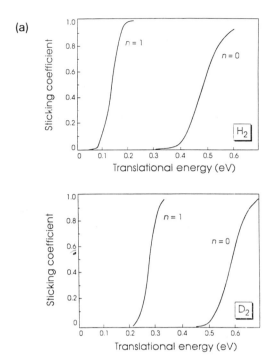

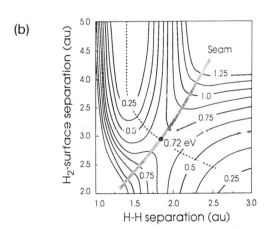

Fig. 10. (a) the effect of vibrational excitation (n = quantum number) on the dissociation of H_2 and D_2; (b) the H_2–Cu interaction potential proposed by Darling and Holloway (1992).

of $\sim 10^{-6}$ to near unity, and this barrier is very approximately 50 kJ mol^{-1} in classical kinetic terms. However, both experiment (Hayden and Lamont, 1989; Berger et al., 1991; Hodgson et al., 1991; Rettner et al., 1992) and theory (Halstead and Holloway, 1990; Darling and Holloway, 1992) have shown quite conclusively that the translational energy requirement is significantly reduced if the hydrogen is excited to the first vibrational level. Thus, vibrational energy is as effective as translational energy in this case and energy transfer occurs between states near the barrier point in the molecular trajectory, which helps the molecule surmount it and dissociate. Since the energy levels of D_2 are closer together than for H_2, then in that case a quantum of vibrational excitation is not so effective in aiding dissociation (fig. 10a). The scheme then for direct dissociation of hydrogen on Cu(110) is as shown in fig. 10b, the molecule approaching the surface with little change until it gets near the energy barrier, dissociation then occurring directly over it. It is clear why vibrational extension aids the dissociation process. Such a model describes the barrier as a "late" one — occurring near the end of the molecule's trajectory to the surface.

3.2.   Precursor-mediated adsorption

A good example of the role of precursor states in adsorption is N_2 dissociation on some W surfaces, especially those based on the (100) plane. In this case King and Wells (1974) showed that dissociation proceeds with a high probability (0.6 at 300 K) and the adsorption shows the classical behaviour characteristic of a precursor state, that is, an initial period of high sticking probability, S, as the coverage increases (fig. 11). This is due to mobility in the precursor layer which enables diffusion *over* filled sites to find empty ones, *before* desorption occurs. If we assume that incoming molecules can only adsorb if they hit empty sites then the sticking coefficient would obey the following relationships for random adsorption.

$$S = S_o(1 - \theta) \qquad \text{for molecular adsorption}$$
$$S = S_o(1 - \theta)^2 \qquad \text{for dissociative adsorption} \tag{8}$$

where S_o is the adsorption probability on a clean surface and θ is the surface coverage. These yield the dependencies shown in fig. 12a. Early on, the surface studies of Taylor and Langmuir (1933) showed that this model is unrealistic, finding a very high sticking coefficient with increasing coverage of Cs on a W foil. They postulated that this was due to the existence of a highly mobile second layer on the surface. Kisliuk (1957) later quantified this behaviour with a simple model which included the possibility of initial adsorption into a molecular state which was a "precursor" to further strong chemisorption/dissociation, which can diffuse over sites while in this

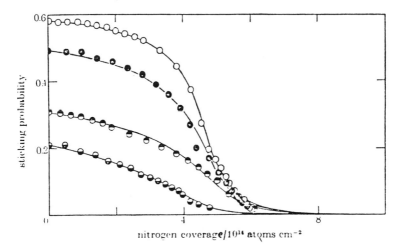

Fig. 11. The coverage dependence of nitrogen sticking on W(100) (from King and Wells, 1974). In order of decreasing sticking probability the crystal temperatures were 300 K, 433 K, 663 K and 773 K.

precursor state and which may find empty sites during its sojourn on the surface. A simple relationship then follows for molecular adsorption.

$$S = S_o \left(1 + \frac{K_p \theta}{1 - \theta} \right)^{-1} \tag{9}$$

where K_p is the so-called precursor state parameter. A curve for this relationship is shown in fig. 12b and illustrates the plateau of sticking probability as the coverage increases. The meaning of the precursor state parameter is illustrated in fig. 12c and in the following relationship.

$$K_p = \frac{k'_d}{k_a + k_d} \tag{10}$$

A low value for K_p means a big precursor effect since k'_d is low, that is the probability of desorption is low and the lifetime in the precursor state is high allowing a wide area of diffusion on the surface and thus a high probability for adsorption into the final chemisorbed state. Approximate "diffusion circles" are shown in fig. 13 based on the following simple Frenkel relationships,

$$\tau_{des} = \tau^o_{des} \exp \left(\frac{E_{des}}{R T_{des}} \right) \tag{11}$$

$$\tau_{diff} = \tau^o_{diff} \exp \left(\frac{E_{diff}}{R T_s} \right) \tag{12}$$

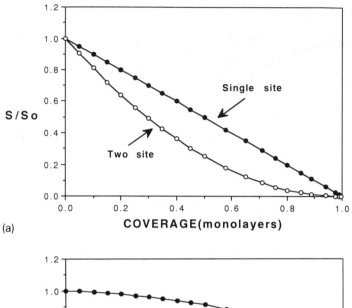

(a)

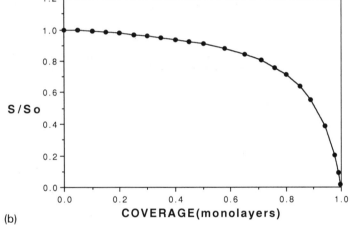

(b)

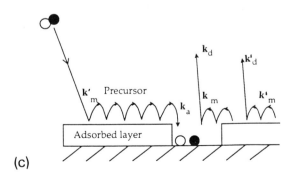

(c)

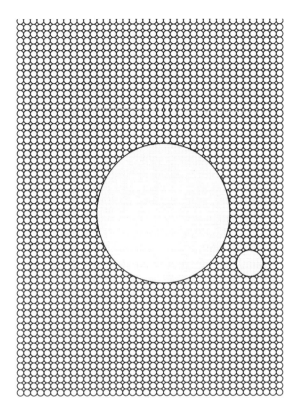

Fig. 13. A model of the extent of molecular diffusion on surfaces, so-called diffusion circles, showing the strong temperature dependence of the number of diffusion events. Circles are shown here superimposed on a (100) lattice for surface temperatures of 400 K and 950 K; at lower temperatures the diffusion circles are much more extensive (see text).

where the subscripts refer to desorption and diffusion. Assuming, for simplicity, that the τ_o values are equal, then the ratio of the lifetime τ is the average number of diffusive events (hops) made on the surface before desorption.

$$N_H = \frac{\tau_{\text{des}}}{\tau_{\text{diff}}} = \exp\left[\frac{E_{\text{des}} - E_{\text{diff}}}{RT_s}\right] \tag{13}$$

Fig. 12. (a) The Langmuir forms for sticking probability dependence on adsorbate coverage. (b) The effect of a precursor on the coverage dependence of sticking. Note the plateau of high sticking. The precursor state parameter value is 0.1. (c) Model of the adsorption process, showing the relevant rate constituents.

As a further simplification let us assume that the energy barrier for the diffusive event, always considerably smaller than the desorption barrier, is $E_{des}/3$, then

$$N_H \sim \exp\left(\frac{2E_{des}}{3RT_s}\right) \tag{14}$$

The size of the diffusion circle is then approximated by the following, assuming completely random directional diffusion.

$$r = \left(\frac{N_H}{\pi}\right)^{1/2} \tag{15}$$

where r is in lattice units. If we assume the precursor state is a physisorbed state with a heat of adsorption of 30 kJ mol^{-1}, then the average number of hops is low at 400 K (300), is higher at 300 K (3000) but very high at 200 K (160,000), where the lifetime of the physisorbed state is high ($\sim 10^{-5}$ s) in terms of the diffusion lifetime. This effect is illustrated in fig. 13.

For adsorption into the stable bound state, there is competition with the process of desorption from the weakly held state. Thus in terms of the process, we can write, at least for temperatures where the lifetime (and therefore steady state coverage) of the precursor state is low

$$A_{2(g)} \xoverset{1}{\rightleftharpoons} A_{2(a)}$$
$$A_{2(a)} \xrightarrow{2} 2A_{(a)}$$

$$\frac{d[2A_{(a)}]}{dt} = k_2[A_{2(a)}] = k_2 K P_{A_2} \tag{16}$$

where K is the equilibrium constant of step 1. This is over-simplistic since it ignores the difference between intrinsic (over empty sites) and extrinsic (over filled sites) precursor states, which, however, may be energetically small. This rate relationship then indicates the following energetic dependence:

$$\frac{d[2A_{(a)}]}{dt} = \frac{A_1 A_2}{A_{-1}} \exp\left\{-\frac{E_1 + E_2 - E_{-1}}{RT}\right\} \cdot P_{A_2} \tag{17}$$

In general, physisorption is non-activated ($E_1 = 0$) and so the energetic term here is a measure of the *difference* in barrier heights for dissociation and desorption from the precursor state. This term can be positive or negative, though for precursor *dominated* adsorption systems it is usually negative (desorption energy higher than dissociation energy) and so there is a negative dependence of initial dissociation rate on sample temperature (fig. 11), exemplified for the dissociation of nitrogen on W(100) where the initial sticking coefficient decreases from 0.6 to 0.2 between 300 K and 770 K (King and Wells, 1974). If the value of ($E_2 - E_{-1}$) becomes negative (that is, a net barrier to dissociation from the gas phase), then there is a positive dependence on substrate temperature, but gas phase temperature and direct dissociation become increasingly dominant as discussed in sections 3.1, above, and 3.3, below.

These kinds of precursor states can have a significant role in catalysis and surface reactions, particularly where "active sites" are sparsely distributed. This will be discussed further in sections 4 and 5 below.

3.3. Mixed adsorption channels

It can be the case that both adsorption channels are important for a particular system. Examples of this are given here for O_2 adsorption on Ag and Cu and for N_2 dissociation on Fe. In these cases we can generalise and say that the precursor mediated route tends to dominate at low substrate and gas temperatures, while direct activated adsorption dominates at high gas temperatures. Furthermore, in all these cases, molecular *chemisorbed* states of adsorption can exist which complicate the pathway of adsorption. A one dimensional potential energy profile is shown in fig. 8 for the case of O_2 adsorption on Ag taken from the work of Dean and Bowker (1988/89, 1989) and of Campbell (1985), although this is likely to be a general representation for this type of adsorption system with other adsorbate/metal combinations.

It appears that, at high gas temperatures, adsorption is dominated by direct dissociation and "hot" molecules go directly over the activation barrier, which in this case exists between physisorbed and chemisorbed *molecules*. This appears to be associated with electron harpooning into the incoming oxygen molecule to form a negative ion state which goes on to dissociate [although with low probability on (111) planes and catalysts (Dean and Bowker, 1988/89; Campbell, 1985)]. Gas temperature variation has a significant effect on such dissociation while substrate temperature variation has little effect. For low substrate temperatures on Ag crystals the negative ion state can be trapped stably on the surface and can be observed spectroscopically (Campbell, 1986). When the surface is heated some of these molecules can dissociate from this state on Ag(110) as shown in fig. 14, but few do on (111) where there is a much higher barrier. This shows that the precursor route is now important, but how important depends on the absolute value of the barrier to dissociation from this state.

A system which shows this dual mechanism even more clearly is oxygen dissociation on Cu(110) where the *net* barrier to dissociation from the gas phase has been measured to be low, at 3 kJ mol^{-1} (Pudney and Bowker, 1990; Hodgson et al., 1993). For this system it was shown that the S_0 value of 0.21 was almost independent of substrate temperature between 300 K and 800 K, but strongly dependent on gas temperature, increasing to 0.48 at $T_g = 850$ (fig. 15). Under these conditions dissociation was dominated by the

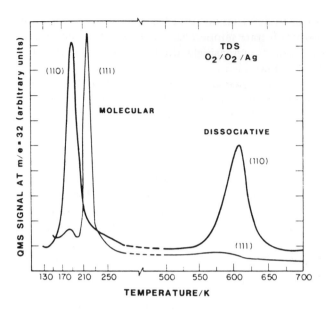

Fig. 14. A comparison of thermal desorption of O_2 from Ag(110) and (111) (from Campbell, 1985). Note the small extent of dissociation (the state desorbing at 600 K) on Ag(111).

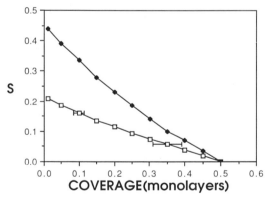

Fig. 15. Sticking coefficient dependence on coverage for oxygen adsorption on Cu(110) at two gas temperatures, 300 K (open squares) and 850 K (solid diamonds), showing the activated nature of adsorption. After Pudney and Bowker (1990).

direct channel. However, at low substrate temperature there was evidence of an increase in S_o and of a plateau in the coverage dependence of the sticking. This was backed up by more detailed work carried out by Hodgson et al. (1993) using seeded beams, who determined a threshold translational energy of ~5 kJ mol^{-1} for transition between the two channels.

N_2 dissociation on Fe crystal planes seems to be an example also of the presence of mixed adsorption channels, which has led to some confusion over the detailed nature of the potential energy surface for this system. Ertl et al. (1982) have claimed that there is a zero net barrier on Fe(111), with adsorption dominated by precursor kinetics, whereas highly activated adsorption is measured in supersonic beam experiments (Rettner and Stein, 1987). This probably relates to the different regimes of measurement, Ertl using low gas temperatures, while Rettner and Stein varied the gas energy.

4. Desorption

Desorption is the reverse of adsorption for a simple adsorption process and is the final step of surface reactions in which products are evolved,

$$A_{(a)} \rightarrow A_{(g)} + S$$

where S is a surface site liberated in the bond-breaking process of desorption. In fact, desorption is a much more widely measured process than adsorption, in particular through the technique of temperature programmed desorption (TPD). Other, less exact, acronyms for this technique are thermal desorption spectroscopy (TDS) and flash desorption. The particular utility of this technique is that it is uniquely adaptable to different materials and conditions. Thus it can be used to measure desorption from single crystals in UHV, or from powdered catalysts in microreactors at high pressure. An example of this is shown in fig. 16 for acetate decomposition and product desorption from Rh.

At the simplistic level the kinetics of desorption can be described by the following standard representation

$$\frac{-\mathrm{d}[A_a]}{\mathrm{d}t} = k_d[A_a]^n = A \exp\left(-\frac{Ed}{RT}\right) \cdot [A_a]^n \tag{18}$$

where n is the desorption order: this can be converted to a temperature dependence in a simple way for a linear heating rate where $\beta = \mathrm{d}T/\mathrm{d}t$.

$$\therefore \frac{-\mathrm{d}[A_a]}{\mathrm{d}T} = \frac{k_d}{\beta}[A_a]^n \tag{19}$$

For a dynamic system in which there is continuous removal of the product it can be shown that the rate $-\mathrm{d}[A_a]/\mathrm{d}t$ is equivalent to P_A, the instantaneous pressure of product measured over the adsorbent, and so represents a relatively facile measurement. The type of desorption seen for such a simple first-order case is shown in fig. 17. The profile shows a peak which is due to the convolution of the rate constant (which increases with increasing

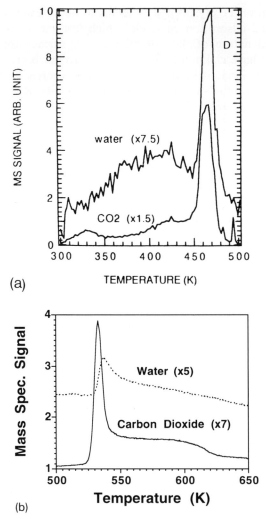

(a)

(b)

Fig. 16. A comparison of acetate TPD after dosing acetic acid on oxygen predosed (a) Rh(110), and (b) Rh/Al₂O₃ catalyst. From Li and Bowker (1993a) and Cassidy et al. (1993).

temperature) and the coverage (which decreases with temperature). The order of desorption strongly affects the desorption lineshape and coverage dependence of the peak temperature, as discussed in greater detail elsewhere (Goltze et al., 1981; Morris et al., 1984; Zhdanov, 1991a).

Examples of the dependence for a fixed rate constant are given in fig. 17. Zero-order desorption shows an increase in peak temperature with coverage and a precipitate drop in desorption rate when all the material is

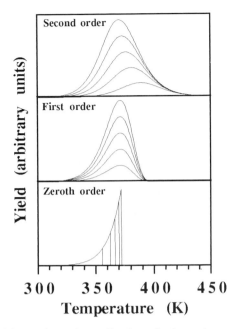

Fig. 17. The effect of desorption order on lineshape for increasing coverages of 0.2, 0.4, 0.6, 0.8, and 1.0 monolayer.

desorbed from the surface. For first-order desorption the peak is asymmetric, with the peak temperature independent of coverage, whereas second-order desorption (typical of atomic recombination) yields a symmetric peak which shifts to lower temperature with increasing coverage.

In this way a great deal can be learned from carrying out a desorption experiment and it could be said that this technique has perhaps the greatest information content of any used in surface science. The basic kinetic parameters can be determined, as also can coverage (by integration of the desorption peak), and this is a parameter not so easily (or accurately) found from most other techniques. However, it is rare for desorption experiments to show simple integral order desorption. More typically the desorption equation should be written as follows.

$$\frac{-d[A_a]}{dt} = R_1 + R_2 + R_3 \ldots$$

where

$$R_x = A_x(\theta) \exp\left(-\frac{Ed_x(\theta)}{RT}\right) \cdot [A_x^a] \qquad (20)$$

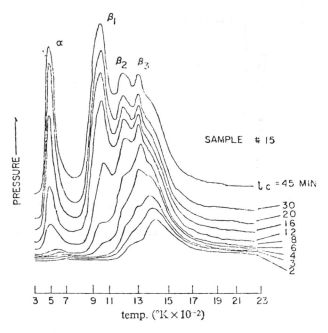

Fig. 18. The complexity of desorption shown in the pioneering work of Redhead (1961) studying CO desorption from polycrystalline W.

Here x represents a number of different adsorption states, associated with different sites on the surface and where the kinetic parameters are strongly coverage dependent. An example of such kinetic complexity is given for an early study of CO desorption from polycrystalline tungsten (Redhead, 1961) shown in fig. 18.

In such a situation it is very difficult to extract kinetic information from the data, although with care it can be done. Nevertheless, qualitative information about relative binding strengths can be determined from the relative peak temperatures of different peaks, a graphical representation of this being given in fig. 19.

The lineshape has been shown in the equations above to be potentially coverage dependent and this is usually due to lateral interactions in the adlayer which result in attractions or repulsions (increased, or decreased adsorption heat) between the species on the surface. Detailed descriptions of the effects of lateral interactions on desorption have been given elsewhere (King, 1978; Goltze et al., 1981; De Jong and Niemantsverdriet, 1990; Zhdanov, 1991a, b).

As described in the previous section, precursor states are significant for adsorption, and they can similarly strongly affect the desorption process, act-

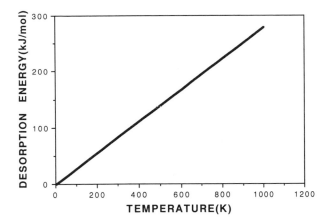

Fig. 19. The linear relationship between desorption peak temperature and desorption energy assuming a pre-exponential factor of 10^{13} s^{-1} and a heating rate of 1 K s^{-1}.

ing as intermediate states between the strongly adsorbed species and the gas phase. They invariably slow up the desorption process since they represent a state from which readsorption can occur. The kinetic consequences of the precursor state on desorption have been described by King (1977), Cassuto and King (1981) and Gorte and Schmidt (1978) and an example is shown in fig. 20. Basically the desorption rate is slowed by a factor F as follows.

$$\frac{-d[A_a]}{dt} = k_d[A_a] \cdot F \qquad (21)$$

where F is complex and coverage dependent and contains the precursor state parameter K_p.

In general, desorption experiments attempt to elucidate, at least approximately, the desorption energy, since this is related to the adsorption heat as shown in figs. 8 and 9 by

$$E_d = E_a + \Delta H_a \qquad (22)$$

In many cases adsorption is not activated ($E_a = 0$) and so the desorption energy is a direct measure of the adsorption heat. As discussed in the previous chapter, however, adsorption is often direct and activated, and desorption methods can be used to determine E_a. Two main types of experiment are used to measure the adsorption activation barrier — angle resolved desorption and time of flight measurements. When desorbing over a net barrier molecules enter the gas phase with excess energy commensurate with the top of the barrier. Since desorption usually involves breaking a surface–molecule bond in the reaction coordinate there is then an enhanced distribution of

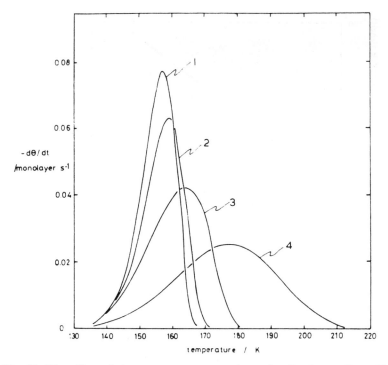

Fig. 20. The effect of the precursor state on desorption lineshapes (from King, 1977). Desorption *1* is unaffected by the precursor state, whereas desorption *4* is most affected.

molecules desorbing from the surface in a near normal direction. An example of this can be seen in the work of Cosser et al. (1981) who measured the angular distribution of N_2 desorbing from a stepped W crystal, W(310), which has a high sticking coefficient for N_2 and from the flat close-packed W(110) which has a low value of $\sim 10^{-3}$ (fig. 21). Commensurate with this, the former shows a normal cosine distribution of molecules desorbing from the surface, indicating a loss of memory of initial bond angles, probably desorbing via the weakly held precursor state. For the (110) surface, in contrast, the distribution is a higher power cosine distribution, highly lobed toward the surface normal. This equates to a high barrier to adsorption of ~ 17 kJ mol^{-1}, which explains the low sticking coefficient on this surface.

Time of flight measurements can yield useful information in a similar vein. The work of Comsa et al. (1980), for instance shows a shift of D_2 desorption from Pd(100) from a Maxwell–Boltzmann, surface thermalised desorbing flux to one with fast D_2 molecules emerging from the surface with a narrow distribution of energies after sulphur is deposited on the surface

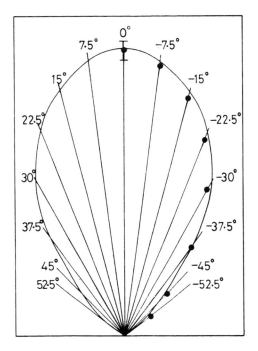

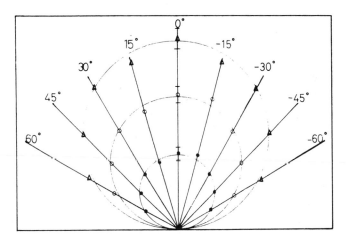

Fig. 21. Angle resolved desorption measurements of nitrogen desorption from W(110), upper figure, and W(310), lower figure.

(fig. 22). This in turn reflects a zero net activation barrier to adsorption on the clean Pd surface to one which is poisoned by S and has a high activation barrier.

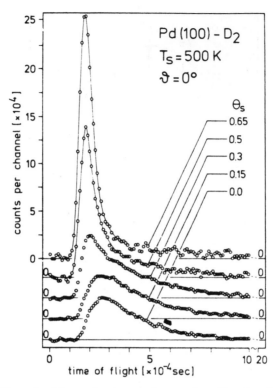

Fig. 22. The effect of sulphur poisoning on the speed profile of molecules detected by their time of flight. In the absence of S, the distribution is essentially Maxwellian showing desorption with no net adsorption activation barrier, while after poisoning the speed distribution is sharp and indicative of fast molecules desorbing over a net activation barrier to adsorption. From Comsa et al. (1980).

The technique of TPD has proved particularly useful for the elucidation of reaction mechanisms in catalysis and in surface reactions more generally. It has been of especial utility for the discovery of the nature of adsorbed intermediates on surfaces, and in the following, two examples of the application of this technique will be given, with further examples described in sect. 6.

The first is from the early pioneering work of Madix and co-workers (Ying and Madix, 1980; Bowker and Madix, 1981b) and concerns formic acid adsorption on Cu(110). If labelling is combined with TPD studies, the mechanism of reaction can be elucidated completely. Figure 23 shows the desorption pattern observed after dosing formic acid on to the surface predosed with oxygen. The steps in the mechanism are as follows

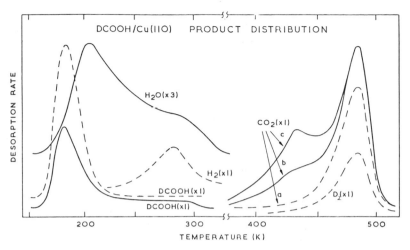

Fig. 23. TPD experiment for deuterated formic acid adsorption on Cu(110) (dashed lines) and on Cu (110) with predosed oxygen (solid lines). From Bowker and Madix (1981b).

$$DCOOH_{(g)} \rightarrow DCOOH_{(a)}$$

$$DCOOH_{(a)} + O_{(a)} \rightarrow DCOO_{(a)} + OH_{(a)}$$

$$2DCOOH_{(a)} + O_{(a)} \rightarrow 2DCOO_{(a)} + H_2O_{(a)}$$

$$H_2O_{(a)} \rightarrow H_2O_{(g)}$$

$$DCOO_{(a)} \rightarrow CO_{2(g)} + D_{(a)}$$

$$2D_{(a)} \rightarrow D_{2(g)}$$

The formate, formed by oxidative dehydrogenation of the acid, is quite stable and doesn't decompose until 480 K. This decomposition is a classical first-order case with a decomposition activation energy of 130 kJ mol^{-1} and a normal value pre-exponential of 10^{13} s^{-1}. The great ability of the TPD technique is the separation of the individual steps in the reaction in temperature. It is clear that the step proceeding over the highest barrier in this case is the formate decomposition, and that in a catalytic oxidation of formic acid the most abundant surface intermediate is likely to be the formate with its decomposition being rate determining.

Another example which is directly related to industrial catalysis is the adsorption and decomposition of propene from a mixed oxide, namely FeSbO$_4$. This material is used for the industrial production of acrolein and acrylonitrile (Yoshino et al., 1971). If the surface is dosed with both propene and ammonia, then all the reaction products in the industrial process are seen to evolve as shown in fig. 24 (Hutchings et al., 1991). Some intact propene desorbs at low temperatures, while the selective ammoxidation

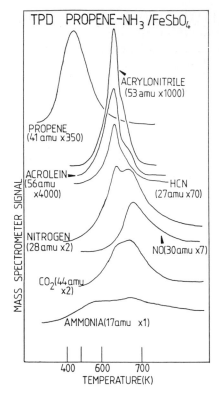

Fig. 24. The application of desorption techniques to an industrial catalytic process, namely propene ammoxidation on an FeSbO$_4$ powdered catalyst. This shows all the significant products.

product, acrylonitrile, evolves at 650 K, just before the further oxidation products CO_2, N_2 and NO. Small amounts of acrolein and HCN can also be seen. From such a simple experiment several important conclusions can be made. Firstly, it is clear that it is surface lattice oxygen which is directly involved in the reaction, not gas phase or molecular oxygen, since no oxygen was dosed. Secondly, it is clear that there is considerable oxygen mobility in the lattice since repeated experiments of this type yield the same product pattern without redosing oxygen, even though significant amounts of oxygen are lost in the process of desorption. Thirdly, defects are very important for the reaction. This is shown because the first desorption from the fresh catalyst surface yields only the products of combustion and selective products are seen only after some oxygen is lost from the surface.

   These two examples are meant simply to illustrate the utility of the TPD technique in surface reactivity and catalysis, and other examples follow in sect. 6. However, TPD is perhaps the most widely used technique in surface studies and the literature on this subject is enormous and could not be reviewed thoroughly in one article. Fuller review papers are available (King, 1975; Menzel, 1975; Zhdanov, 1991a).

5.   Surface diffusion

Mobility at the surface is very important for determining reaction rates, and the kinds of mobility involved are illustrated in fig. 25 and act as the basis for subdividing this section. These three subsections relate to diffusion in the weakly held layer (1), diffusion in strongly chemisorbed layers (2) and mobility of the surface atoms themselves (3).

5.1.   Precursor state diffusion

The concept and significance of diffusion in weakly held layers has already been described in sect. 3, in which its importance for affecting adsorption kinetics was highlighted, and was briefly discussed in sect. 4 in relation to the desorption process. Thus little further will be added here on this subject. However, it is worth noting some early work using field emission microscopy, which nicely illustrates the role of diffusivity in the adlayer. In this experiment a field emission microscope was immersed in liquid hydrogen and oxygen was admitted from one side of the tip (Gomer and Hulm, 1957). It condensed there forming a boundary and a shadow area with no adsorbate. Upon warming, but still at low temperatures, they found a "moving boundary" kind of diffusion, like an unrolling carpet as the physisorbed species diffused over the chemisorbed layer. At the edge of the boundary the oxygen transferred from a physisorbed to a chemisorbed state, as it encountered clean surface. This was a particularly clear demonstration of this kind of diffusion.

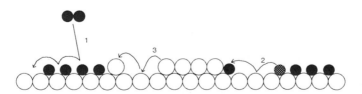

Fig. 25. A schematic diagram of surface diffusion processes: (1) diffusion in a weakly held precursor layer, (2) diffusion of a chemisorbed atom or molecule, and (3) diffusion of surface atoms of the solid.

5.2.   Diffusion in the chemisorbed state

As illustrated in fig. 2, diffusion occurs in the chemisorbed state and its rate is a strong function of temperature in the usual way, being an activated process. Generally the activation barrier to diffusion is something less than half the desorption activation energy, but can be very low indeed.

The effectiveness of such diffusion in aiding adsorption is clearly seen in the work of Singh-Boparai et al. (1975), who studied N_2 dissociation on stepped W surfaces. The N_2 dissociation rate on W(110) is very low, whereas on W(320), a surface with mostly flat (110) terraces and approximately 20% of the surface atoms at (100) step sites, it is high. The importance of diffusion in the precursor state has already been shown for this system, but it is clear from fig. 26 that diffusion in the chemisorbed state is also very significant, at least on the atomic scale, since the coverage of the adsorbate goes to completion — including the (110) planes which are inactive for direct adsorption. This is effected by diffusion as summarised in fig. 27. Without these combined diffusion processes the adsorption would be very inefficient in two ways: (i) the dissociation probability would be approximately 20% of that measured, since it would occur only at the active step sites, and (ii) the saturation coverage would be reduced also to ~20% of that observed.

An STM study of oxygen adsorption on Al (Brune et al., 1992) demon-

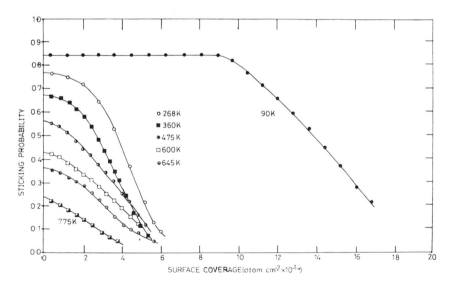

Fig. 26. The adsorption probability for nitrogen on W(320) (from Singh-Boparai et al., 1975). The lowest temperature data is mainly molecular adsorption whereas the others are exclusively dissociative.

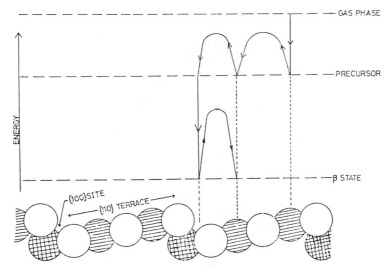

Fig. 27. Model for nitrogen adsorption on W(320) showing diffusion in a precursor state, dissociation at active (100) steps and diffusion of the atoms onto otherwise inactive (110) terraces. From Singh-Boparai et al. (1975)

strates atomic diffusivity at the microscopic level. Diffusion can occur directly over an activation barrier or, sometimes, as may be the case for N_2 dissociation above, the adsorbed atoms can retain some energy from the initial bond breaking process and in losing energy to the lattice may diffuse over a large number of sites. This seems to be the situation for O_2 dissociation on Al, since the atoms eventually reside in particular sites on the surface, but well away from the site of dissociation; essentially, the atoms appear to "fly apart" on the surface, and disperse themselves.

In a similar vein a plot of the sticking coefficient of oxygen on Cu(110) was shown in fig. 15 and has an unexpected shape for dissociative adsorption, showing close to a $(1 - \theta)$ rather than $(1 - \theta)^2$ dependence (Pudney and Bowker, 1990). There is little sign of a precursor state involvement in this particular adsorption since the dissociation probability decreases at low coverage and further, there is no substrate dependence of the sticking, at least above 300 K. The reason for this dependence is oxygen atom diffusivity on a surface with growing islands of a dense oxygen 0.5 monolayer $p(2\times1)$ phase, the oxygen adsorbing and dissociating on the clean areas of the surface and rapidly diffusing to the island edge. The adsorption rate is then simply proportional to the amount of inter-island surface which is proportional to $(1 - \theta)$, where θ is the surface averaged coverage. It must be noted that Cu metal atom diffusion is also significant for the structural changes occurring here during adsorption, as discussed below in sect. 5.3.

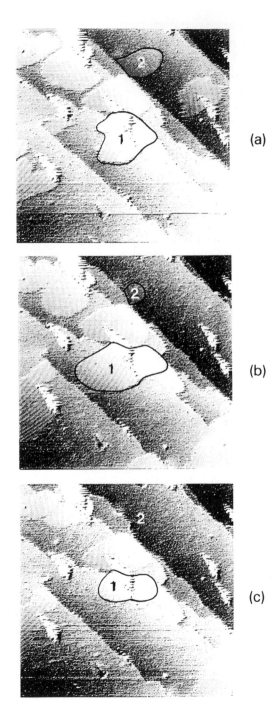

Figure 28 clearly shows the importance of diffusion within a chemisorbed layer to surface reaction processes (Leibsle and Bowker, in prep.). In this series of STM images the surface methoxy species on Cu(110) is decomposing (evidenced by the loss of total area of methoxy islands), but diffusion is taking place between islands since big islands get bigger at the expense of smaller ones, which eventually disappear. This kind of diffusion phenomenon can be classified as surface mediated Ostwald ripening.

5.3. Substrate atom diffusion

The surface has often been considered to be a "checkerboard" of fixed sites on which reactions take place. However, in recent years our ideas of the surface have changed partly due to LEED analysis, but perhaps more significantly due to the advent of the atomically resolving techniques of field ion microscopy (FIM), high resolution electron microscopy (HRTEM) and especially scanning tunnelling microscopy (STM).

FIM has shown unusual forms of diffusion of metal atoms on a surface, including correlated motion between separated atoms (Tsong, 1993). Such diffusion is usually strongly anisotropic, on (110) surfaces for instance it occurs along the close-packed (110) direction which has only a weak potential corrugation compared with the orthogonal direction.

If we return to consider oxygen adsorption on Cu(110), STM has given us considerable extra insight into the mechanism of this process It is now clear as a result of the work of Ertl and colleagues (Coulman et al., 1990) and of others (Jensen et al., 1990; Wintterlin et al., 1991), that this process, which results in a p(2×1) structure of oxygen at the Cu surface, proceeds by the formation of "added rows" of Cu–O units, as shown in fig. 29. These are formed by the diffusion of Cu atoms onto the (110) terraces from steps on the surface, and these Cu atoms join growing Cu–O strings which initially grow in the (110) direction, followed by later agglomeration of such strings to make narrow islands. Diffusion of these Cu–O strings is shown very nicely by the work of Besenbacher and Stensgaard (1993).

The important point here in relation to surface reactions is the timescale of the diffusion. If the surface atoms diffuse once in an hour, the surface can effectively be considered as a checkerboard (and since good images of surface atoms are often observed in STM at ambient temperature, diffusion must be slow), whereas if it occurs once every microsecond it may not be considered rigid with respect to surface reactions taking place on that kind of timescale.

Fig. 28. Sequential STM images of methoxy islands on a Cu(110) surface showing the loss of methoxy as it decomposes. Between images *a* and *b* the island labelled *1* has increased in size, while that labelled 2 has decreased; in *c* island 2 has gone altogether.

Fig. 29. STM image showing long islands of oxygen covered Cu (dark areas) on Cu(110) separated by clean surface (bright areas). From Jensen et al. (1990).

Another example of this for a surface reaction is shown in fig. 30 for a system discussed above, namely, decomposition of CH_3O units on a Cu(110) surface to yield H_2CO in the gas phase (Wachs and Madix, 1978; Bowker and Madix, 1980). This reaction mechanism is discussed in more detail in sect. 6.2.2 below. Associated with the CH_3O structure are added Cu atoms and their diffusion *back* to steps is clearly seen in the figure, since when the methoxy has gone a nearby step edge has changed shape and has extended due to the addition of the Cu atoms originally associated with the methoxy island.

6.   Surface reactions

The surface reaction is very often the step in the network shown in fig. 2 which is rate limiting for any conversion, including catalysis. Thus, for instance, in methanol synthesis the rate determining step is thought to be

$$HCOO_{(a)} + H_{(a)} \longrightarrow I \xrightarrow[-O]{+2H_{(a)}} CH_3OH_{(g)}$$

where I is a hydrogenated version of the formate, HCOO, with one of the CO bonds either broken or intact, the exact situation being uncertain. The nature of the surface involved in bonding such an intermediate as the

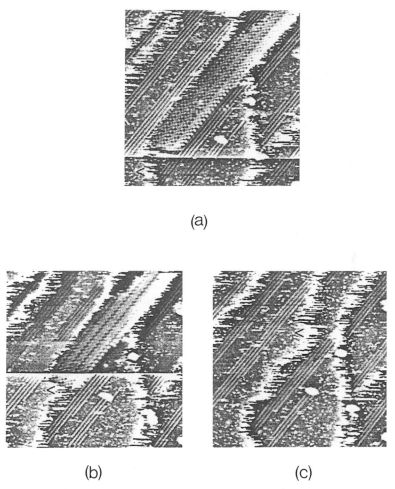

(a)

(b)   (c)

Fig. 30. STM images showing a large island of methoxy (in the centre of image *a*) and the enhancement of a nearby step (marked by the arrow) as it decomposes (*b* and *c*). From Leibsle et al. (1994).

formate is then crucial in dictating the pathway of further reaction; the composition and structure both strongly affect the kinetics of the decomposition process. Thus a Cu/Pd alloy, albeit with zero Pd in the top layer, decomposes formic acid 20 times faster than monometallic Cu with the same surface structure (Newton et al., 1991, 1992), while acetate on Rh(110) decomposes below 300 K (Li and Bowker, 1993a), but, on Rh(111), it decomposes at 360 K (Li and Bowker, 1993b). It is perhaps useful to begin a discussion of surface reactions with a brief and simple consideration of the kinetics relating to them.

6.1.   *Surface reaction kinetics*

6.1.1.   *Simple monomolecular reactions*

If we consider the simplest reaction as follows

$$A \rightarrow B$$

with the following mechanism at the surface.

$$A_{(g)} \overset{1}{\Longleftrightarrow} A_{(a)}$$
$$A_{(a)} \overset{2}{\longrightarrow} P_a + Q_a \ldots$$
$$P_a, Q_a \overset{3}{\longrightarrow} P_g, Q_g,$$

If step 2 is the rate limiting step then the rate equation is simply.

$$R_p = k_2 \theta_A \tag{23}$$

If we further represent the pre-equilibrium of A-adsorption by the Langmuir isotherm, then we obtain the so-called Langmuir equation (also known in enzyme kinetics as the Michaelis–Menten equation);

$$R_p = \frac{k_2 K_a P_A}{1 + K_a P_A} \tag{24}$$

here K_a is the adsorption equilibrium constant and the form of this equation is given in fig. 31, which shows first-order behaviour in A at low pressure and zero-order behaviour at high pressures. In practise all the rate constants are

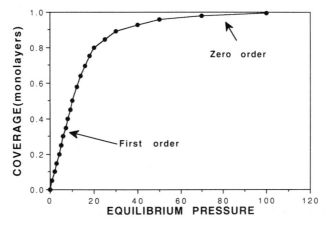

Fig. 31. The form of equilibrium coverage dependence on pressure from the Langmuir equation.

strongly coverage dependent and so will affect the detailed shape in fig. 31 quite strongly.

6.1.2. Bimolecular surface reactions

In principle a surface reaction can result in a multiplicity of products. As shown above two reactants A and B can produce product C (the desired product) and D, which is the non-selective product. If we consider first the situation when we have only one product, the reaction scheme can be written as follows

$$A_g \xleftrightarrow{\ 1\ } A_a$$
$$B_g \xleftrightarrow{\ 2\ } B_a$$
$$A_a + B_a \xrightarrow{\ 3\ } P_a$$
$$P_a \xrightarrow{\ 4\ } P_g$$

If the surface reaction is rate limiting, then the rate is as follows.

$$R_p = k_3 \theta_A \theta_B \tag{25}$$

that is, first order in both surface coverages, second order overall. This is the simplest form of equation for the reactants and in principle the rate "constant" k_3 is likely to be a strong function of both θ_A and θ_B. Ignoring this dependence for the sake of brevity, the equation can be written in terms of gas phase pressures, and is called the Langmuir–Hinshelwood equation

$$R_p = \frac{k_3 K_A K_B P_A P_B}{(1 + K_A P_A + K_B P_B)^2} \tag{26}$$

where K_A and K_B are the equilibrium constants associated with the adsorption of A and B, both forward and reverse steps proceeding at very high and equal rates. For all the simplicity of this relationship many experimental results for simple surface reactions at least exhibit the general trends, as shown in fig. 32 for CO oxidation on Rh(110) (Bowker et al., 1993b). Here the rate goes through a maximum as the initially dosed oxygen coverage is reduced and the CO coverage increases, in accordance with eq. (25).

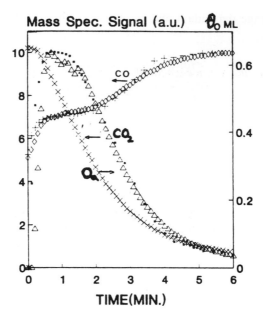

Fig. 32. Data for CO oxidation on Rh, showing the maximum in CO_2 production as a function of time which is characteristic of a Langmuir–Hinshelwood type of mechanism. From Bowker et al. (1993b).

6.2.   Surface intermediates

In this section the nature of a series of simple intermediates will be considered. In catalysis and surface reactions in general it is usually organic species which form these intermediates and the following discussion will itemise a few of the main types of species which have been well studied by surface science methods. The bonding and stability of such species is often crucial for the efficiency of heterogeneous catalysis.

6.2.1.   Carboxylates

Carboxylic acids tend to absorb on surfaces and lose the acid hydrogen function to form a carboxylate. The stability of carboxylates generally relates to their basicity — the more basic, the more stable. Thus on Cu(110) the acetate intermediate decomposes at 600 K (Bowker and Madix, 1981a), whereas the formate decomposes at 480 K (Ying and Madix, 1980; Bowker and Madix, 1981b). It is also the case that species have the greatest stability on the least reactive surfaces; on Cu (Ying and Madix, 1980; Bowker and Madix, 1981b) and Ag (Barteau et al., 1980) the formate decomposes at 480 K and 410 K, respectively (shown in fig. 23), whereas on Pd(110) for

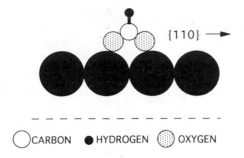

Fig. 33. A model of formate binding on a (110) fcc surface.

instance (Aas et al., 1991) it is at 240 K. These carboxylates tend to be bound in a bidentate fashion to the surface as shown by IRAS (Lindner et al., 1987) and XPS (Bowker and Madix, 1981b). On (110) surfaces they bind with the molecular axis parallel to the close-packed direction (fig. 33) as shown by Woodruff et al. for formate on Cu(110) (Crapper et al., 1986) and by Newton et al. (to be published) for acetate on the same surface.

The stability of such intermediates can be strongly affected by alloying and recent work with Cu/Pd alloys referred to above, has shown destabilisation of the formate by the influence of Pd which is not present in the top layer, but is in the second layer (Newton et al., 1991, 1992).

STM and LEED show that the simplest carboxylate, the formate, forms several surface structures on Cu(110), depending on the mode of adsorption. For formic acid adsorbed on a Cu(110) surface with 0.25 ml of adsorbed atoms the oxygen is titrated off leaving 0.5 monolayers of formate in a well-ordered (2×2) structure (fig. 34; Leibsle and Bowker, in prep.). At high oxygen predosed coverages, mixed layers of oxygen and formate are produced. Decomposition of the formate proceeds by dehydrogenation to yield CO_2 and H_2, as shown in fig. 23.

6.2.2. Alkoxides

Alkoxides can be formed by reaction of alcohols with surfaces or with oxygen-treated surfaces

$$ROH \rightarrow RO_{(a)} + H_{(a)}$$
$$2ROH + O_{(a)} \rightarrow 2RO_{(a)} + H_2O$$

On metals to the right hand side of the transition series the alkoxides are reasonably stable. They decompose in a partial oxidative fashion on IB metals to produce formaldehyde and an example of temperature programmed desorption from methanol adsorbed on copper with predosed oxygen is

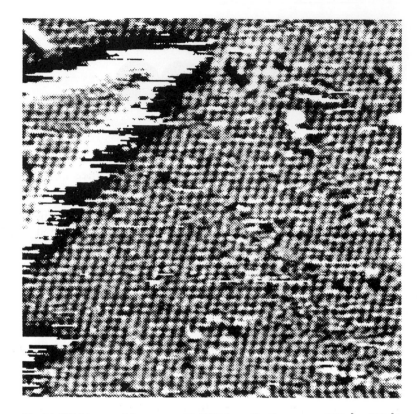

Fig. 34. STM image of formate on Cu(110) in a c(2×2) structure (200 Å × 200 Å).

shown in fig. 35. The coincident evolution of H_2CO, CH_3OH and H_2 indicates that all these products evolve from the common intermediate, the
methoxy. Thus the methoxy decomposes in the following way

$$CH_3O_{(a)} \rightarrow H_2CO_{(a)} + H_{(a)}$$
$$CH_3O_{(a)} + H_{(a)} \rightarrow CH_3OH_{(g)}$$
$$2H_{(a)} \rightarrow H_{2(g)}$$

Indeed, Ag metal catalysts are utilised for the industrial partial oxidation of
methanol to produce formaldehyde, which is used largely to make adhesive
resins (Davies et al., 1989).

On the group VIII metals the methoxy is formed, and is fairly stable
on Ni, for instance, decomposing at around 450 K (Johnson and Madix,
1981), whereas on Pt(111) the decomposition temperature is ca. 300 K
(Abbas and Madix, 1981). On all these metals the molecule is completely

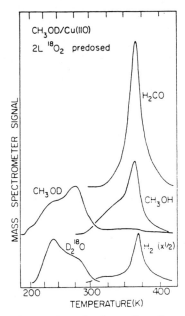

Fig. 35. The reactive desorption spectrum for the methanol/oxygen reaction on Cu(110). The products at 370 K are indicative of the presence of methoxy. From Wachs and Madix (1978).

dehydrogenated to yield CO and H_2 in the gas phase. On more reactive metals further to the middle of the transition series the molecule tends to be completely split. For instance, on W(100) this is observed, at least at low coverage until the atomic binding sites are saturated (Ko et al., 1980a, b).

On oxides the product pattern is strongly dependent on the nature of the oxide surface. The methoxy tends to be stable, but can be further converted to the formate species, as on ZnO (Bowker et al., 1981) and SrO (Pringle et al., 1994) for instance. The reaction is then completed by heating above 500 K,

$$CH_3O_{(a)} \rightarrow H_2CO_{(a)} + H(a)$$
$$H_2CO_{(a)} + O_{(a)} \rightarrow H_2COO_{(a)}$$
$$H_2COO_{(a)} \rightarrow HCOO_{(a)} + H_{(a)}$$

to decompose the formate which then releases hydrogen and carbon monoxide into the gas phase.

$$HCOO_{(a)} \rightarrow CO + O_s + H_{(a)}$$
$$2H_{(a)} \rightarrow H_2$$

where O_s is a surface lattice oygen. On oxides such as TiO_2 and $SrTiO_3$ the main product in such an experiment is that of deoxidation, namely CH_4 (Pringle et al., 1994). Other oxides are good H_2CO producers and the most recently developed commercial catalysts for this reaction are oxides, based on Fe, for instance $FeMoO_4$ (Pearce and Patterson, 1981).

6.2.3. Hydrocarbon intermediates

As already described (sect. 3.1) alkanes are very inert molecules, very active surfaces are needed to attack the C–H bonds in such molecules, as a result of which subsequent bond breaking will readily occur leading to significant dissociation of the whole molecule. However, hydrocarbons are produced in CO hydrogenation experiments on metals such as Fe (Fischer–Tropsch catalysts) and are formed from the hydrogenation of a variety of chain lengths of hydrocarbon intermediates. Alkyl and alkylidine intermediates are likely to exist, but are highly unstable species with respect to hydrogenation/dehydrogenation and C–C bond fission. Alkenes, on the other hand, are much more reactive with surfaces due to the functionalisation of the molecule, especially their π-donor and π^*-acceptor capabilities. If we use ethene as an example, one of the major stable intermediates which this forms is the ethylidyne ($M–C–CH_3$) which is formed by hydrogen transfer. Evidence of this as the major intermediate first came from EELS carried out in the Ibach group (Ibach and Mills, 1982). This is the case for (111) type surfaces, for others the situation is not quite so clear. The ethylidyne decomposes by dehydrogenation and the formation of polymeric residues on the surface, which, if the surface is heated enough, dehydrogenate further to form graphite, which is difficult to remove from the surface. For aromatic molecules, if we here use benzene as an example, the molecules tend to adsorb flat with the π ring interacting strongly with the surface (Van Hove et al., 1983). π donation and π^* backbonding from the surface tends to result in weakening of the aromatic bonding and the molecule splits up into acetylene like entities which form intermediates like acetylene does, as evidenced mainly by EELS (Koel et al., 1986).

6.3. Substrate dependence of reactivity

The nature of the electronic and geometric structure of the surface is usually crucial in determining the reaction rate. The effect of surface morphology will be considered in the next section, but here we will exemplify the effect of global variations in electronic structure by considering a particular catalytic reaction, namely CO hydrogenation, and how the choice of different transition metals affect the selectivity of the product pattern. This is illustrated in fig. 36 and shows that metals in group VIII tend to be the Fischer–Tropsch

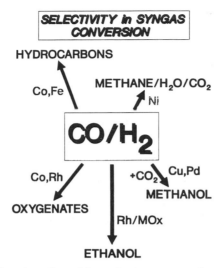

Fig. 36. The range of products formed in synthesis gas conversion over various metal catalysts.

materials, producing a wide range of hydrocarbons, whereas group IX produces methane, some higher hydrocarbons and oxygenates. Group X tend to be methanation catalysts, although Pd can produce methanol. Cu is a very selective metal for methanol synthesis from CO and hydrogen. A metal such as Rh can have unusual behaviour in that it can be very selective to the production of oxygenates and especially ethanol (Bowker, 1992). The understanding of this reactivity pattern rests in a knowledge of the variation in the nature of bonding of CO to the metals involved.

   In this respect the nature of the thermodynamic quantities involved is important, regarding which some discussion was given in sect. 2 above. Figure 37 shows the variation in the heat of adsorption of CO across the transition series and reflects the surface energy of the metals which can be taken to relate to the cohesive energy of the host lattice, which in turn is manifested in such bulk quantities as the latent heat of vapourisation. For refractory metals like W, which has both the highest heat of vapourisation, melting point and therefore surface energy, CO has a very high heat of adsorption, whereas for low melting point solids like the 1B metals it is very low at $\sim 1/10$ of that for W. There is a fairly smooth variation in adsorption heat across this series. Figure 37 also shows, in an approximate way, the ability of metals to dissociate CO. Metals up to group VIII dissociate CO easily (W has a dissociation probability per impinging molecule near unity), whereas after this the predominant adsorption form is molecular, that is, the barrier to dissociation increases significantly between groups

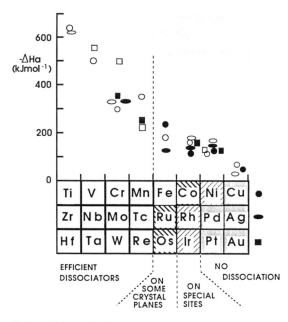

Fig. 37. A compilation of heat of adsorption/desorption data for transition and 1B metals. From Bowker (1992).

VII and IX. Note that although this is true, pure metals like Ni and Rh do dissociate CO since in CO hydrogenation at high pressure they are essentially methanation catalysts with Ni being the metal of choice for such a reaction. It is the case however that the dissociation probability under such circumstances of high adsorbate coverage are very low (estimated as $\sim 10^{-12}$; Bowker, 1992) and attempts to measure their value on Rh(110), a low surface coordination, high reactivity plane for that metal, indicate a value for the clean surface of $< 10^{-6}$ (Bowker and Joyner, unpublished result). In this case, however, recombination is very easy as evidenced by the low desorption peak temperature (400 K) of CO which results from oxygen adsorption after carbon deposition (Bowker, 1992). From experiments of this kind an enthalpy plot shown in fig. 38 can be drawn which illustrates why CO dissociation is so difficult on Rh(110). The enthalpy of adsorption is only slightly negative on the absolute scale and dissociation with respect to gas phase CO is *endothermic*, proceeding over a very high activation barrier, estimated to be ~ 100 kJ mol^{-1} both from experiment (Bowker, 1992; Bowker and Joyner, unpublished result) and the calculations of Van Santen (Bowker, 1992; De Koster and Van Santen, 1990). The relatively low barrier to $C + O$ recombination can be seen from the figure and explains why this occurs at a high rate at a low temperature of 400 K.

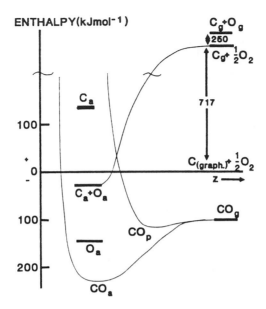

Fig. 38. A proposed enthalpy plot as a function of reaction coordinate for CO adsorption and dissociation on Rh. From Bowker (1992).

From these simple thermodynamic considerations it is interesting to note that the dissociation on metals like W proceeds over a zero activation barrier, therefore $C + O$ is adsorbed *exothermically* with respect to gas phase CO. This is a result of the enhanced binding of O and C pulling down the enthalpy of the adsorbed product. In the case of Rh it is not the binding of O which causes the difficulty of CO dissociation, since O_2 itself is easily and stably dissociated, and upon heating recombination does not occur until ~ 1000 K (at least at low coverage). The problem is in the thermodynamics of surface carbide formation, as fig. 38 demonstrates the C is adsorbed in a highly *endothermic* fashion, by approximately $+200$ kJ mol^{-1} on the absolute scale. However, the *change* in dissociation ability across the series is probably more related to the weakening of the metal–oxygen bond, since the diatomic carbides have similar bond strengths. This can be seen in fig. 39 which shows the dissociation energies of diatomic carbides and oxides in the gas phase (Lide, 1992), which again shows a similar trend to that of CO dissociation on the bulk metals.

This adsorptivity for CO is the cause of the broad reactivity pattern which is illustrated in fig. 36. For the 1B metals Cu is a good methanol synthesis catalyst ($\sim 99\%$ selective) because it produces little methane due to its inability to dissociate CO. Pd is most like a 1B metal in that it has an almost filled d-band, with an occupancy of ~ 9.5. It too can produce methanol, though

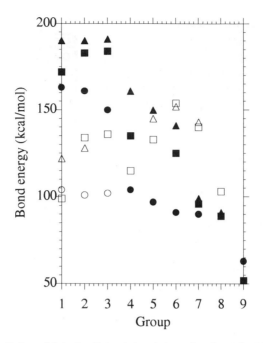

Fig. 39. A compilation of data for diatomic bond strengths of metal oxides (solid data points) and carbides (open points) against valence electron number in the transition series. Circles for first row elements, squares for second row and triangles for third row.

with not such good selectivity as Cu (Ryndin et al., 1981). Ni is the metal of choice for methanation carrying out the following hydrogenation reactions,

$$CO + 3H_2 \rightarrow CH_4 + H_2O$$
$$2CO + 2H_2 \rightarrow CH_4 + CO_2$$

the latter being known as "dry methanation" (Bowker et al., 1993a). The reasons for this efficiency are that: (i) it dissociates CO at a low rate, and (ii) the carbon is weakly bound and is hydrogenated fast. Both of these properties result in low steady state coverages of carbon at reaction temperatures (at higher temperatures in the reverse reactions, steam reforming and dry reforming, C build up can be a problem), and limited C–C forming reactions. The latter result in higher hydrocarbons on a metal like Fe, where there is a high steady state coverage of carbon and carbonaceous intermediates which are hydrogenated off as a range of hydrocarbons. Further to the left of the transition series the metals are inefficient catalysts because they tend to form very stable oxides and carbides which are not easily hydrogenated off, which in turn results in surface blockage and reaction poisoning. In effect they are converted into surface compounds with much lower surface energy and relative catalytic inactivity. This kind of behaviour is often discussed in terms of the "volcano plot" which describes why a maximum in catalytic turnover is seen for most metal catalysed reactions at some group in the transition series. This is due to a trade-off between the ability of a surface to bind reactants and yet to leave active sites available for reaction. In a simplistic manner this can be expressed in the following way

$$R = k\theta_A \cdot \theta_B \cdot (1 - \theta_A - \theta_B) \tag{27}$$

On the left of the transition series the product $\theta_A\theta_B$ is very high, but the vacancy term, which is essential for the reaction to proceed, is very small, near zero, due to site blockage. On the right hand side of the volcano plot the adsorbates are much more weakly bound and the opposite is the case, namely the $\theta_A\theta_B$ term is very small, and there are many vacancies on the surface. In between these two extremes there is a good distribution of adsorbate on the surface with vacancies for adsorption.

Similar considerations to these apply to the range of catalytic reactions and materials. The strength of binding and ease of dissociation of molecules are crucial determining factors. On oxide surfaces there is often a scarcity of "active sites", these usually being associated with defects of one kind or another — anion vacancies for instance. In these cases the nature of binding and reaction is strongly affected by the distribution of these vacancies and by the morphology of the surface. Oxides are often used not for their activity, but for their selectivity for particular reactions. Thus metals are generally totally unselective for hydrocarbon oxidation reactions — they tend to result in the combustion products CO_2 and H_2O. As an example, a range of oxide materials can be used to selectively oxidise propene to acrolein

$$O_2 + C_3H_6 \rightarrow C_3H_4O + H_2O$$

an example being $FeSbO_4$. When this material is in a fully oxidised state, the catalyst has low selectivity to the desired product and predominantly burns the propene (Allen et al., 1991). As the surface is reduced it becomes more selective to the partial oxidation product, largely due to the lack of available oxygen for the high oxygen requirement of combustion.

7. Structure dependence of reactivity

It is often the case that surface reactions can depend more strongly on surface structure than on the substrate atomic number itself. If we consider as examples the sticking coefficients for N_2 and O_2; the former varies significantly at 300 K, between W(100), with a value of 0.6 (King and Wells, 1974) and W(110) with a value of $\sim 10^{-3}$ (Pfnür et al., 1986; Tamm and

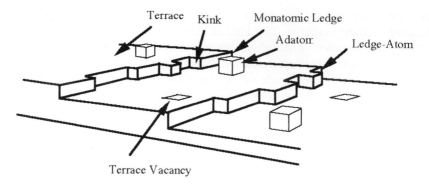

Fig. 40. The TLK model of surfaces. From Somorjai (1991).

Schmidt, 1971), whereas for Mo(100) it is high. The sticking coefficient of oxygen on the open planes of metals up to and including Pt is high, whereas on Pt(111) the dissociative adsorption is activated with a probability of only 0.1 (Rettner and Mullins, 1991).

The surface has often been pictured in terms of the terrace–ledge–kink model shown in fig. 40, although it is clear from recent STM studies that surface atoms move around at a rather significant rate [on Cu(110), for instance (Besenbacher and Stensgaard, 1993)]. This model shows surface sites of very different coordination, from the highly coordinated terrace atoms through to low coordination steps and very weakly coordinated adatoms. Clearly such a model represents a very anisotropic surface with sites which would be expected to have very variable reactivity; and so it is found to be by adsorption and desorption measurements. Two specific examples are for CO and H_2 adsorption on stepped and stepped–kinked surfaces, carried out by Somorjai and his colleagues. We see that H_2 desorption reveals a low temperature (350 K) desorption state on flat Pt(111) (fig. 41a). A higher temperature state at $\sim$450 K is associated with the steps present on a (552) surface, while an even higher temperature state ($\sim$550 K) desorbs from a surface with kinks present within the steps (Somorjai, 1991). In a like manner, CO desorption from a stepped Pt surface shows that the higher energy, low coordination sites bind the CO more strongly and are the preferential sites for adsorption at low coverages (fig. 41b; Somorjai, 1991).

Marked examples of crystallographic anisotropy in adsorption are found in adsorption of nitrogen on W and Fe. The strong structural dependence is clear, and in the case of N_2 dissociation on W and Mo, the morphology effect is much bigger than the difference between the two metals as stated earlier. The variation for Fe is significant in relation to ammonia synthesis since that metal is the material of choice (when in a promoted state) for industrial

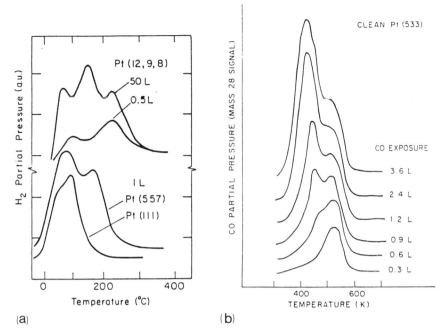

Fig. 41. (a) Hydrogen desorption from plane, stepped and kinked Pt surfaces. (b) CO desorption from a stepped surface, showing filling of the step sites first at low gas doses, followed by terrace adsorption. After Somorjai (1991).

nitrogen fixation — the so-called Haber process for ammonia synthesis. The rate of this process carried out near industrial conditions of high temperature and pressure shows a similar dependence to the nitrogen dissociation rate, implying the latter to be the rate determining step in the process. Clearly then, all other things being equal, it would be important to try to favour (111) like surfaces in a real catalyst and there is some evidence that these are indeed predominant on the catalyst surface (Strongin and Somorjai, 1991).

It is usually the case that activated, direct dissociations are strongly dependent on surface structure and another nice example of this is oxygen dissociation on Ag. Here oxygen adsorbs directly over a large barrier on Ag(111), the close packed fcc plane, whereas on (110) there is a much smaller barrier. This is reflected in thermal desorption experiments carried out by Campbell (1985; fig. 14). On the (110) surface recombination of dissociated oxygen atoms is observed at 600 K while the molecularly chemisorbed state is much more weakly bound and desorbs at 200 K. A similar molecular state is seen on Ag(111), but the dissociated state can hardly be seen at all.

Somorjai and his colleagues have spent several years investigating the effects of surface morphology on simple organic reactions at surfaces and a particularly illuminating example is given in fig. 42; a detailed review of this kind of work has been produced by Davis and Somorjai (1982). Here they compare the dehydrogenation of cyclohexane and its hydrogenolysis to *n*-hexane (Blakely and Somorjai, 1976), both over a range of Pt surfaces with varying step and kink densities. It is clear that the dehydrogenation reaction is structure insensitive ("undemanding"), whereas the other reaction is very

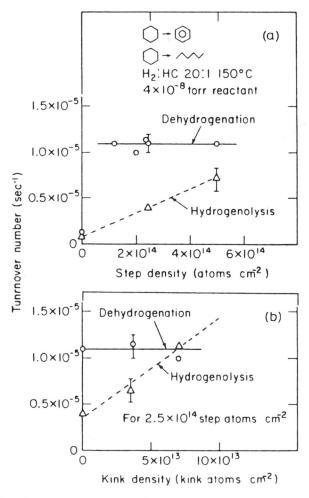

Fig. 42. Showing the structure dependence of hydrogenolysis and structure independence of dehydrogenation of cyclohexane on Pt. From Blakely and Somorjai (1976).

sensitive ("demanding"), the rate for a high density of steps and kinks being some 20 fold greater than for a flat Pt(111) surface.

8. Modification of surface reactivity: poisoning and promotion

Many industrial processes, especially those using catalysts, modify the behaviour of the surfaces involved by doping with additives of one form or another; table 1 shows a list of some of the more well-known industrial processes and these all use promoters, for a variety of reasons, but usually to improve the time-yield efficiency of a particular reaction.

In beginning to understand the effects of additives on surface reactivity we should first consider the effect on the distribution of surface sites. When an additive which is non reactive itself is placed on a surface it has a primary effect of blocking sites on the surface, but may have a secondary effect of activating (or deactivating) adjacent sites by electrostatically modifying the solid, or by altering the site geometry. Figure 43a shows the effect of adding a promoter to the distribution of unactivated and activated sites assuming only those adjacent to the promoter are affected (Bowker, 1988). This is based on the following relationship.

$$\theta_U = (1 - \theta_P)^{n+1}$$
$$\theta_A = 1 - \theta_P - \theta_U$$

where θ_U, θ_A and θ_P are the coverage of unpromoted, promoted and promoter sites, respectively, and n is the size of the affected ensemble of

Table 1
Some promoted industrial catalytic reactions

Process	Basic catalyst	Promoter	Parameter promoted
Ammonia synthesis ($N_2 + 3H_2 \rightarrow 2NH_3$)	Fe/Al_2O_3	K_2O	Activity
Fischer–Tropsch ($xCO + 2xH_2 \rightarrow C_xH_{2x} + xH_2O$)	Fe/SiO_2	K_2O	Product distribution
Methanation ($CO + 3H_2 \rightarrow CH_4 + H_2O$)	Ni/Al_2O_3	Alkalis	Activity/ lifetime
Water–gas shift ($CO + H_2O \rightarrow CO_2 + H_2$)	Iron oxide/Al_2O_3	Alkalis	Activity/ lifetime
Ethylene epoxidation ($C_2H_4 + \frac{1}{2}O_2 \rightarrow C_2H_4O$)	Ag/Al_2O_3	K, Cs	Selectivity
Propene ammoxidaton ($C_3H_6 + \frac{3}{2}O_2 + NH_3 \rightarrow C_2H_3CN + 3H_2O$)	BiMo oxide	K_2O	Selectivity

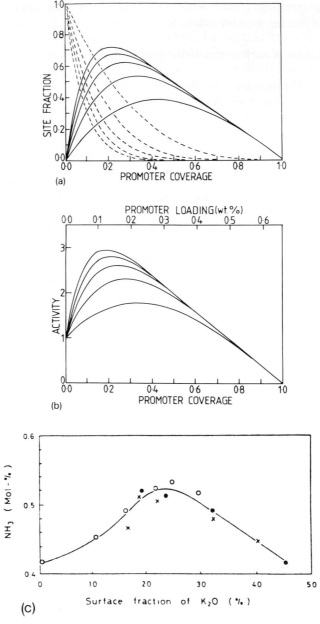

Fig. 43. (a) The dependence of the fraction of promoted (solid lines) and unaffected sites (dashed lines) upon promoter coverage for ensembles of 2, 4, 6, 8 and 10 promoted sites. (b) The activity dependence on promoter coverage/loading for a promoted catalyst (from Bowker, 1988). (c) Experimental data for ammonia synthesis dependence on promoter loading (from Krabetz and Peters, 1965).

sites. Promoters which in themselves are inactive always deactivate at high coverage due to blockage of active sites. It is clear that a general rule for optimised active site distribution is that the promoter coverage should be no more than approximately 1/3 of a monolayer. The effect of this distribution on activity for a hypothetical situation in which the activated sites are 4 times as active as an unpromoted site is shown in fig. 43b. Again it is clear that low promoter coverages are essential. Addition of greater than 0.7 monolayers of promoter results in a less active catalyst. Figure 43c shows some experimental data for the activity of an ammonia synthesis catalyst, which shows a similar dependence (Krabetz and Peters, 1965).

The simplest interpretation of these effects is in terms of local electronic redistribution, and this will be illustrated in relation to CO bonding. Because the $4s$ level of potassium is above the Fermi energy of metals the alkali atom will autoionise upon adsorption (at least at low coverage). The region immediately around the alkali centre has enhanced electron density and so in the case of molecular CO adsorption for example, there can be greater $d \rightarrow 2\pi^*$ electron donation, weakening of the CO bond and easier dissociation. The presence of promoter species has similar effects to those of steps on desorption from otherwise flat surfaces. Alkalis often have a marked effect on sticking probabilities for systems where s is low, for example N_2/Fe (Ertl et al., 1982), O_2/Ag (Dean and Bowker, 1989; Kitson and Lambert, 1981). Promoters can alter the selectivity to a product, a well-defined example being CO hydrogenation on Ni crystals as demonstrated by Goodman (1982). As shown in fig. 44, and as is well-known, Ni is an excellent methanation catalyst giving high selectivity to methane alone; indeed it is widely used in industry and academia for removing small amounts of CO from gas feeds by hydrogenation. When the Ni crystal is promoted with K, however, its characteristics change to those more akin to a Fischer–Tropsch catalyst like Fe, making higher alkanes/alkenes in much greater abundance (fig. 44).

Thus, in a gross catalytic sense, the effect of promoters is to shift behaviour more towards metals left of the promoted metal in the transition series.

The most marked effect of alkali promoters is a local one at adjacent sites, with any delocalised effects being very slight indeed, although most studies have concentrated on alkali coverages too high to determine non-local perturbations. Theoretical calculations confirm mainly local effects. Thus Feibelman and Hamann used slab calculations for a Rh(100) surface doped with an ordered array of Li atoms; as fig. 45 shows, there is a significant enhancement of electron density at the adjacent site in the [001] direction, but the next site (which is equidistant from two Li atoms) has little increased density. The effect of promotion on activated dissociation processes is nicely illustrated by the calculations of Tomanek and Benneman (1983) who used a cluster approach to calculate the barrier for CO dissociation on Ni.

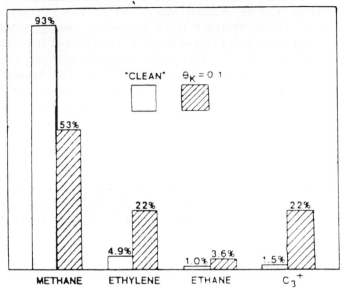

Fig. 44. The effect of promotion on the syn gas reaction products on Ni(100) (after Goodman, 1982).

Figure 46 shows that the promoter *reduces* the net barrier, thus aiding dissociation and tying in nicely with the catalytic work shown in fig. 44.

Although the commonly accepted description of promotion is in the terms given above, that is, enhanced back donation from the metal, Holloway et al. (1987, 1984) have proposed a more fundamental cause of stabilisation. This is due to the very strong electrostatic field which exists normal to the surface at sites adjacent to the promoter atom. These fields are such as to lower the energy of the molecular orbitals enabling extra back-bonding into the CO $2\pi^*$ orbital. A particularly illuminating demonstration of such field effects in catalysis at surfaces was given by Chauh et al. (1989) using the technique of pulsed field desorption. They showed that methanol decomposition on a Rh field emission tip proceeds in the expected fashion at low field strength yielding the products of total dehydrogenation, CO and H_2, while at strengths of 20 V/nm [in the range of fields due to promoters (Holloway et al., 1987)], the decomposition changed to yield formaldehyde, a partial dehydrogenation reaction more characteristic of Cu (Wachs and Madix, 1987; Bowker and Madix, 1980), as described earlier (sect. 6.2.2).

Poisons can act in a variety of ways in a catalytic sense. They can alter the selectivity of the reaction, or reduce the number of active sites by adsorption or by enhancing sintering rates. Surface reactivity is reduced

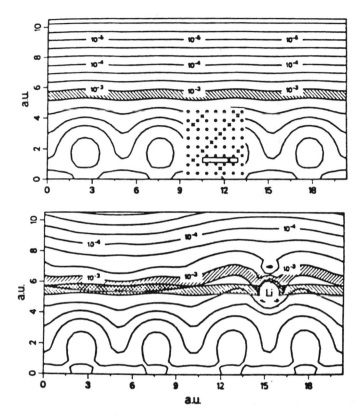

Fig. 45. The work of Feibelmann and Hamann (1985) showing increased electron density at sites adjacent to an added Li promoter atom on a Rh(100) model surface.

simply by site blockage by inert elements (e.g. S, C, Cl), but such poisons often preferentially block the most active sites on a catalyst and therefore cause a greater than $(1 - \theta)$ or $(1 - \theta)^2$ detriment to the activity (Kelley and Goodman, 1982). Furthermore, like promoters, the poisons (usually electronegative species) can cause electronic effects in the adjacent region which can deactivate the material even further, inducing fields of opposite sign to those of promoters, which in turn tend to raise the energy of unfilled orbitals and stabilise molecular states. Thus, fig. 46 shows an increased barrier to CO dissociation on Ni in the presence of Cl.

Poisons can be used to preferentially block a non-favourable pathway. Such selective poisons are often called reaction modifiers. An example of such a system is ethylene epoxidation catalysis. Above a certain coverage of chlorine the ethylene combustion reaction is severely deactivated, whereas the effect on the selective route is less. As soon as the EDC is introduced

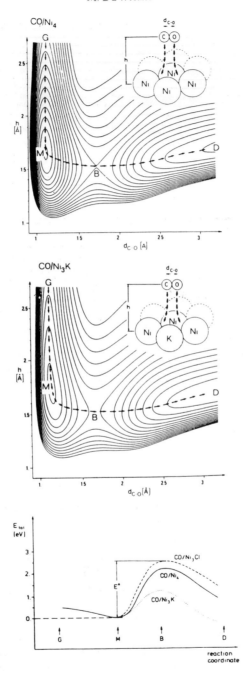

Fig. 46. Theoretical models and calculated results for CO dissociation on a small Ni cluster and the effect of a promoter, K, and a poison, Cl. From Tomanek and Benneman (1983).

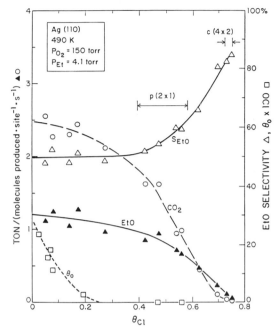

Fig. 47. Effect of surface coverage of chlorine on ethylene oxidation. EtO stands for ethylene oxide, the product of partial oxidation. From Campbell and Paffett (1984).

into a reactive mixture it is clear that the selective product yield is almost unchanged, whereas the combustive product, CO_2, is significantly decreased (Law and Bowker, 1991; Campbell and Paffett, 1984). Figure 47 shows the single crystal results of Campbell and Paffett (1984) which show a similar trend to the catalyst data (Law and Bowker, 1991) which confirm the single crystal findings. In this case it is thought that these result from the nature of the transition state geometry. The combustion reaction requires a bigger transition state (bigger site geometry) than does the selective oxidation because the former needs to be oxidised by more than one O atom.

Such selectivity changes by a poison are shown very nicely by the work of Madix and co-workers on Ni(100) which showed that clean Ni totally dehydrogenates methanol, whereas with S poisoning the partial dehydrogenation to formaldehyde is favoured (fig. 48; Johnson and Madix, 1981). Thus the effect of the selective poisoning is to make the Ni surface behave more like a copper surface, and reflects the kind of behaviour described above for Rh in the presence of a high field.

The preferential poisoning of strong adsorption sites is clearly shown by the work of Goodman and co-workers for the effects of S, P and Cl on CO

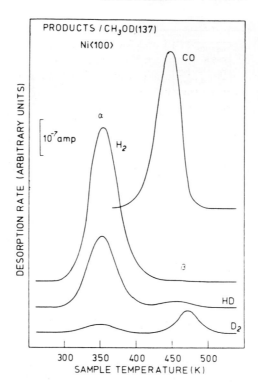

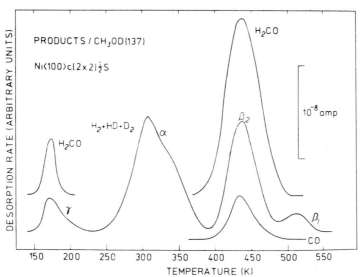

Fig. 48. Effect of the presence of sulphur on the Ni(100) surface on methanol decomposition. Total dehydrogenation occurs on the untreated surface, while that with S yields the partial dehydrogenation product, formaldehyde. From Johnson and Madix (1981).

binding — all weaken it in a similar way (Kelley and Goodman, 1982). In all cases the coverage is reduced and the reduction is most marked for the strongest adsorption states on the unmodified surface.

These examples show clearly that poisoning has the gross effect of shifting the reactivity behaviour of the surface more towards that of elements to the right of that which is poisoned, the opposite effect to promotion.

9. Conclusions

The nature of surfaces and the study of the gas–solid interface is of crucial importance in a number of technological areas, perhaps the most important of these being catalysis. Over the last twenty years the methodology and understanding of surface science has gone some considerable way to improving our understanding of the microscopic properties underlying catalysis, and in particular, the relationship between surface structure and reactivity.

Of course there are still some significant gaps in our understanding. For example, many real catalysts consist of very small diameter metal particles of very high surface free energy, whose electronic properties can differ significantly from a macroscopic single crystal of low surface : bulk ratio. It is the case, however, that many catalysts consist of large particles (≥ 1 nm radius), and then model single crystals mimic better both the average surface coordination and electronic properties. Another gap in knowledge exists because of the lack of truly in-situ technologies available for the study of catalytic reactions under the real industrial conditions of high pressure and temperature which often prevail. The advent of novel techniques using X rays from synchrotrons may go some way towards improving this situation. More importantly perhaps, the advent of STM technology, some applications of which were described above, may be further developed to study reactions under these conditions, with atomic resolution. The hope would be then that the "Holy Grail" of catalysis would have been found, that is the ability to "see" the so-called "active site" during catalytic turnover. Such developments are likely to occur over the next 10 years.

References

Aas, N., Y. Li and M. Bowker, 1991, J. Phys Cond. Matter **3**, 5281.
Abbas, N., and R.J. Madix, 1981, Appl. Surface Sci. **7**, 241.
Adams, D.L., H. Nielson, J. Andersen, 1983, Surface Sci. **128**, 294.
Allen, M., R. Betteley, M. Bowker and G. Hutchings, 1991, Cat. Today **9**, 97.
Bao, S., G. Liu and D.P. Woodruff, 1988, Surface Sci. **203**, 89.
Barteau, M., M. Bowker and R.J. Madix, 1980, Surface Sci. **94**, 303.
Becaerle, J.D., Q. Yang, A. Johnson and S. Ceyer, 1987, J. Chem. Phys. **86**, 7236.
Becaerle, J.D., Q. Yang, A. Johnson and S. Ceyer, 1989, J. Chem. Phys. **91**, 5756.
Beebe, T., D.W. Goodman, B. Kay and J.T. Yates, 1987, J. Chem. Phys. **87**, 2305.

Berger, H., M. Leisch, A. Winkler and K. Rendulic, 1991, Chem. Phys. Lett. **175**, 425.
Besenbacher, F., and I. Stensgaard, 1993, in: The Chemical Physics of Solid Surfaces, eds. D.A. King and D.P. Woodruff (Elsevier, Amsterdam), ch. 15.
Blakely, D.W., and G.A. Somorjai, 1976, J. Catal. **42**, 181.
Bowker, M., 1983, Vacuum **33**, 669.
Bowker, M., 1988, Appl. Catal. **45**, 115.
Bowker, M., 1992, Catal. Today **15**, 77.
Bowker, M., and R.W. Joyner, unpublished result.
Bowker, M., and R.J. Madix, 1980, Surface Sci. **95**, 190.
Bowker, M., and R.J. Madix, 1981a, Appl. Surface Sci. **8**, 299.
Bowker, M., and R.J. Madix, 1981b, Surface Sci. **102**, 542.
Bowker, M., M. Houghton and K.C. Waugh, 1981, J. Chem. Soc. Faraday I **77**, 3023.
Bowker, M., T. Cassidy, A. Ashcroft and A.K. Cheetham, 1993a, J. Catal. **143**, 308.
Bowker, M., Q. Guo and R. Joyner, 1993b, Surface Sci. **280**, 50.
Brass, S.G., and G. Ehrlich, 1987, Surface Sci. **187**, 21.
Brune, H., J. Wintterlin, J. Behm and G. Ertl, 1992, Phys. Rev. Lett. **68**, 624.
Campbell, C., 1985, Surface Sci. **157**, 43.
Campbell, C., 1986, Surface Sci. **173**, L641.
Campbell, C.T., and M. Paffett, 1984, Appl. Surf. Sci. **19**, 28.
Cassidy, T., M. Allen, Y. Li and M. Bowker, 1993, Catal. Lett. **21**, 321.
Cassuto, A., and D.A. King, 1981, Surface Sci. **102**, 388.
Ceyer, S., 1990, Science **249**, 133.
Chauh, G.K., N. Kruse, W. Schmidt, J. Block and G. Abend, 1989, J. Catal. **119**, 342.
Copel, M., T. Gustafsson, W. Graham and S. Yalisave, 1986, Phys. Rev. B. **33**, 8110.
Comsa, G., R. David and B. Schumacher, 1980, Surface Sci. **95**, L210.
Cosser, R., S. Bare, S. Francis and D.A. King, 1981, Vacuum **31**, 503.
Coulman, D., J. Wintterlin, R. Behm and G. Ertl, 1990, Phys. Rev. Lett. **64**, 1761.
Crapper, M.D., C. Riley, D.P. Woodruff, A. Puschmann and J. Haase, 1983, Surface Sci. **171**, 1.
Darling, G., and S. Holloway, 1992, J. Chem. Phys. **97**, 734.
Davies, P., R. Donald and N. Harbord, 1989, in: Catalyst Handbook, 2nd ed., ed. M.V. Twigg (Wolfe, London), ch. 10.
Davis, S., and G.A. Somorjai, 1982, in: The Chemical Physics of Solid Surfaces, eds. D.A. King and D.P. Woodruf, Vol. 4 (Elsevier, Amsterdam), ch. 7.
De Jong, A. and J. Niemantsverdriet, 1990, Surface Sci. **223**, 355.
De Koster, and R.A. Van Santen, 1990, Surface Sci. **233**, 366.
Dean, M., and M. Bowker, 1988/1989, Appl. Surface Sci. **35**, 27.
Dean, M., and M. Bowker, 1989, J. Catal. **115**, 138.
Debe, M.K., and D.A. King, 1979, Surface Sci. **81**, 193.
Ertl G., 1991, in: Catalytic Ammonia Synthesis, ed. J.R. Jennings (Plenum, New York, NY), p. 109 and references therein.
Ertl, G., S. Lee and M. Weiss, 1982, Surface Sci. **114**, 527.
Feibelman, P., and D. Hamann, 1985, Surface Sci. **149**, 48.
Goltze, M., M. Grunze and W. Hirschwald, 1981, Vacuum **31**, 697.
Gomer, R., and J. Hulm, 1957, J. Chem. Phys. **27**, 1363.
Goodman, D.W., 1982, Surface Sci. **123**, 413.
Gorte, R., and L. Schmidt, 1978, Surface Sci. **76**, 559.
Halstead, D., and S. Holloway, 1990, J. Chem. Phys. **93**, 2589.
Hayden, B., and C. Lamont, 1989, Phys. Rev. Lett. **63**, 1823.
Heinz, K., and K. Muller, 1982, in: Structural Studies of Surfaces, Springer Tracts of Modern Physics, Vol. 91 (Springer, Berlin) p. 1.

Hodgson, A., J. Moryl and H. Zhao, 1991, Chem. Phys. Lett. **182**, 152.

Hodgson, A., A. Lewin and A. Nesbitt, 1993, Surface Sci. **293**, 211.

Holloway, S., and J.K. Nørskov, 1984, J. Electroanal. Chem. **161**, 193.

Holloway, S., J.K. Nørskov and N. Lang, 1987, J. Chem. Soc. Farad. 1 **83**, 1935.

Hutchings, G., M. Bowker, A. Crossley, R. Betteley and M. Allen, 1991, Catal. Today **10**, 413.

Ibach, H., and D.L. Mills, 1982, Electron Energy Loss Spectroscopy and Surface Vibrations (Academic Press, New York, NY), pp. 326ff.

Jensen, F., F. Besenbacher, E. Laegsgaard and I. Stensgaard, 1990, Phys. Rev. **B41**, 10233.

Johnson, S., and R.J. Madix, 1981, Surface Sci. **103**, 361.

Kelley, R., and D.W. Goodman, 1982, in: The Chemical Physics of Solid Surfaces, Vol. 4, eds D.A. King and D.P. Woodruff (Elsevier, Amsterdam), ch. 10 (a review of Goodman's work on poisoning of the methanation reaction).

King, D.A., 1975, Surface Sci. **47**, 384.

King, D.A., 1977, Surface Sci. **64**, 43.

King, D.A., 1978, Crit. Rev. Sol. State Mater. Sci. **7**, 167.

King, D.A., and M.G. Wells, 1974, Proc. Roy. Soc. Lond. **A339**, 245.

Kisliuk, P., 1957, J. Phys. Chem. Solids **3**, 95.

Kitson, M., and R. Lambert, 1981, Surface Sci. **109**, 60.

Ko, E., J. Benziger and R.J. Madix, 1980a, J. Catal. **62**, 264.

Ko, E., J. Benziger and R.J. Madix, 1980b, J. Catal. **64**, 132.

Koel, B., J. Crowell, B. Bent, C. Mate and G.A. Somorjai, 1986, J. Phys. Chem. **90**, 2709.

Krabetz, R., and C. Peters, 1965, Angew. Chem. **77**, 333.

Law, D., and M. Bowker, 1991, Catal. Today **10**, 397.

Lee, M., Q. Yang and S. Ceyer, 1987, J. Chem. Phys. **87**, 2724.

Leibsle, F., and M. Bowker, to be submitted.

Leibsle, F., S. Francis, S. Haq and M. Bowker, 1994, Surface Sci. **318**, 46.

Lennard-Jones, J.E., 1932, Trans. Farad. Soc. **28**, 333.

Li, Y., and M. Bowker, 1993a, J. Catal. **142**, 630.

Li, Y., and M. Bowker, 1993b, Surface Sci. **285**, 219.

Lide, D.R. (ed.), 1992, Handbook of Chemistry and Physics, 73rd ed., (CRC Press, Boca Raton, FL), pp. 9-129–9-132.

Lindner, T., J. Somers, A. Bradshaw and G. Williams, 1987, Surface Sci. **185**, 75.

Luntz, A., and J. Harris, 1992a, J. Chem. Phys. **96**, 7054.

Luntz, A., and J. Harris, 1992b, J. Vacuum Sci. Tech. **A10**, 2292.

Menzel D.., 1975, in: Interactions on Metal Surfaces, ed. R. Gomer (Springer, Berlin), pp. 101–142.

Morris, M., M. Bowker and D.A. King, 1984, in: Comprehensive Chemical Kinetics, Vol. 19, eds. C. Bamford, C. Tipper and R.G. Compton (Elsevier, Amsterdam), ch. 1.

Murray, P., F. Liebsle, C. Muryn, H. Fisher, C. Flipse and G. Thornton, 1994, Phys. Rev. Lett. **72**, 689.

Newton, M., S. Francis and M. Bowker, 1992, in: Catalysis and Surface Characterisation, eds. T. Dines, C. Rochester and J. Thomson (Royal Society of Chemistry, Cambridge), p. 165.

Newton, M., S. Francis and M. Bowker, to be published.

Newton, M., S. Francis, Y. Li, D. Law and M. Bowker, 1991, Surface Sci. **259**, 45.

Nørskov, J.K., A. Houmouller, P. Johansson and B. Lundqvist, 1981, Phys. Rev. Lett. **46**, 257.

Pearce, R., and W. Patterson, 1981, in: Catalysis and Chemical Processes (Wiley, New York, NY), p. 263.

Pfnür, H., C. Rettner, J. Lee, R.J. Madix and D. Averbach, 1986, J. Chem. Phys. **85**, 7452.

Pringle, T., N. Aas and M. Bowker, 1994, J. Chem. Soc. Faraday Trans. **90**, 1015.

Pudney, P., and M. Bowker, 1990, Chem. Phys. Lett. **171**, 373.

Redhead, P.A., 1961, Trans. Faraday Soc. **57**, 641.

Rettner, C., and C. Mullins, 1991, J. Chem. Phys. **94**, 1626.

Rettner, C., and H. Stein, 1987, J. Chem. Phys. **87**, 770.

Rettner, C., D. Auerbach and H. Michelson, 1992, Phys. Rev. Lett. **68**, 1164.

Rettner, C., H. Pfnür and D. Auerbach, 1985, Phys. Rev. Lett. **54**, 2716.

Rettner, C., H. Pfnür and D. Auerbach, 1986, J. Chem. Phys. **84**, 4163.

Ridler, D., and M.V. Twigg, 1989, in: Catalyst Handbook, ed. M.V. Twigg (Wolfe, London), ch. 5.

Root, T., L. Schmidt and G. Fisher, 1983, Surface Sci. **134**, 30.

Ryndin, Y.A., R. Hicks, A.T. Bell and Y. Termakov, 1981, J. Catal. **70**, 287.

Schlogl, R., 1981, J. Catal. **70**, 19.

Singh-Boparai, S.P., M. Bowker and D.A. King, 1975, Surface Sci. **53**, 55.

Somorjai, G.A., 1991, Catal. Lett. **9**, 311.

Strongin, D., and G.A. Somorjai, 1991, in: Catalytic Ammonia Synthesis, ed. J.R. Jennings (Plenum, New York, NY), p. 133ff.

Sun, Y.K., and W.H. Weinberg, 1990, J. Vacuum Sci. Tech. **A8**, 2445.

Tamm, P., and L.D. Schmidt, 1971, Surface Sci. **26**, 286.

Taylor, J., and I Langmuir, 1933, Phys. Rev. **44**, 423.

Tománek, D., and K. Benneman, 1983, Surface Sci. **149**, L111.

Tsong, T., 1993, Surface Sci. **299/300**, 153.

Van Hove, M.A., R. Koestner, P. Stair, J. Biberian, L. Kesmodel, I. Bartos and G.A. Somorjai, 1981, Surface Sci. **103**, 189 and 218.

Van Hove, M.A., R. Lin and G.A. Somorjai, 1983, Phys. Rev. Lett. **51**, 778.

Wachs, I., and R.J. Madix, 1978, J. Catal **53**, 208.

Winters, H., 1975, J. Chem. Phys. **62**, 2454.

Wintterlin, J., R. Schuster, D. Coulman, G. Ertl and R. Behm, 1991, J. Vacuum Sci. Tech. **B9**, 902.

Ying, D., and R.J. Madix, 1980, J. Catal. **61**, 48.

Yoshino, T., S. Saito and B. Sobukawa, 1971, Japanese patent 7103438.

Zhdanov, V.P., 1991a, Elementary Physicochemical Processes on Solid Surfaces (Plenum, New York, NY), p. 194 ff.

Zhdanov, V.P., 1991b, J. Chem. Phys. **95**, 2162.

AUTHOR INDEX

SUBJECT INDEX